Teodora Guenkova-Luy

Multimedia Networking

Teodora Guenkova-Luy

Multimedia Networking

Coordination of Multimedia Services in Next Generation Mobile Networks

VDM Verlag Dr. Müller

Bibliographic information by the German National Library: The German National Library lists this publication at the German National Bibliography; detailed bibliographic information is available on the Internet at http://dnb.d-nb.de.

Contact: info@vdm-verlag.de
Cover image: www.purestockx.com
Publisher: VDM Verlag Dr. Müller e. K., Dudweiler Landstr. 125 a, 66123 Saarbrücken, Germany
Produced by: Lightning Source Inc., La Vergne, Tennessee/USA
Lightning Source UK Ltd., Milton Keynes, UK

Bibliografische Information der Deutschen Nationalbibliothek: Die Deutsche Nationalbibliothek verzeichnet diese Publikation in der Deutschen Nationalbibliografie; detaillierte bibliografische Daten sind im Internet über http://dnb.d-nb.de abrufbar.

Kontakt: info@vdm-verlag.de
Coverbild: www.purestockx.com
Verlag: VDM Verlag Dr. Müller e. K., Dudweiler Landstr. 125 a, 66123 Saarbrücken, Deutschland
Herstellung: Lightning Source Inc., La Vergne, Tennessee/USA
Lightning Source UK Ltd., Milton Keynes, UK

ISBN: 978-3-8364-5147-5

Preface

The Internet architecture is flourishing through its ability to provide broad variety of electronic services like electronic messaging, telephony, or audio/video conferencing and on demand services. Multiple services display common characteristics, especially in cases of coding of audio and video media, in order to cover the requirements of different users (consumers and providers of media contents) and of various access-network technologies. Micro/macro IP mobility and wireless IP technologies pave the way towards the integration of today's Internet with the next generations of mobile telephone systems into All-IP communication architecture. In order to achieve harmonization, future generations of networks and services need to address the increasingly important challenges of cross bridging of networks, wireless access, mobility management, Quality of Service (QoS) provisioning for different kinds of applications and multimedia-supporting services. Current protocols and mechanisms for providing QoS cover predominantly single facets of the global QoS management. However, QoS is a system aspect that crosses all components and layers of a distributed multimedia system. Consequently, the control mechanisms for service management turn to be increasingly complex with the growing system heterogeneity and the service users and providers should be prepared to dynamically exchange and evaluate possible system configurations in different network environments and scenario situations. The complexity of the system configuration however shall not be delegated to the human users as the average user of multimedia applications lacks the knowledge to perform such configurations or is unwilling to spend the time for manual system adjustments. Furthermore, network and service providers wish to have possibly exact management of the services which can be guaranteed only having suitable automatic controls. Hence, automatic negotiations of system configurations and scenario-dependent configuration decisions can enable the comfortable deployment of novel multimedia services on the Internet that have the flexibility to adjust themselves without human intervention to different environment and application scenarios. In addition to providing an advanced description-model that covers the different facets of the multimedia communications and scenarios, an optimisation of the control-data and its successful transfer is also a significant research task. The minimization of the response times of the services under adaptation conditions through appropriate signalling represents an important aspect of the overall QoS for the multimedia applications. Current investigations address predominantly the adaptation of the human-perceivable media as it represents the greater load for the services. However, the suitable adaptation of the control data and its appropriate transfer over a wide variety of networks represents an additional challenge for the next-generation services, as the proper QoS for the control information can improve further the adaptation behaviour of the multimedia services, introducing also alternative adaptation possibilities that cannot be covered by the media streaming alone, e.g. through network content adaptors or through session- and service-mobility scenarios.

The results presented in this work were motivated through several research projects. Of special importance for the presented investigations were the EU-funded projects MIND and DAIDALOS and the DFG-funded project AKOM. However, the views expressed in this book are those of the author and are not necessarily representative for the projects.

The major topic of this work is the specification and the implementation of the End-to-End Negotiation Protocol (E2ENP). The protocol represents a novel and flexible approach to manage and coordinate multimedia services in mobile and heterogeneous network environments, different from the currently existing static service-coordination methods on the Internet. Significant parts of this work are devoted to the optimisation of the control signalling for distributed multimedia services. This contribution summarises also studies of previous works on multimedia QoS controls in heterogeneous networks and consequently derives requirements for a description model for session controls. The model reflects system management issues relevant for both the media consumers and the network and application providers. Additionally, considerations about the optimisation of the control signalling are discussed, regarding also recommendations for successful and efficient negotiations stemming from other then the pure technical backgrounds. Furthermore, alternative scenarios for service adaptation were developed and investigated to achieve other adaptation possibilities then the currently available sender-receiver adaptation. The resulting description language and control operations of E2ENP support different negotiation modes and phases that cover the requirements of different service and network environments. Detailed contributions stemming from the E2ENP description language and the E2ENP User Agent implementations are the possibility to introduce pre-configuration of the systems through pre-negotiations and through recollection of previous negotiation rounds; the introduction of protocol-specific referencing of memorised data that enables the application of short negotiation rounds to optimise the control-signalling procedures at re-negotiation of parameters of running services; and the direct provider interaction with the protocol to enable exact controls of the services within the access networks of the users. In addition, E2ENP enables the distributed management of service resources integrating a resource-reservation-coordination process within the negotiation roundtrips. Furthermore, integrations of the protocol with standard technologies like Session Initiation Protocol (SIP), eXtensible Mark-up Language (XML) and some meta-data specifications of Moving Picture Experts Group (MPEG) are achieved.

The concept of the protocol is proven through the implementation of the E2ENP User Agent and through the evaluation of the agent's performance in different negotiation roles to estimate the gains of the short E2ENP phases in comparison with the implementation of full descriptions and negotiations, as typical for the currently existing session-management protocols. The test environment for the protocol considers the application of different access-network technologies and the possibilities to provide QoS for the signalling data. Comparisons of E2ENP based on different SIP-carrier implementations are evaluated. Results stemming from E2ENP measurements on different end-devices and within different emulated networks are presented and commented. In order to emphasize the profits of E2ENP distinguishing it from the current state of the art, E2ENP is compared with simulations of the Session Description Protocol next generation (SDPng). Since both of the protocols are similar in their expressiveness and application, their comparison confirms the efficiency of E2ENP.

Acknowledgements

Before all, I would like to thank my scientific supervisor Prof. Dr. Franz J. Hauck for his encouragement and guidance, and for the fruitful discussions we had on telecommunication, middleware and on further special subject matters of my research that resulted in this work.

My special thanks go also to Dr. Andreas Kassler for guiding my work at the beginning of my research time and later on for providing suggestions for enhancements and improvements of my investigations.

I thank also Prof. Dr. Michael Weber and Prof. Dr. Peter Schulthess for helping to improve the quality of this work and for their help in organisational matters.

Further thanks go to my colleagues as paper co-authors and scientific-project partners from the DFG and EU supported projects AKOM, MIND and DAIDALOS. Especially important were the conjoined contributions again with Prof. Dr. Franz J. Hauck and Dr. Andreas Kassler and with Dr. Davide Mandato, Andreas Schorr, Holger Schmidt, Dr. Bernhard Feiten, Ingo Wolf, Dr. Christian Timmerer, Miguel Gomes, Pedro Ruiz and Prof. Tomas Robles.

I would like to thank also my students who contributed to the software debugging and the measurements. Especially, Markus Nosse, Goran Petrovic and Michael Mpeka.

I would like to very, very much thank my husband Johann and my mom Assia and dad Nikolay for their patience and their encouragements.

Teodora Guenkova-Luy

Table of Contents

List of Figures

List of Tables

1 Introduction

The prosperity of the Internet architecture is based on a wide variety of provided electronic services (among many others – electronic messaging, telephony, or audio/video conferencing and on demand services). Not only sedentary but also nomadic users can profit from the application of these. Micro/macro IP mobility and wireless IP technologies pave the way towards the integration of the Internet with the next generations of mobile telephone systems. In order to achieve harmonization, future generations of IP and telephone networks and services need to address the increasingly important challenges of cross bridging of networks, wireless access, mobility management, Quality of Service (QoS) provisioning for different kinds of applications, and multimedia support.

High-demanding IP-based multimedia applications (especially real-time audio and video) that operate in mobile environment experience frequent Quality of Service disturbances, due to packet loss, packet malformation, delay and delay variations. The reasons for these are found in the supporting network, like signal fluctuations in the radio link, handovers between different wired or wireless network technologies and network congestion. Ubiquitous utilization of distributed multimedia services in next-generations mobile systems involves application of heterogeneous end-devices (e.g. Mobile Phones, Portable PCs, etc.), which vary significantly in their capabilities to support multimedia streaming (i.e. different codecs, memory size, processing units, operating systems, etc.). Hence, such participants' heterogeneity also influences the QoS performance of the services. In addition to the pure technical reasons for different QoS capabilities and possible QoS variations of the services, mobility within provider-owned networks can also affect end-to-end system management, as the subscriber management can restrict network-resource utilization in accordance with pre-agreed user-provider contracts and/or pre-defined levels of application services. Because of all these circumstances and restrictions, the enforcement of end-to-end Quality of Service throughout all layers of a distributed multimedia system is a complex process incorporating multiple tasks and requiring the coordination of these tasks.

Current protocols and mechanisms for supporting QoS cover predominantly single facets of the global QoS management:

- The Real-Time Protocol suite (RTP/RTCP) [1][2] is used for multimedia data transfer and receiver-quality feedback
- The Resource Reservation Protocol (RSVP) [3][4], the architecture for Differentiated Services (DiffServ) [5] or the Multiprotocol Label Switching Architecture (MPLS) [6] provide different mechanisms for network resource management and their reservation
- Common Open Policy Service (COPS) [7] manages policy exchange between network entities that shall apply common information management system for QoS
- Forward Error Correction (FEC), Automatic Repeat Request (ARQ), path diversity and other redundancy techniques give the possibility to successfully deliver data in lossy networks [8] – [10]
- Filtering mechanisms are applied to adapt running multimedia streams to specific network or terminal resource requirements [11][12]

However, QoS is a system aspect that crosses all components and layers of a distributed multimedia system. QoS management should therefore incorporate mapping between different classes and types of QoS parameters, orchestrating respectively various system facets [13] – [16]. Due to the heterogeneity of the novel networks and services and due to the required flexibility of the service support for different kinds of clients and for various scenarios, it is no longer expectable that the service users and providers can rely on fixed configurations of their applications. Therefore, the service users and providers should be prepared to dynamically exchange and evaluate possible system configurations in different network environments and scenario situations. However, the technical burden for such data negotiations and processing should not be delegated to the human clients, as the average user of multimedia applications lacks the knowledge to perform such configurations or is unwilling to spend the time for manual system adjustments. Hence, automatic negotiations of system configurations and scenario-dependent configuration decisions can enable the comfortable deployment of novel multimedia services on the Internet that have the flexibility to adjust themselves automatically to different environment and usage scenarios.

1.1 Motivation

There is a multiplicity of QoS architectures and proposed management approaches for these architectures [16] – [32] that treat the orchestration of the quality of the electronic services and applications. Hence, the motivation of this work is not to provide another architecture for the multimedia service support under conditions of Quality of Service, but rather to study the existing approaches and develop a mechanism for architecture-independent QoS management enabling quality of service applications even in heterogeneous terminal-, network- and provider-environments.

The deployment of novel QoS-aware multimedia services faces a major dilemma concerning the users and providers relations. The users would like to have ultimate control over the system whenever accessing, utilizing and personalizing multimedia services. On the other side, the network operators and service providers require strict control over the network and application service resources, in order to balance resource distribution between users (based on user contracts and/or current system load), to charge the users for the resource consumption and finally to optimise their profit. This trade-off raises the requirement for highly flexible admission-control mechanisms at the operator/provider and sophisticated network resource-reservation and charging techniques. This requirement could lead to user-driven service models, unlike the currently existing inflexible provider-driven service models. [33][34]

An adequate QoS coordination mechanism should consider how the user evaluates and experiences system performance. Consequently, the configuration mechanism of the application as a whole should take into account the signalling of applications capabilities and performance restrictions (due to technical or contractual reasons), and be able to adequately react to different system situations and changes, e.g. handovers or resource-availability changes that result in QoS disturbances of different kinds and seriousness. In such situations, applications typically have to adapt and reconfigure themselves. The applications' configuration signalling must be efficient enough to provide all the necessary configuration data at service establishment and to minimize the service disruption at QoS disturbance and adaptation situations [14]. Current mechanisms for application adaptation (typical example is the Offer/Answer Model with SDP [35][36]) treat in a similar way the establishment and the adaptation of the sessions in their application management logic and do not take care about optimising the application signalling at service adaptations (see [17] – [19], [35] – [44]). Such approaches might bring the distributed application in an inconvenient situation, where the application's operations disrupt since the single parts of the service cannot express the adaptation efficiently in their signalling or cannot quickly handle the adaptation situation resulting in user-driven or adaptation-logic-driven interruption of the service. Hence, a study on successful negotiation of application features (e.g. capabilities, QoS, etc.) based on concepts stemming from human negotiations principles [45] shall be provided, aiming to propose an optimisation approach for the application signalling. Furthermore, different possibilities for taking decisions [46] under different application knowledge conditions [47] are to be considered in order to achieve appropriate expressiveness of the application signalling and to provide efficient interaction between the signalling protocol language and the coordination and decision engines of the application software.

1.2 Negotiations – Where, When and How?

The term *negotiation* is usually associated with the law sciences [48]. The negotiation process is an essential part of the coordination of team actions and of the resolving of disputes. *Wikipedia* provides the following definition of negotiation [49]:

> *Negotiation is the process whereby interested parties resolve disputes, agree upon courses of action, bargain for individual or collective advantage, and/or attempt to craft outcomes which serve their mutual interests.*
>
> *Given this definition, one can see negotiation occurring in almost all walks of life, from parenting to the courtroom.*

The propagation of distributed computing [50] – [52] in everyday life and the higher demand for user-centric services [33][34] make it necessary that also the distributed peers of a service shall be able to perform real teamwork attempting to reach collective advantage for their customers. To this extent, the applications shall be able to represent the opinions of their clients (i.e. application users, application providers and intermediary support-entities) and negotiate positions for desired exploitation of the services with stable and long-term applicable results for the service. This requirement is also reflected in the technical definition of the term *negotiation* that is stemming from the area of artificial intelligence [47][53]:

> *Negotiation is a communication process used to achieve coordination and resolve conflicts.*

In distributed systems, the means of communication between the electronic peers and the single parts of the distributed software are protocols. A protocol is specified as [50][54][55]:

> *A protocol consists of the rules for message exchange between components belonging to the same protocol layer. Every single layer is an abstraction of some connectivity functionality. The specified rules govern the syntax, the semantics and the synchronization of the communication.*

Considering this definition, it is clear that the mere exchange of messages and the rules for sending-receiving and ordering of messages alone are much too less for a real negotiation. Flexible configuration and coordination of distributed services in form of negotiation implicates much more the contents of the messages rather then the way they are exchanged. This sense is also expressed in a specific technical definition of QoS negotiation [13]:

> *Negotiation – two or more distributed agents that try to reach an agreement on the quality-of-service that they will attempt to provide to each other*
>
> *The agents communicate their requirements and their provisions*

The following two introductory sub-sections present an overview of the negotiation process in general, consider some associations of the general negotiation process in terms of its technical application and discuss to what extent the protocol arts applied in the distributed systems can be considered as negotiations.

1.2.1 Negotiation Backgrounds

This sub-section discusses some concepts and definitions of the human-typical negotiation process. The roles of the entities (e.g. in the human negotiations the entities are persons or professions) that may be involved in a negotiation are presented and some general issues concerning the negotiation mechanisms are introduced.

The participants in a negotiation process can be divided into two groups according to their *personal* interest in the results of the negotiation. These are:

- The entities that are directly interested in and affected by the outcomes of the disputed topic and
- The entities that may steer the negotiation into certain direction and manipulate the results of the dispute without being affected themselves by these results.

The first category of entities is named *peers* or *partners*. The partners can be in position to resolve directly their dispute or they may take the advantage of using the mentoring services of the second category of negotiating entities. The mentoring entities are separated into three types according to the fact how strong is their influence on the outcome of the negotiation:

- *Conciliator* is an entity that may discover and reveal to the mentored peers certain facts and constraints concerning the scope of the discussion about which existence the partners are not aware. The conciliator does not enforce actively any constraints, as within the conciliation process it is up to the peers how they apply the revealed facts and constraints in order to reach an effective agreement. [56]
- *Mediator* is an entity that actively processes the inputs of the partners. It prepares the inputs in a way that certain contradictory positions are dropped and not presented to the partners when decisions shall be taken. [57]
- *Arbiter* is an entity that actively processes the inputs of the discussion and takes a decision for its outcome on behalf of the partners [58]

The negotiation process itself follows one of the two principles *win-lose* or *win-win* [45][49]. The *win-lose* negotiation mechanisms leads to an agreement where one of the partners wins because of having or claiming to have a better position with respect to the outcome of the discussion or no agreement is reached at all as the negotiating parties do not have a *common language* about the dispute. The *win-win* approach aims primarily to discover the common gains of the partners before taking a decision about the results of the dispute. The major difference between the two approaches is in the way the partners prepare their inputs for the negotiation. The efficiency of the negotiation, the fairness of the taken decision and the stability and long-term applicability of the achieved results depend on the fact to what extent the delivered inputs are comparable [59] (i.e. the *common language*), if the inputs concern absolute or relative values [46] (i.e. only absolute values are a basis for a *common language*, as relative values can also be interpreted relatively) and if the inputs can rely on previous agreements and common knowledge [46][47].

The above described roles and mechanisms can be applied when analysing the behaviour of user terminals and provider services in a telephone or computer network system in order to discover what active or passive control functions they may take in an automated negotiation process that aims the coordination of the distributed parts of an application. Furthermore, the proper application of the negotiation roles may serve the optimisation of the automated negotiation in consideration with the specific features and constraints of the distributed environment and with respect to different establishment and management scenarios of the applications.

1.2.2 Application of Negotiations in Automated Systems

This sub-section gives an overview on current techniques and approaches for service and session signalling in a distributed system in general, considering in particular the application signalling for Quality of Service (QoS) coordination. The QoS signalling encompasses multiple protocols and abstraction levels of the application [13] thus it can be considered an appropriate example for demonstrating the negotiation and decision taking area of interest and its respective problems. Furthermore, a discussion is presented that shows the advantages/disadvantages of the current signalling techniques in terms of their applicability for QoS negotiations.

1.2.2.1 Service/Session/QoS Signalling

The signalling complexity for service, session and QoS management depends on three features of the application (see also [60]), namely:

- Applied signalling protocol/-s for global Service/Session/QoS coordination
- Applied protocols for network reservations and network resource-management
- Protocols/techniques for enforcing provider restrictions to the reservation process and to the resources consumption (i.e. evaluation of resource-monitoring data and enforcement of Authentication, Authorisation, Accounting, Auditing, and Charging (A4C) rules in form of Service Level Agreements (SLA) or of user-subscription information)

These different techniques can be applied independently of each other or the management protocols can be interleaved and applied in parallel. Additionally, these protocols differ in the way they traverse the network and on their ability to support alternative service/session configurations, redundant information and fault tolerance.

The QoS signalling conveys application- and/or network-performance requirements and parameters. The types of QoS signalling are defined in accordance with the way the QoS information is delivered [24][25], i.e. if the signalling information is a part of the associated data traffic or not and if the signalling information traverses the same or a different network path compared with the data traffic:

- In-band – In this case the signalling information is delivered together with the data traffic. Some control fields of such protocols serve the QoS-relevant signalling (e.g. for filtering means like with DiffServ [5])
- Out-band – In this case, the QoS information is delivered in a separate control protocol (e.g. Real-Time Control Protocol (RTCP) [1] or Session Initiation Protocol (SIP) [62])
- Path-Coupled – The signalling information belongs to a protocol which traverses the same network path as the data packets. Hence, the path-coupled protocols can either

be in-band (e.g. RTP [1] with DiffServ Type-Of-Service field [61]) or out-band (e.g. RTCP [1] or RSVP [3][4]) protocols.

- Path-Decoupled – The signalling information is carried through different network paths compared with the data packets (e.g. SIP [62]). It is evident that the path-decoupled protocols are always out-band protocols.

Independently of the protocol type, all the above kinds of application signalling can be applied in a way that they provide a negotiation pattern. The different negotiations patterns are defined in the next section.

1.2.2.2 Signalling Scenarios

The complexity of the session signalling and of the resource reservation processing depends on the intelligence of the applied protocols, i.e. if one or many session/system configurations can be exchanged with a single request/response, if the response to a request can detail the reason for having exactly such response and how exact and detailed is the description format for defining the session/system configurations. The demonstrated signalling scenarios are developed according to the Offer/Answer paradigm [35] as a typical example of a negotiation-like approach where one of the parties makes a proposal (i.e. Offerer) and the addressed party replies with a decision or a counter-proposal (i.e. Answerer).

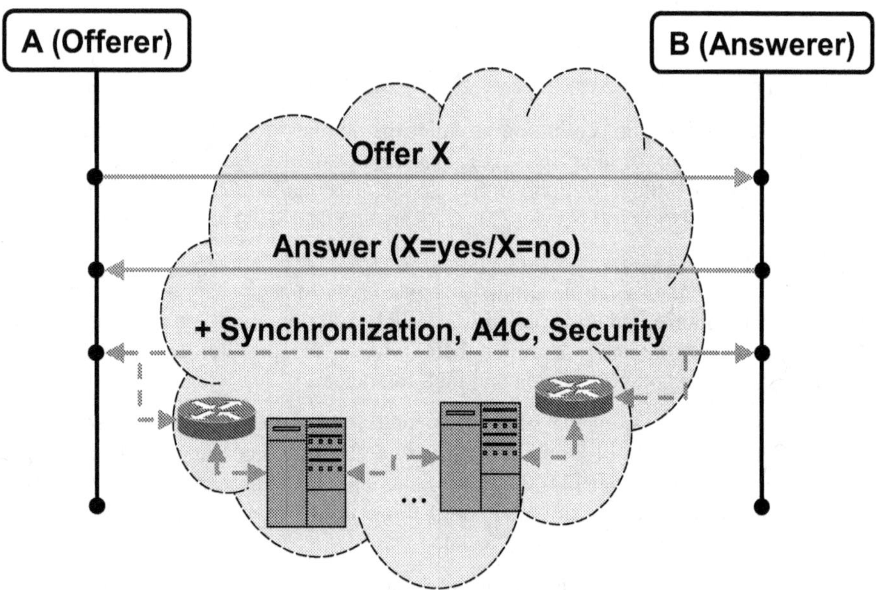

Figure 1: Yes/No signalling – single configuration delivery

1.2.2.2.1 Yes/No Signalling

In this scenario (Figure 1) End-system A (Offerer) asks End-system B (Answerer) if the configuration X (just a single configuration) can be supported by the End-system B. The answer of End-system B can be Yes (End-system B can support the configuration proposed by End-system A) or No (End-system B cannot support the configuration proposed by End-system A). The interactions marked with dashed lines (Figure 1) are interactions to provide protocol synchronisation and/or interactions with network internal services to provide A4C and security for the protocol.

The *Yes/No* scenario is an example for a *win-lose* type of negotiation. As previously discussed (see Section 1.2.1), such kind of negotiation might easily lead to no decision and no result of the negotiation. This scenario provides the smallest negotiation overhead compared with the following ones, as only one configuration is sent per negotiation roundtrip and the answer thereof is short (i.e. *Yes/No* or *ACK/NACK*). However, such scenario requires a great amount of static pre-configuration of the applications, as the protocol itself is much too simple for expressing reasons for the specific answers. Furthermore, in a highly flexible and heterogeneous network environment the maximum number of roundtrips to reach an agreement is infinite, since when receiving a decline the Offerer may repeat the negotiation roundtrip with new parameter.

Such protocols are usually applied for tracking and filtering packets in the manner DiffServ [5] does it or in the manner how RSVP [3] provides resource reservations on the network. However, these protocols alone cannot provide means for managing complex multimedia sessions and services as they have no notion of the different parts of the service and of the different abstraction models the applications require. Hence, the protocols of the *Yes/No* type alone are not enough for performing successful QoS negotiations and achieving flexible QoS management.

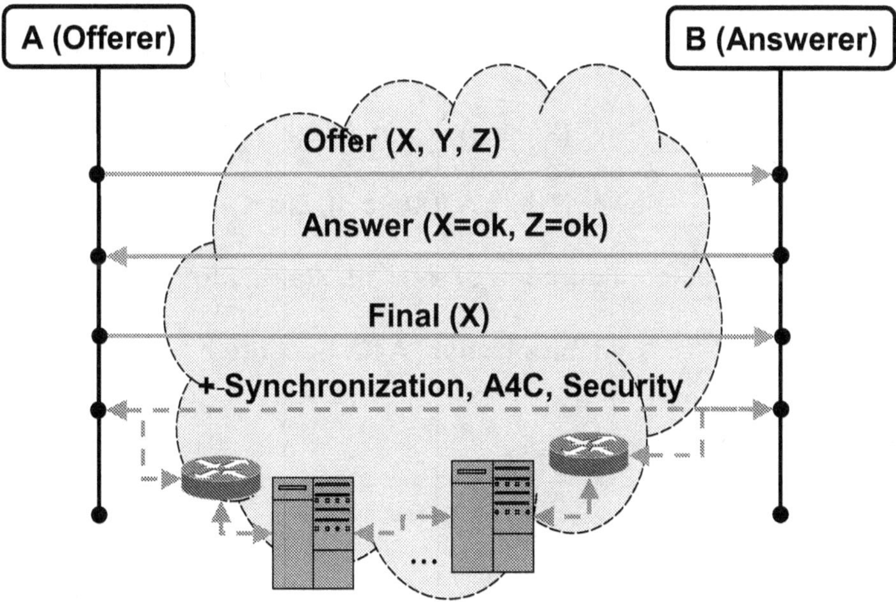

Figure 2: Raw granularity signalling – configurations level signalling

1.2.2.2.2 Raw Granularity Signalling

In this scenario (Figure 2) End-system A (Offerer) proposes several standard configurations (i.e. X, Y, Z) to End-system B (Answerer) so that End-system B can chose one or multiple of those. In its answer, End-system B indicates which are the possibly supportable configurations and finally End-system A indicates which configuration would be used to establish a session or to configure the service. The interactions marked with dashed lines (Figure 2) are interactions to provide protocol synchronisation and/or interactions with network internal services to provide A4C and security for the protocol.

Several currently existent methods for service/session/QoS configuration correspond to the model depicted in Figure 2. These standard methods for multimedia-session establishment and adaptation, typically exchange a complete set of session-configuration information every time a session should be adapted (e.g. the Offer/Answer Model with SDP [35], 3GPP signalling [22], ITU H.245 [63] or FIPA Nomadic Application [64]). The application of such methods might be useful for broadband networks, where the volume of the end-to-end application/session signalling-impact does not play an important role and where it is not expected that frequent handovers and radio signal fluctuations shall take place. However, highly heterogeneous networks with mobility application require on one side flexible definition and reasoning of the provided offers and answers, on the other side the minimization of the signalling traffic is an important problem to solve [14][65]. Since, QoS is an issue that concerns many levels of a QoS framework, the format for describing QoS configurations should be flexible enough to detail the multiple features of the QoS framework, to describe different negotiation scenarios, to be able to depict novel system characteristics, even if they are not standardised and to be able to explicitly express constraints concerning the actual situation within the system (i.e. specific resource or provider restrictions). These requirements lead to the third model of service/session/QoS signalling (see Section 1.2.2.2.3 below).

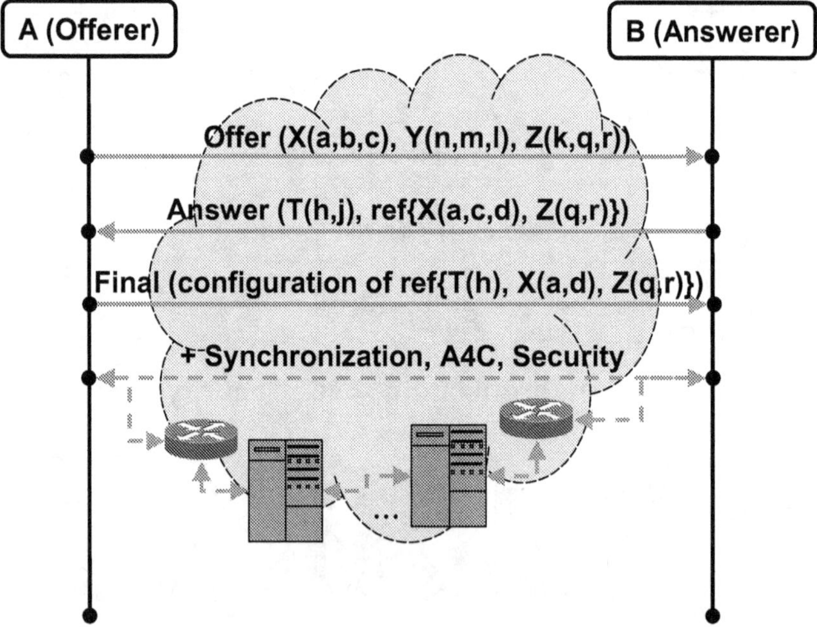

Figure 3: Fine granularity signalling – parameter level signalling

It is evident that the raw granularity signalling of service, session and QoS already shows typical features of *win-win* negotiations (see Section 1.2.1), such as the ability to make multiple proposals when searching for a common gain. However, as the protocol language in this case does not provide possibilities to express situation-specific restrictions this leads to unstable decisions that might need to be renewed through re-negotiation if a new situation shall occur. Furthermore, the lack of specific restriction expressiveness in the negotiation language leads to the necessity to apply mediating entities (e.g. like in the 3GPP architecture [22]), which might increase the impact on the signalling as the mediating entities also demand decision time on one side, on the other side they restrict the peers' own decisions in case of a different situation that is outside of the responsibility and the knowledge of the currently applied mediating entity (e.g. at vertical handovers between two provider domains).

1.2.2.2.3 Fine Granularity Signalling

In this scenario (Figure 3), End-system A (Offerer) proposes several parameterised configurations to End-system B (Answerer). The End-system B adapts these configurations to its own needs and answers to End-system A with a new configuration or a set of configurations. End-system B could deliver new information about itself (i.e. T (h, j) in Figure 3). The Answerer can either repeat the configuration-descriptions from the Offer in its Answer and/or can only reference the already known information sent by the Offerer (e.g. by using some identifier of the information). The Offerer creates a final configuration out of the common knowledge of End-system A and End-system B, and both end-systems use this configuration for session establishment and later on at session adaptation.

This kind of service/session/QoS signalling leads to the application of a very complex configuration description model. However, compared with the two previous models it can express greater variety of configurations and system parameters, thus this model could provide higher possibility that the negotiating peers meet an agreement on their own without the involvement of mentoring entities.

The current state of the art (independent and individual studies, but also Internet Engineering Task Force (IETF) research [13], [17], [60], [66] – [70]) predominantly expresses requirements that such protocols shall be developed stating that QoS requires higher precision of the description model to cover the different facets of the QoS-aware network environment. Some currently developed solutions that describe QoS parameters trying to fulfil the requirements for flexible QoS definitions are:

- OWL (Web Ontology Language) [71] is the language of the *Semantic Web* that represents a service-description and management framework for automatically collecting and processing information about services on the Web. However, the scenarios currently discussed in the OWL specification [71], consider predominantly the search for appropriate services, but not the dynamic automatic configuration of suitable services. Consequently, the current work on OWL is concentrated on service description and discovery but not on service negotiation and optimised configuration for applications.
- QML (Quality of service Modelling Language) [18] is a proprietary definition language using its own syntax and semantics. QML is applied for defining QoS parameters as an internal configuration of a terminal or network entity. A negotiation process using this language is not defined.
- MPEG-21 DIA (Moving Picture Experts Group specification 21, Digital Item Adaptation) [72] provides system and technology independent description

specification that also includes QoS features for network and application QoS definitions. The descriptions of MPEG-21 DIA utilize the Extensible Markup Language format (XML) [73]. However, being technology independent MPEG-21 does provide neither signalling mechanism, nor negotiation patterns for QoS.

- QuO (Quality Objects) [38][39] utilize the CORBA technology [74] and enhance it with definitions of QoS features (i.e. the so-called Quality Description Languages – QDL). However, the CORBA technology still faces some problems with respect to device mobility [75]. Furthermore, the QDL contain just specifications of QoS as utilization policies (i.e. the Contract Description Language – CDL [39]) and only some requirements for a QoS-negotiation process are being currently developed [40][41]. The negotiation process developed for QuO [40][41] is just a policy exchange that does not fulfil the requirements for session and device mobility [76].
- HQML (Hierarchical QoS Markup Language) [16][19] is an XML-based solution suited to the requirements of a specific QoS Broker (described in [19]). The QoS Broker specification [19] provides requirements for QoS negotiations, but the proposed solution for the definition language mixes system-dependant and system-independent QoS features and the proposed *negotiation* process is actually a policy-exchange process between the single parts of the QoS broker rather then a real negotiation between independent parts of a heterogeneous network environment.
- Quality of Service Ontology Specification [77] of FIPA (Foundation for Intelligent Physical Agents) [78] describes QoS features using the Agent Communication Language (ACL) [79]. This language can be applied for negotiations together with the nomadic agent negotiation pattern [64] (further information on nomadic and mobile agents is given in [80]). The specifications of FIPA consider, however, only the network aspect of the QoS. Respecting these ACL features, it is evident that ACL might not be applicable in non-typical agent environments like the current applications of multimedia streaming (e.g. VoD or VoIP), as these real-time applications have different performance requirements as the FIPA agents.
- Session Description Protocol next generation (SDPng) [81] – SDPng is a meta-protocol that describes other protocols and configuration mechanisms of applications. Usually, the SDPng description is carried inside other session layer protocols (like SAP [82], SIP [62], RTSP [83], MGCP [84] or Megaco [85]), to exchange for instance configuration information on RTP-based multimedia streams. However, not all combinations of SDPng and a session layer protocol can enable QoS negotiations. Only the Session Initiation Protocol [62] using the communication pattern of the Offer/Answer Model [35] has the ability to carry out negotiations, as the other carriers defined for SDPng are fixed to one of the first two signalling models described above (see Sections 1.2.2.2.1 and 1.2.2.2.2). Currently, the basic specification of SDPng treats only the adaptation of its predecessor SDP [86] to the XML format of SDPng [69][81]. SDPng has specified different containers for different information purposes in its basic specification [81] together with a package for Real-time Transport Protocol formalized description (based on the specifications [1] and [2]). Furthermore, some initial definitions of audio and video capabilities of the services are provided that are by no means complete considering some other description languages for system capabilities (e.g. MPEG-21 DIA [72]). Outside of the basic SDPng specification, a package for network QoS parameters was proposed [87], however the development of this package was not considered further within the SDPng evolution process [81].
- H.245 "Control protocol for multimedia communication" [63] is the ITU equivalent to the session configuration and control mechanisms in IETF (e.g. SIP [62] and RTSP

[83] for transport of the control messages and SDP [86] and SDPng [81] for the session descriptions). It is able to describe and deliver audio, video and transport capabilities for multimedia sessions. However, its signalling logic is similar to SDP and SDPng. Hence, H.245 is not in position to provide control-data optimisation and adaptation and to flexibly communicate such information between the negotiating peers.

All the above description languages and logical solutions for QoS negotiations have more or less the ability to specify system capabilities of a service and respective constraints, however, none of them has up to now taken care of optimising the negotiations in a way that a following re-negotiation can address previous negotiation results. Furthermore, only few works (in the area of QoS Brokers [17][19][20] and research on requirements for future signalling protocols [36][60][70]) take into account the coordination of the negotiation and the resources management process of a QoS-aware service. Considering the rules for successful automated negotiations [47] (see also Section 1.2.1) the parties need to have possibly best information about the available resources within the scope of the disputed topic to avoid inconsistent statements during the negotiation process.

1.3 Optimising the QoS Signalling for Multimedia Applications

Requirements for optimised QoS signalling and a theoretical definition of a methodology for automated QoS negotiation/re-negotiation process for service establishment and adaptation were developed in studies [17] and [70]. These works serve as a motivation for the development of an advanced set of definitions for an end-to-end negotiation protocol for QoS, including resource coordination at the terminals and on the network [14][20][1]. The contributions [14], [17], [20] and [70] represent the major theoretical backgrounds for this book. The ideas and concepts presented in these works are evaluated, proved and improved through the implementation and the tests of the research of this contribution.

1.3.1 Goals of Study

The goal of this study is to develop a negotiation language for QoS of multimedia applications that can express different abstractions of the applications, including the possibility for coordination of various kinds of resources used by the multimedia services. This language shall exhibit typical features of a *win-win* negotiation (see Sections 1.2.1 and 1.2.2) in terms of:

- *Architecture independence* – ability to evaluate and to apply different levels of abstractions of QoS as different parts of a specific architecture may see it (e.g. the way the terminals and the network/service provider see the configuration information)
- *Heterogeneity support* – ability to evaluate and to utilize different technologies of a specific distributed architecture, like different protocols of various levels (see [60]) or different resource-reservation mechanisms
- *Stability of the negotiated information* – ability to exchange, evaluate, remember and re-use multiple proposals of possible system configurations in consideration with:

---

[1] The author of this contribution has participated in the cited EU IST project MIND (Mobile IP Based Network Developments, IST-2000-28584)

- o *Prediction and planning* – possibility to exchange information about the configurations of a multimedia service before a service shall be actively used (e.g. before a multimedia session takes place) in order to plan effectively common gains for the service users and be able to optimise negotiation interactions at the start of the service and during the service utilization
- o *Flexible information management*
  - *Re-negotiation* – ability to request additional information through re-negotiation, whenever the available configurations are not enough for taking an appropriate session establishment or adaptation decision
  - *Validation* – ability to validate sets of information against the available system resources and to invalidate sets of information that are no longer applicable for the decision taking of the service (e.g. for session establishment or for session adaptation)
  - *To-the-point expressiveness* – ability to structure the negotiated information in an optimised way in order to minimize the internal language processing at the terminals and at the network entities

- *Efficiency of the negotiation* – ability to interact both in peer-to-peer mode and within a network infrastructure, and to utilize mentoring services for performing a successful negotiation whenever the own knowledge of the partner entities does not enable an efficient decision about the service establishment or an adaptation process of the multimedia application
- *Flexibility of adaptation* – ability of being transparent to the different means of adaptation of a service (e.g. to provide adaptation alternatives different then the pure sender-receiver adaptation)
- *Multi-application management* – ability to administrate the access of the applications to a shared signalling environment, thus, ensuring resources coordination between different services and enabling priorities-based management between applications

The specification of the negotiation language incorporates studies of existing capabilities and QoS descriptions like SDP/SDPng [81][86][87], MPEG-21 DIA [72] and FIPA QoS ontology [77]. The QoS negotiation language specification is used to implement a negotiation protocol and protocol agents as software for supporting negotiations between different kinds of peers utilizing QoS-aware multimedia applications in heterogeneous terminal and network environment. Furthermore, the negotiation process is tested with different signalling carriers for performing remote calls (e.g. RMI [88] or SIP [62]) and within different network environments to prove the efficiency of the solution.

1.3.2 Contributions, Results and Benefits

The contribution of this work addresses the following research areas:

- *Domain and interaction models*

  Considering the fact that there is a multiplicity of QoS architectures (see Section 1.1), this work provides a study on what are the common features of these architectures in order to identify proper general negotiation entities in form of negotiation roles (domains). A domain in the domain model identifies specific decision and negotiation responsibility area. The specification of the domain model includes also possible negotiation interactions at different levels of abstractions. It is used to identify interaction scenarios for the negotiations in different kinds of situations and to study protocol features with respect to authentication, authorisation and security of the proposed solution. The development of a domain and interaction model is related with the decision taking and the optimisation of the adaptation for the multimedia services (see below). This investigation influences the specification of the protocol description language and it is applied to define preconditions for the utilization of the negotiation protocol.

- *Descriptions of QoS, description formats and optimisation of the descriptions*

  There are various approaches how QoS is defined (see Section 1.2.2.2), within this work these approaches are studied to identify a set of QoS descriptions at different levels of abstractions, appropriate for the negotiation of multimedia-application configurations in comprehensive manner and in consideration with different existing techniques for application management. The identified sets of information are independent enough to provide a common vocabulary for the multimedia applications of different kinds and to allow multimedia system interactions in different network and provider environments and under conditions of applying heterogeneous terminal devices. The study on QoS definitions leads to the specification of a hierarchical description model that is used by the terminals to provide and negotiate different possible technical specifications of their services and to choose the most appropriate configuration for their current environment of operation. The terminals are enabled to associate different QoS abstractions like the definitions of application and transport QoS and the specifications of various QoS management levels for the service in form of QoS contracts and contexts. Furthermore, the access-network providers apply the hierarchical model to associate restrictions upon the services for the purpose of resources management in the current network of access of the terminals. In addition, the hierarchical QoS specification is a part of the concept for optimisation of the signalling for service establishment and adaptation (see below). The enhancements to the state-of-the-art mechanisms provided through the protocol description language are the possibilities to define univocal naming of the components of the hierarchical QoS model and to directly and modularly address these components by their names. This is an improvement to the current art of making negotiation statements, as the existing mechanisms expose only restricted modularity and have to completely filter negotiation statements in order to recognise what are the changes to the configurations as proposed by the communication partners.

- *Optimisation of the signalling for service establishment and adaptation*

  Current application signalling mechanisms are not in position to differentiate between session management activities (e.g. session pre-configuration, establishment and adaptation) upon the contents of the exchanged messages and treat similarly session establishment and session adaptation procedures (see [35][36][81][86]). Consequently, they cannot optimise the session signalling through appropriate arrangement and exchange of negotiation information. However, the optimisation of the signalling in connection with the optimisation of description format leads to several effects for the negotiation protocol. On one side the greater amount of information exchange will take place before or at the very beginning of a multimedia session, where the real-time effects and time constraints of the media adaptation still have little or no influence on the service. Furthermore, the possibility to memorize decisions at different levels of abstractions of the multimedia service allows the usage of *short signalling phases* (i.e. just referring to already exchanged information) during the media streaming when the time constraints for the media delivery are more stringent and when the application signalling and the media transport have to share the connectivity medium (e.g.

hardware network device, the network band and link, etc.) and processing power (e.g. CPU, memory, etc.). Thus, the process of service signalling, resource reservation and media adaptation is optimised. Additionally, the exchange of multiple configuration sets before the session begins will enable optimised decision taking at adaptation time and the changes in the media will be signalled through the media-transport protocol alone without applying the more expensive application signalling (see [1], [2] and [89]). In such cases, where the resources consumption stays in tolerable boundaries the application signalling will remain inactive, promoting the simpler and the less time-consuming adaptation mechanisms of the application. These considerations lead to the specification and the implementation of different modes and phases of the negotiation protocol. The efficiency of application of different modes and phases in accordance with different session establishment and adaptation scenarios is proven through the implementation and the evaluation of the performance of the protocol agent software of the End-to-End Negotiation Protocol (E2ENP). The performed measurements confirm the benefits of E2ENP in comparison with state-of-the-art protocols.

Current solutions for service establishment and adaptation (e.g. contributions [17], [19] and [35]) consider predominantly multimedia adaptation through adaptation of resources at the media sender and/or receiver terminals, and/or of router forwarding-policies in the network. However, adaptation of media contents can occur also at some intelligent media-management services inside the network (i.e. so-called media adaptors [90][91]) or through migration of the software sender/receiver to another hardware entity [92] (typical examples for such behaviour is known from the area of mobile software agents [80]). The study of such scenarios leads to developing of a model on how to negotiate adaptation also in cases that third parties shall support the negotiation and/or the media adaptation process. The session migration scenario as an alternative to the pure sender-receiver adaptation is integrated within the logic of the E2ENP protocol agent.

- *Decision taking*

  A study of possibilities for decision taking in a heterogeneous network environment and for multimedia service application is provided. The scope of this study in connection with the domain and interaction models allows the specification of a validation/invalidation algorithm for processing the negotiation information transported via the negotiation protocol. The goal of such algorithm is to optimise the local decision taking at each entity of concern by filtering the decisions starting from more general ones and proceeding further till a specific decision can be taken. The validation algorithm specified in this work is a recommendation for the applications of E2ENP on how to optimally filter their decisions in consideration with local terminal resources and in accordance with provider restrictions for the services (see also above the definition of the protocol language and the optimisation of the signalling).

- *Resources management and coordination of multimedia applications*

  Different kinds of resources in a heterogeneous terminal and network environment expose different availability and priority [19]. The way and the order these resources are reserved and hold by the media producer and consumer entities may influence the efficiency of the complete multimedia service. Thus, a requirement that the resources shall be managed in an *economic way* [14][20] holds true also for the negotiation protocol for QoS. This work provides a model and an implementation thereof on *economical* resource coordination of multimedia applications in heterogeneous and mobile environment. For the purpose of managing resources *economically*, the coordination of their reservation is integrated within the negotiation process of the protocol.

The typical distributed systems [50] – [52] have the requirement for consistency of the view upon the coordinated resources for all participants of the service. This is also a requirement when taking an appropriate decision in a negotiation process [45][47][59]. This contribution presents a model and an implementation thereof to provide consistent information for the multimedia-application peers at the end of every negotiation round. However, as the multimedia applications are very user-centric, some of the typical assumptions valid for distributed operating systems do not hold true in cases that user preferences and utilization habits shall be considered. User behaviour might not always be reasonable [45][46][47], nevertheless, the service shall be able to react adequately also in cases that user preferences shall be considered (like forcible negotiation or service interruptions in favour of other communication partners or other automatic services). The negotiation model considers the possibility that the negotiation rounds are not atomic and the fact that intermediary results might be still re-usable. For providing consistency of the E2ENP-managed information, the protocol language implements associations between the different statements exchanged among the peers in a negotiation process and the protocol agent implements a procedure for prioritising negotiation calls and a mechanism (in form of a cache memory) for saving, retrieving and removing of protocol information. The cache memory is also associated with the ability of the protocol to provide negotiation memory for the applications and thus to optimise the negotiation procedure (see above – optimisation of the signalling and decision taking).

The utilisation of the signalling for resources coordination and for guaranteeing the consistency of the negotiated data should consider the existing concurrency of applications in QoS-aware environments. Consequently, this work provides a model and implementation thereof on how to extend the existing signalling/negotiation environments in order to enable multi-application and priority-based management of services.

The application of QoS and resources management for the multimedia services involves also the QoS guarantees for the media signalling. This work provides a study and an emulation of QoS guarantees for the media signalling as a part of the global resource management for the services. The provisioning of QoS for the media signalling is associated also with the optimisation of the signalling (see above) and it represents an innovation to the current approaches for multimedia management that consider QoS guarantees only for the user-perceivable media but not for the control process for the media.

- *Integration with standard technologies and enhancements to standards*

  The integration of existing standard technologies within the negotiation protocol is important in order to achieve harmonisation between the services within a heterogeneous environment. This work provides a study on standard technologies and integrates several standard mechanisms within the negotiation language and within the implementation of the protocol agent.

  Standard definitions of application- and transport-performance parameters are reused in the protocol language to achieve inter-operation of different multimedia applications and network types.

  XML [73] is currently a widespread technology that is used in several other technologies of interest for the development of the negotiation protocol like MPEG-21

DIA [72] and SDPng [81]. Due to the extensibility of the XML descriptions through implementation and integration of various XML schemas [93][94] into a single description format, it is possible to integrate the proposal of this work into other technologies that use XML. The reverse integration of already existing XML-based description formats into this work is also possible. An example for integration of QoS descriptions stemming from different specification sources is shown in [95] – [97]. The specification of the E2ENP language is made in XML and it reuses concepts and descriptions stemming from XML, MPEG and SDPng.

The optimisation of the description format for the purposes of a negotiation protocol requires memorizing and referencing of different parts of the protocol payloads and cross-referencing between protocol payloads. The XML technology provides a document referencing mechanism through the so-called linking technology (XLink [98], XPath [99] and XPointer [100]). However, the XML referencing concentrates predominantly on named and persistent documents and on the physical structure of the documents, on the other side protocol payloads are usually not named and are transient. The study on the XML referencing leads to a proposal on how to treat internal and cross-referencing of the negotiation protocol payloads. The implementation of the E2ENP negotiation language provides the possibility to make logical links between negotiated information, unlike the only-physical links of the XML linking technology. The logical links represent a novel approach on how to associate negotiation information contained in different negotiation statements. Furthermore, E2ENP integrates the XML linking technology in order to enable provider-specific marking of the negotiated information. Such marking is not possible with current approaches like SDP [86] or SDPng [81] and is applied to enhance existing negotiation mechanisms like the Offer/Answer Model [35][36] (see below).

Current mechanisms like the Offer/Answer Model consider at negotiation only the role of the communication partners, omitting the possibility that mentoring entities may influence the negotiation process. E2ENP enhances the state-of-the-art mechanisms with the possibility that provider and third-party entities interact with the negotiation protocol in order to dynamically enforce environmental constraints or to propose environment enhancements (e.g. through alternative adaptation like transcoding or session migration) of which the peers are not aware. Consequently, E2ENP enables additional adaptation scenarios for the applications.

E2ENP reuses the Session Initiation Protocol [62] for transferring the negotiation messages. However, the language of E2ENP enables the application of SIP in different modes and phases that are not available through SIP alone. This approach leads to the optimisation of the signalling (see above).

The innovations of the negotiation protocol E2ENP presented in this book in comparison with the state-of-the art mechanisms can be summarised as follows:

- E2ENP enables *win-win* negotiations through multiple proposals (actual, potential configurations) and optimises the control signalling through modular definition of configurations and through recollection of and referencing to past negotiations
- E2ENP facilitates negotiations with mentoring entities thus providing higher configuration flexibility of the multimedia applications and makes their utilisation within heterogeneous operation environment possible.

1.4 Organisation of the Work

This work contains four major parts organised to provide as first introduction to the negotiation topic; secondly the theory and methods for performing automatic negotiations for quality of service are considered; as third topic the concepts and the contributions of the End-to-End Negotiation Protocol (E2ENP) are presented in detail and finally evaluation of the protocol and summary and conclusions of this work are discussed.

This first chapter already provided an overview on the negotiation backgrounds. Chapter 2 discusses the technical basis and the theoretical possibilities for application of automatic negotiations of quality of service. Furthermore, Chapter 2 represents the major discussions about the state-of-the-art negotiation techniques and how they can be improved using E2ENP. The scenarios and the functional interactions affecting and related to the End-to-End Negotiation Protocol are presented in Chapter 3. The negotiation language of the protocol is described in Chapter 4. The logic of the negotiation engines of E2ENP and the architecture of the corresponding software modules (i.e. the architecture of the E2ENP User Agent) are presented in Chapter 5, followed by Chapter 6 that includes an evaluation of the E2ENP User Agent (UA) for supporting of the Offer/Answer Model with network-reservation coordination. The provided measurements estimate the performance of the UA components; evaluate the UA operation in different network environments and under conditions of application of different E2ENP phases and finally compare E2ENP with Session Description Protocol next generation. The last Chapter 7 presents the conclusions to this contribution.

The work contains five Annexes that describe topics related to this contribution. Annex A provides a short overview on the projects where E2ENP-relevant contributions were developed. Annex B presents a model for definition of provider profiles reflecting specific transport and application constraints a provider may enforce. Annex C shows the complete XML examples that were used for the measurements presented in Chapter 6. Annex D presents classification schemes for audio and video codecs similar to those in MPEG-7 MDS and used within the scope of the E2ENP configurations (see Chapter 4). Annex E presents recommendations to the applications of E2ENP in terms of possible interactions of E2ENP with the resources-reservation entities of the applications (see Chapter 5). Finally, the Index of this work indicates important terms and abbreviations used in this contribution. The Index provides references to the respective pages where these topics are handled.

For the reader of this work it is important to note that the text of this book is strongly cross-referenced. The cross-references represent threads of related topics. Some of the cross-references pointing to later chapters and sections of this work are important for the deeper understanding of the related topics, however, for the first-time readers, it is not necessary to follow such forward-pointing references for comprehending the contribution.

2 Negotiating Quality of Service

The negotiation of quality of service between the distributed parts of a multimedia application covers three main principles, developed here according to concepts presented in general works about negotiations and their specific features [45] – [47], [53], [59]; in IETF research on requirements and negotiation patterns for signalling protocols [14], [35], [36], [60], [66], [101] – [106] and in studies and standardisation on future internet architectures developed from the Third Generation Partnership Project [21][22] and the International Telecommunication Union [63][107]. These three principles are defined in this work for the purpose of clarification of the similarities of the existing technologies and architectures. Furthermore, they are used for the analysis of the automatic configuration of a distributed multimedia environment in order to propose an optimisation for the negotiation process of the applications. The specified principles are:

- The first principle is the one of the *involved talkers*. This principle concerns the identification of the relevant partners and mentors for the negotiation (see Section 1.2.1). The end-points in a distributed application (e.g. terminals or automated multimedia sources like Video-on-Demand servers (VoD)) are not the only devices that may determine the results of the negotiation and their consequence for the QoS of the application, the network- and the service-provider intermediary entities may also affect the negotiation process by assisting and/or restricting it. Hence, the relations between the terminals and the intermediary services needs to be studied in order to determine how different entities may influence the negotiation process and how this influence relates to the *common vocabulary* and the *communication arrangement* principles of the negotiation (described below).

- The second principle is named *common vocabulary*. It specifies what to negotiate, what is the common knowledge of the distributed parts of an application and is this knowledge sharable between different applications.

  The general definition of QoS according to [13] and [108] includes any possible parameters of the applications that may be configurable or negotiable. However, the definition of a QoS language that covers any configurable parameters of the application is not practicable taking into account the complexity of definition of such a language as a single piece of information. A simpler solution is the grouping of features relevant for a specific type of applications. A *common language*, concentrated only on specific characteristics of the applications, is named *profile*. The application of one single *profile* covers only one single aspect of QoS for the service, but the application of different *profiles* covers different aspects of QoS, thus enabling the support of services with different complexity with respect to their QoS understanding. Similar definitions of a *profile* belonging to a *common language* of the multimedia applications are described within the scope of SDPng [81] and MPEG-7 [109]. The following research will concentrate on the relevant parameters for estimation of a *common vocabulary* for streaming multimedia applications under consideration that the user end-terminals are heterogeneous and they may attach to different kinds of access technologies to enable application mobility with the end-terminal or between end-terminals. Since the streaming multimedia applications are the most demanding ones among all the multimedia services, they are most appropriate for demonstrating negotiations with fine granularity signalling (Section 1.2.2.2.3). Hence, this work provides a special consideration of such services.

The ability of the terminals to attach to different technologies is associated with the relation of the *common vocabulary* and the *involved talkers* principles. This is because of the fact that different networks of attachment may belong to different providers which may like to express their opinion about the availability of the offered services with the terminology of the common language. Hence, the *profile* for multimedia communication in heterogeneous networks shall contain also the possibility for expressing provider views about the end-to-end exchanged configuration data of the terminals.

- The last principle is the *communication arrangement*. It defines the way and the order how the partners exchange their proposals. In order to make a proposal and to meet a common agreement the partners have to consult the state of their resource availability, so that they can reach a durable agreement about their configuration/-s after the negotiation (see also Section 1.2.1). The resources in case of distributed multimedia applications are the resources of the involved terminal-devices and the resources of the network connecting the terminals. Hence, the principle of the *communication arrangement* is also associated with the so-called *economy principle* [19][70] that defines the order of resources reservations in the terminal and the network in cases that guaranteed services shall be established. With this respect, the negotiation procedure includes both the exchange order of negotiation messages and decision points where the negotiating peers may consult and approve the state of their resources in order to guarantee their services to their partners (see also [13]).

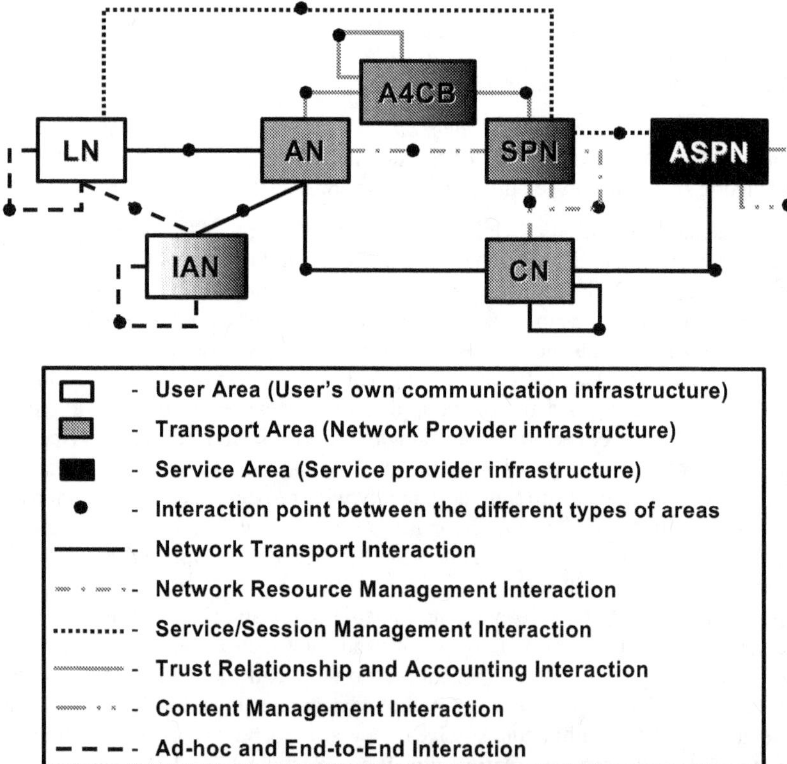

Figure 4: Domain Model - a class diagram of the service roles

The *communication arrangement* is associated with the *involved talkers* principle in a way that the *communication arrangement* specifies when the intermediary *involved talkers* are allowed to interfere the negotiation procedure and when such involvement is not appropriate. The relation between the *involved talkers* and the *communication arrangement* principles leads to a definition of different peer-to-peer and multi-peer communication scenarios.

These three principles are detailed in the next sections and utilized for the specification of a negotiation protocol for QoS termed End-to-End Negotiation Protocol (E2ENP). This protocol is applied between end-terminals with involvement of intermediary network and provider entities. The next sections present theoretical concepts and backgrounds that are used for deriving of requirements and for designing solutions for the negotiation protocol for service and session control.

2.1 Domain Model for Identification of the Concerned Entities in the Negotiation

This section details the *involved talkers* principle (See the beginning of Chapter 2). Multiple works describe end-to-end negotiation models [13], [16] – [18]. However, they omit the fact that intermediary entities on the path of the negotiation may influence or manipulate the negotiation. The only architectural scenario where the provider interacts with the peer-to-peer negotiation is the IP Multimedia Subsystem (IMS) of the Third Generation Partnership Project (3GPP) [21][22]. A concept similar to the 3GPP scenario is also expressed within the scope of ITU H.323 [107]. However, the actual application mentoring of the negotiation process is left open in H.323. In addition to the provider mentoring of the negotiations, advanced scenarios where terminal devices may also take a mentoring role as third-party call-control entities, need to be considered as an optimisation factor for the multimedia communication. Conceptually similar scenarios, where third parties assist some management actions of the applications are described within the scope of the mobile agents architectures [64][80], however, specific applications for multimedia call control are not considered there. In the following, a model for identifying the involved entities in a negotiation is developed and explained, taking into account existing concepts about peer-to-peer negotiations assisted from the network or from third-party entities and enhancing these concepts where necessary and appropriate.

Requirements for clear separation of responsibilities when providing end-to-end services are defined in [110] – [112]. These requirements serve here for specifying a model for analysing the management functionalities for end-to-end negotiations of QoS for multimedia applications. The objective of the Domain Model (DM) (Figure 4) is to provide a generalized description for any network architecture that may include multimedia services without giving some specific example architecture and without considering specific technologies. The DM developed in this work is used to analyse what kinds of management roles may appear within the scope of a distributed multimedia application and how these roles may interact with each other. The DM can be applied for different analysis purposes, like for example, the analysis of the architecture's security issues (as done in [20], where some components of the DM presented in this work stem from). Within the scope of QoS negotiations for distributed multimedia applications, the Domain Model is used to specify what are the concerned entities in a negotiation and what are their relations. Furthermore, a deeper analysis of the interactions identified via the Domain Model allows the specification of concrete technologies that are applicable in terms of a specific architecture and its services. Such analysis is necessary to discover what technologies are appropriate for negotiation purposes in their original form and to recognise missing features with which existing technologies shall be enhanced to meet the requirements for automated QoS negotiations.

2.1.1 Domains, Relations and Activities

The Domain Model in Figure 4 identifies three kinds of areas of responsibility:

- The *User Area* includes the users' own communication infrastructure. These are the terminal devices (e.g. mobile phones, portable or stationary PCs, etc.). Devices within the *User Area* can be end-points in the consumer-producer chain of the multimedia contents (see also [17]), or the devices can be content forwarders or content transcoders for other end-point consumers and/or producers in ad-hoc scenarios [20][65]. In the second case (ad-hoc scenarios), devices belonging to the *User Area* take over functionalities that are specific for the two remaining areas described below.
- The *Transport Area* includes the infrastructure of the network providers that is responsible for forwarding data and for supplying transportation guarantees for the data within the global network in the manner of technologies like IntServ [3][4] or DiffServ [5] (see also [25]).
- The *Service Area* provides higher management logic and added value services. The logic of the *Transport Area* alone can provide only best-effort services (as in the current Internet). The *Service Area* monitors and controls the *Transport Area* in order to provide separation between different kinds of services and service qualities. This separation between the management responsibilities in the provider part of the network can improve the performance and the scalability of the network as shown with examples in works about telecommunication and multimedia architectures [24], [25], [65], [111] – [113]. Furthermore, the *Service Area* includes the automatic multimedia sources and multimedia processors that may exist on the network (e.g. VoD or conference servers, media transcoders, etc.).

The exact domains within the Domain Model are defined as follows (see Figure 4):

- **Leaf network** (LN) – The LN depicts one or many end-terminals/-s, which may or may not be mobile devices and can belong to different users. The single devices belonging to LN can establish point-to-point connection with each other or LN devices can establish a connection to the global network using either the IAN role (in ad-hoc scenarios) or AN role (in mobile or sedentary scenarios where provider infrastructure is involved). IAN and AN represent the respective access points and access services for LN. The LN can utilise the IAN/AN services by following the IAN/AN registration and AAA (Authentication, Authorisation, Accounting) rules and mechanisms. These rules and mechanisms can be established and provided by SPN that administrates the end-user information in sense of a network provider (e.g. within initial terminal configuration with static data like the data of a SIM chip [114] or at registration procedure of the terminal when it attaches to a new network/service like in the case of SIP [62]). Within the scope of the QoS negotiation, the LN role is the role of a peer as it generates the QoS requirements of the terminal application (according to the QoS requirements of the user – see also Section 2.3.2.2.2) and communicates these requirements either to other devices belonging to the LN role (e.g. in telecommunication scenario, like Voice-over-IP (VoIP)) or to automatic services within the ASPN role responsibility (e.g. VoD).
- **Intermediary Access Network** (IAN) – The IAN is subscriber-owned equipment that supports its own user with application services and may provide external users with

such services (e.g. acting as multimedia transcoding service or as a mentoring entity in third-party call-control scenarios – see Section 3.3). Furthermore, IAN may act as an access to the global network in ad-hoc scenarios. From the point of view of negotiation of QoS, the IAN incorporates both peer (similar to LN) and mentoring functionality (similar to SPN). In case of a mentoring functionality, the IAN incorporates also the ASPN role in order to express the mentoring logic in application-specific way. The IAN can be configured by the respective network provider (the configuration procedure is similar to LN – see above), what credentials it is allowed to accept, in case IAN is the connection point to the global network. If no network provider is involved (i.e. formation of just spontaneous network), it is the responsibility of the user of the IAN-capable device to determine what credentials of other users the terminal is enabled to accept.

- **Access Network** (AN) is the combination of network devices (e.g. routers, gateways, etc.) for accessing the global network. In the sense of a negotiation the AN provides the communication medium for transferring the negotiation data through the network. The AN may influence the negotiation process (e.g. how long the negotiation process takes place considering the capacity of the access network) and the results of it (e.g. how the access network treats resources reservations for providing service guarantees). AN does not manage the negotiation information by itself but it may provide rules for influencing the negotiation information as AN provides the physical network resources. The actual and direct interaction with the negotiation information is carried out applying the higher network logic represented by the SPN role.
- **Core Network** (CN) is the combination of network devices (e.g. routers, gateways, etc.) interconnecting different ANs. As CN also represents the physical network resources, it can also influence the negotiation process in terms of transport and resource granting, but again the SPN enforces the resource granting requirements of CN within the scope of a negotiation of QoS.
- **Service Provider Network** (SPN) – The SPN is the user-profile administration role that defines the subscription rules to the network and regulates the service utilization. This role can be separated in two actual roles, i.e. for network management and for services management. However, the current Internet and mobile telephony systems do not separate between these two roles (e.g. the Internet *dial-in* providers grant automatically access to the Internet services or mobile telephony providers grant access to additional added value services like Short Message Service (SMS) [115]). Furthermore, also novel protocols and architectures for system access (e.g. 3GPP IMS [21], SIP [62] or H.323 [107]) do not differentiate between network and service registrations. Hence, regarding the current state of the art, the network and service registrations are not considered as different kinds of registrations. In such case, the *user profile* means the aggregated information about the allowed network and service access for the user (see also Section 2.2.1). SPN is responsible for maintaining this user-specific information. SPN uses the services of the A4CB within the *user profile* management process. Within the scope of a negotiation, the SPN represents a management role that enforces the network constraints upon the services. Furthermore, the SPN takes care for the bridging between different AN and CN roles that belong to different technologies and providers.
- **Application Service Provider Network** (ASPN) – The ASPN is responsible for the application of the protocol- and of the service-specific rules. The ASPN provides protocol-specific expressiveness of the negotiations together with precise AAA and security mechanisms of the respective technologies. Within the scope of the

negotiation, the ASPN is the role that takes care for understanding the negotiated information between the peers and for representing the network and the service constraints in application-specific manner (Section 2.2.1). For example, an ASPN role can be fulfilled within a Video-on-Demand Service that negotiates as a peer or within a Content Adaptation Service or a Provider Negotiation Mentoring Service that may negotiate as third-party call-control entities.

- **Authentication, Authorisation, Accounting, Auditing and Cashing Broker** (A4CB) – The A4CB provides trust relationship establishment between different AAA-relevant services (see also [116] – [118] for further details on accounting services). Such functionality is important when establishing relations between different users, respectively different network and service providers. This functionality has no direct influence on the negotiation. However, within the scope of the decisions *what to negotiate* the role of the A4CB may provide important rules on how the peers authenticate each other for the negotiation and within the negotiation process. The role of the A4CB is important when analysing the security of the negotiation procedure. The auditing and cashing actions of the A4CB deal with the composition of bills for the user when one or multiple networks and services are applied. These functions can be used for providing additional rules for the decision taking within the negotiation process, as providers might apply different QoS management rules based for example on time-of-the-day or roaming information.

Within the Domain Model (Figure 4) six major kinds of interactions are identified:

- The *Network Transport Interaction* corresponds to the actual transportation of information through the network (e.g. the wired or radio network technologies as transportation medium associated with the first two layers of the OSI model [54], the IP packets [119] – [121] as art of data fragmentation for transport purposes, TCP [122] and UDP [123] for end-to-end data transfer or RTP [1] for streaming data transfer). The *Network Transport Interaction* is the basic interaction in the Domain Model that serves as a foundation of all the other *logical* interactions in the DM.

- The *Network Resource Management Interaction* corresponds to management actions for granting and/or maintenance of network resources. Network guarantees and reservation technologies like RSVP [3][4], DiffServ [5] and MPLS [6] or redundancy techniques like FEC, ARQ or path-diversity [8] – [10] belong to this type of interaction.

- The *Service/Session Management Interaction* corresponds to actions for session establishment and management with protocols like SIP [62] or RTSP [83] that may include additional information about the services and the sessions using description formats like SDP [86] or SDPng [81].

- The *Trust Relationship and Accounting Interaction* corresponds to actions for establishing trust relationships between different users and different providers. Such interaction can be implemented with protocols like RADIUS [124], DIAMETER [125] and COPS [7] (see also [126]).

- The *Content Management Interaction* corresponds to processing actions for different kinds of formatting and presentation of the media content. Protocols belonging to this kind of interaction are MGCP [84] for controlling of media gateways, MPLS [6] for network-data presentation adaptation on the boundaries of different technologies or RTP [1][89], which is the actual carrier of the contents and which can be manipulated to produce different QoS levels from the user's point of view [90][91].

- The *Ad-hoc and End-to-End Interaction* includes all the above interactions in case of spontaneous network formation. Furthermore, at the level of network formation protocols like MANET [127] or extensions to it [65] are applied.

The considerations about the Domain Model presented in this section are used to specify an example architecture for end-to-end interactions between the LNs (i.e. the user-terminals) and between LN and ASPN (i.e. automatic services) reflecting also the influence of the Service and Transport Areas upon the end-to-end negotiations (see Section 2.1.2.1). The example architecture provides an overview about the interplay between resources management, decision taking and negotiations of the applications in the specific case of multimedia services (Section 2.1.2). A refinement of the example architecture is used also later for definition of a QoS Parameters Model (see Section 2.2).

2.1.2 Aspects of the QoS Provisioning

This section presents the different aspects of QoS from the point of view of the different participants in the service (i.e. the users and the providers) in consideration with different abstractions that represent the relevant service guarantees for different kinds of interconnection scenarios (i.e. mobile/sedentary, homogeneous/heterogeneous, etc.).

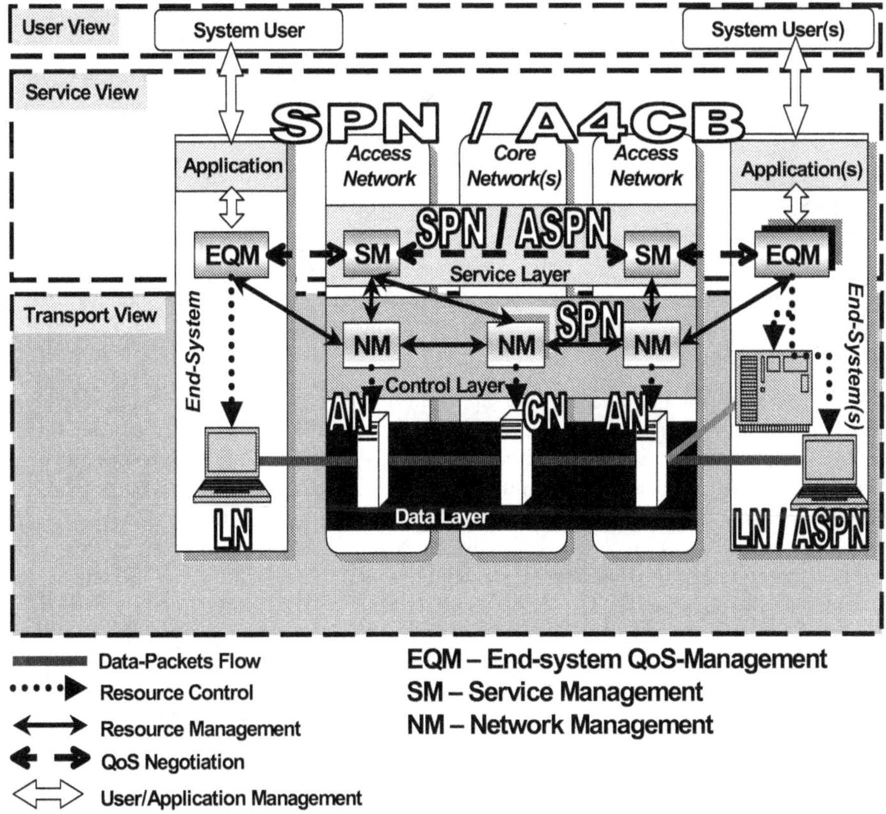

Figure 5: Architecture for end-to-end application interactions

2.1.2.1 User, Service and Transport Views

A generalized architecture for distributed multimedia services (Figure 5) was developed and is presented in this work to distinguish between two important features of the streaming applications and their end-to-end performance, i.e. packet forwarding and service management. This architecture represents a summary of the current research on architectures for multimedia services considering QoS Brokers [16][17], of studies and standardization on telecommunication and multimedia networks [21][24][25][107][111][112][128][129] and of the initial expedience with the End-to-End Negotiation Protocol [130] – [133]. The model (Figure 5) is used to analyse different negotiation and decision roles within the architecture for multimedia provisioning and serves as a basis for defining a strict differentiation between QoS-relevant system facets. The roles of the different devices (e.g. terminal, router, security/authentication services, etc.) involved the QoS decisions and negotiations are characterized via a separation into two views upon the services, namely the Transport View and the Service View. Furthermore, the model includes also a User View in order to complete the picture of the QoS relevant system features. The generalized architecture (Figure 5) shows in addition how the roles of the Domain Model (Section 2.1.1) are associated with the architectural layering.

The User View corresponds to the user interaction with the multimedia services as end-consumer of the services or as producer respectively supplier of multimedia contents.[2] The User View is not directly involved in the automatic decision and negotiation process but this view is important when designing user interfaces towards the application that enable the proper preparation of the user data (e.g. the QoS preferences and the payment requirements) for the automated negotiation process of the application (see also Section 2.3.2.2.2).

The Transport View represents the system responsibility for QoS-aware packet forwarding and contains all management functions that provide QoS for IP packets. Examples are resource reservation via RSVP [3][4], packet marking and filtering in DiffServ domains [5][61], etc. The Transport View represents the restrictions of the network that influence the service performance. The parameters corresponding to the Transport View are thus performance constraints for the application that can be expressed during the negotiation.

The Service View considers the application performance as a whole. Hence, it represents the service aspects of multimedia applications like evaluation of the perceivable features of the multimedia (e.g. quality of the audio sound or of the motion pictures, see also [134] – [136]), decision taking and negotiation of session descriptions and capabilities for session setup; Authentication, Authorization, Accounting, Auditing, Charging (A4C); security features of a service, etc. (see also Section 2.1.2.2). It is the Service View that provides the means for automated negotiations and coordination of the distributed multimedia system.

Within the generalized architecture (Figure 5) a service abstraction for the application is provided by the end-system resource management, which is typically implemented within the terminal's operating system (OS) or within a middleware, and it is denoted here

---

[2] It is important to differentiate between contents producer and contents supplier as this differentiation leads to different multimedia communication scenarios (e.g. VoIP or VoD) that have different requirements on the negotiation and the media transfer process with respect to time constraints, e.g. pre-recorded data is usually only supplied and has less stringent play-out requirements compared with live communication.

as End-system QoS-Management (EQM). EQM informs the application about available resources and capabilities to handle multimedia data (type of audio/video hardware devices, multimedia codecs, etc.) and for controlling network transport (type of network devices, currently available access networks and signal quality thereof, etc.). The application can flexibly configure/re-configure itself in accordance with the available resources (both locally and on the network) by using monitoring data provided from the resource management (i.e. the monitoring is performed within the components for Network Management, NM, and is delivered to the local management EQM). NM nodes are also responsible for controlling the admission to network resources in accordance with: provider policies (originally defined over the Service Management (SM) components of the system), results of monitoring of network resources, packet-routing rules (which may also be specified from SM), etc. SM and NM represent the implementation of the management separation in terms of service management (i.e. SM) and packet forwarding (i.e. NM).

SM can participate actively or passively in the negotiations of end-systems when establishing or adapting a multimedia session by listening in and/or actively participating in the negotiation process. SM can instruct the NM entities how to apply provider policies by uploading such policies in the NMs. Respectively the NMs provide the SMs with monitoring information, which is used in the SMs decision taking when participating in the negotiation of end-terminals. The listening in the negotiation process allows the SMs to reserve network resources on behalf of those end-system which cannot or by definition are not allowed to reserve resources on the network or the SMs instruct the NMs how to handle the traffic in networks where no explicit reservations are performed, e.g. packet-marking and -filtering in a DiffServ [5] or MPLS [6] domains.

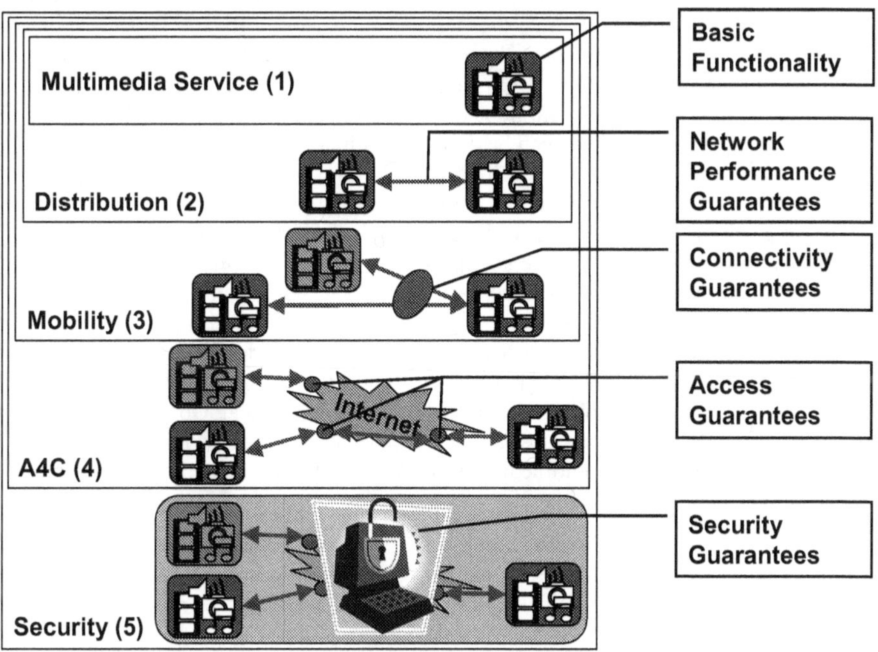

Figure 6: Quality of Service aspects of the multimedia application

The presented model (Figure 5) identifies as negotiating entities within the end-systems the EQMs and as mentors within the provider system the SM entities, i.e. the software modules that implement the negotiation procedure. Note that the EQM entities can also perform mentoring in cases of third-party call control and session redirection scenarios (e.g. in cases that the terminals decision logic considers that it alone cannot perform appropriate service adaptation for the user and delegates the call to another available terminal or intermediary adaptation service – see also Section 3.3). Furthermore, the separation of the views on the QoS is used to identify appropriate parameters for the negotiation as shown in Section 2.2 (see also Section 2.1.2.2).

2.1.2.2 QoS Provisioning Guarantees

A multimedia service in a provider infrastructure (like the one shown in the previous Section 2.1.2.1) displays five major aspects characterizing the necessary guarantees for the performance of the service in consideration with different application scenarios of the service (see Figure 6). These aspects characterise the association of the *involved talkers* and the *common vocabulary* principles (see beginning of Chapter 2) and are developed in accordance with concepts for distributed systems management presented in [128][129]. The considered aspects are:

- The first aspect is the *Basic Functionality* of the service in case of service performance on a single device and service utilization from a single user. The *Basic Functionality* is associated on one side with the user view on the service (see also Section 2.1.2.1), i.e. how the user configures the service according to her/his understanding for service quality[3], on the other side with the device's hardware (i.e. device type – PC, PDA, mobile phone, etc.; multimedia processing hardware – audio/video boards; play-out equipment – displays, audio outputs; etc.) and supporting software (i.e. device drivers, multimedia codecs, etc.). Thus, the *Basic Functionality* of a service represents the lower and the upper bounds of QoS that the service may support. Furthermore, the *Basic Functionality* covers continuous or discontinuous spaces of QoS levels (e.g. the playout frequency of a video can be defined continuously like *5-20 frames per second* or may have a discontinuous specification like *(5-10 frames per second) and (15-20 frames per second)*). Furthermore, the QoS levels may display different granularity (e.g. considering the colour of a video, the specification *greyscale, colour at 16 bit* exposes lower granularity as the specification *greyscale, colour at 8 bit, colour at 16 bit*). The separation of the QoS performance of the service into levels and layers is applied when developing tests for user perception of QoS (see [134] – [138]) or when deploying QoS adaptation environments also in consideration with the rest of the service aspects (see below and [16], [17], [70], [139] – [141]). The concept of layering of QoS is applied in this work for defining a hierarchical QoS specification for identifying the manageable abstractions of the applications within the scope of a negotiation protocol for multimedia service coordination (see Section 2.2.2.2 and [17], [70], [132]).
- The *Distribution* aspect covers on one hand the basic performance of the network (as idealized network that serves a single service)[4] in case that the application is

[3] Note that there can be various user understandings of service quality (see [17][135]). However, in a real application these understandings can be restricted through the provided user interface for configuring the application.

[4] The idealized performance of the network is the performance of the network under test conditions as specified for standard network technologies [142] – [145].

distributed between multiple devices, on the other hand it characterizes the performance of the devices in a typical consumer-producer paradigm [52][146], where the actual performance of the service depends on how effectively the producers generate the contents and how effectively the consumers may utilize these contents. The *Distribution* aspect determines the necessity of more stringent upper bounds for the performance of the multimedia application compared with the *Basic Functionality*. In this case, the producer or the consumer of the service, which has the lowest efficiency, determines the uppermost quality of the service. A profound study on the *Distribution* aspect of the application is provided in [17]. The *Distribution* aspect of the application is used here for studying the network constraints on the application performance that shall be applied within a negotiation process (see Sections 2.2).

- The *Mobility* aspect considers the fact that the multimedia consumers and producers may navigate through different networks of the same technology and/or attach to networks with different technologies during the process of multimedia contents utilization/generation. The *Mobility* aspect is associated with the fact that short-term service disruptions may occur or that sudden service up-/down-grading is possible (e.g. in the case when the devices switch between network technologies with different throughput). The *Mobility* aspect is used for defining requirements for the negotiation process in terms of how sensitive is the reaction of the negotiation process towards QoS variations (see Section 2.3.2), i.e. if the decision about a service adaptation can rely on already negotiated information or if a new round of negotiation of the application performance is necessary for taking a decision about QoS adaptation at QoS variation of the service environment.
- The *A4C* aspect of the service covers the management of multiple services that share the same execution environment. The *A4C* is associated with additional deployment-specific guarantees for the services, i.e. how many users in terms of media sources and sinks can simultaneously utilize the execution environment; how much the human users executing the media sources and sinks are willing to pay for sharing the environment[5]; how the providers of the application environment optimise the environment through deployment of different control entities inside it, etc. The *A4C* aspect introduces further restrictions to the performance of the services (compared with the *Distribution* aspect) in terms of management of user profiles that represent artificial separation of the usage environment according to the user's willingness to pay for the environment and respectively provider-side treatment of the service (e.g. indicating services with profiles like gold, silver or bronze [147] – [149]). The *A4C* aspect is in a way similar to the *Mobility* aspect as the application of *A4C* can also result in sudden service up-/down-grading. However, in the *A4C* case the QoS variations are due to the sharing of the usage environment by different services. The *A4C* aspect is used to study different management cases of the negotiation also in consideration with application of intermediary control entities for the negotiation (see Sections 2.3, 3.2 and 3.3).
- Finally, the *Security* aspect considers the fact how the different entities that share the execution environment can guarantee fair negotiations and execution to the other

[5] For the user the service price is defined not only in terms of money, but also in terms of optimal utilization of the local usage environment (i.e. the resources of the user's terminal). This means that the users may specify certain priorities for their local and distributed services running on the user-owned equipment (see Section 2.1.1). In this case, the prize is also the level of utilization of the terminals resources, e.g. CPU, memory, play-out devices and network connectivity devices.

entities in the environment. This aspect is used to define trust relationships for the users and the providers of the execution environment in terms that the statements the different entities make in a negotiation process are provable for their originality, integrity and authenticity. The *Security* aspect includes the study of security mechanisms (see [50], [52], [150]) that shall be used in the negotiation protocol to guarantee the integrity of the negotiation process (see Section 3.2).

The service aspects defined in this section are applied later for identifying requirements (Section 2.2 and 2.3) and performance scenarios for the negotiation protocol (Chapter 3) and for specifying the exact structure of the negotiation protocol (Chapter 4).

2.2 QoS Parameters Model

This section details the *common vocabulary* principle (see the beginning of Chapter 2). In order to identify relevant parameters for the negotiation of QoS for mobile multimedia applications it is necessary to identify first which are the entities that really participate in the negotiation and secondly what are their actual interests (see [45][59]). The Domain Model presented in Section 2.1.1 shows that the end-terminals (user terminals or automated services) are the final users of the results of the negotiations. Hence, they represent the peer role in a negotiation (see Section 1.2.1). The actual implementation of the peer functionality is thus located within the EQM in the example architecture (see Section 2.1.2.1). Furthermore, the provider services in the network may play some mentoring role in the negotiation. However, it is necessary to distinguish which provider entities and devices can be actually involved in the negotiations of the terminals as mentors since the identification of the Service Provider Network and Application Service Provider Network (see Section 2.1.1) as mentoring role is too general. The specification of the provider mentoring devices uses the general architecture for end-to-end application interactions (Section 2.1.2.1) and the respective aspects concerning the QoS provisioning guarantees (see Section 2.1.2.2). The general architecture is simplified to identify the actual negotiation entities in the network. The simplification is based on some observations about the current development of the Internet and the mobile communications:

- The applied network technologies (inclusive radio technologies, which allow device mobility) have higher throughput in the core as in the access part of the network (see the respective works on next generations of telecommunication networks [142] – [145], [151]), thus the resulting bottlenecks for the distributed application performance are concentrated in the access network.[6] Furthermore, access networks based on radio technologies suffer higher instability due to the lesser predictability of the behaviour of the radio links. Hence, the management of the access is more sensitive compared with the management of the core (i.e. the stationary parts of the network).
- The exact and precise manageability of the networks is shifted more and more to the network boundaries, i.e. the vicinity of the terminals [34]. A typical example of such application is the combination of IntServ and DiffServ for network reservation purposes in the access and in the core [23][113], i.e. exacter management of the access network provides higher predictability of the behaviour of the core thus allowing the application of more generalized management in the core (i.e. technologies based on

[6] This statement considers idealized error-free networks. Of course, bottlenecks can appear also in the core in cases of gateways or routers malfunctions in the core. However, the treatment of errors in the network is out of scope of this work. The only consideration with respect to network errors in this book is how the applications may adapt as result of appearing network errors.

generalization of the traffic types and not on management of single services [5], [147] – [149]).

- The manageability of the access networks is based on centralized data collection that includes registration of the terminals to the network and to its services (e.g. H.323 Gatekeeper [107], 3GPP Call Session Control Function [21][22] or SIP Proxies/Registrars [62]). Note that centralized data collection does not necessarily mean centralized data management, i.e. there is a single responsibility for the data collection in the network, which may be actually implemented as a single or distributed entity, e.g. like the registration management in the 3GPP IMS and the SIP architectures. The centralized data collection gives the access network consistent overview on the network resources in the access and in the adjacent core. One of the tasks of the centralized data collection includes the understanding of the application management protocols of the services used from the terminal devices. This task corresponds to the ASPN role identified in Domain Model (see Section 2.1.1).

Because of these observations, the identified device in the access network that corresponds to the provider mentoring of the application's negotiations is a service proxy that understands the application management protocols. Examples for such functionality in existing architectures are 3GPP CSCF [21][22] or SIP Proxy [62]. However, there are additional requirements that have to be regarded in order to make these entities mentors in a negotiation procedure (see Sections 2.3 and 2.4). In the example architecture (Section 2.1.2.1), the functionality of the service proxy in covered through the SM responsibility in the access network.

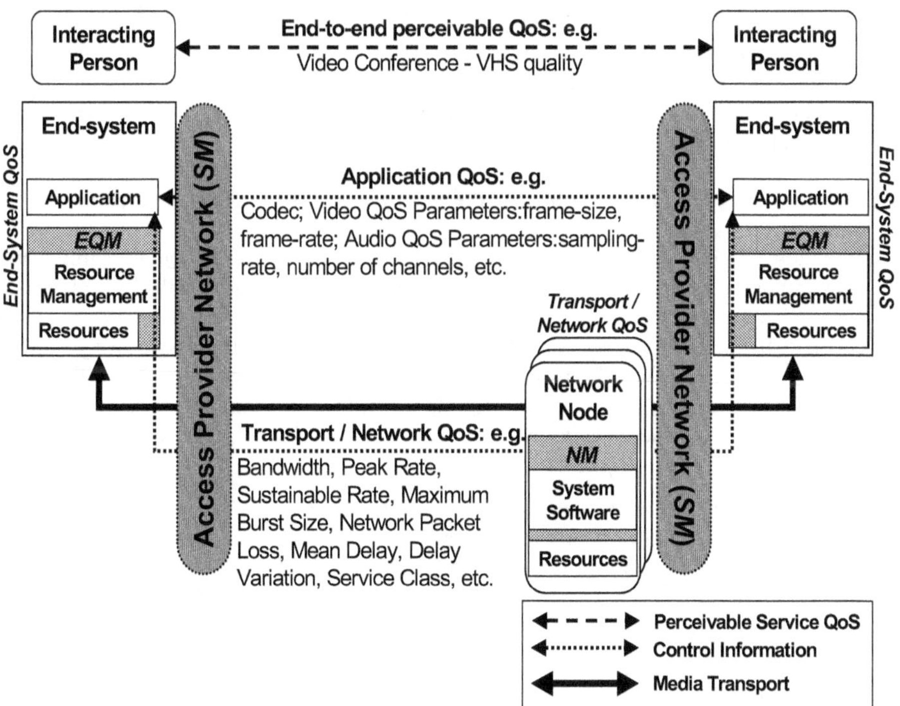

Figure 7: A reference model for identifying the relevant QoS parameters

Consequently, the actual implementation of a peer role for the negotiation is situated in the EQM, which can be a part of the application or of a middleware implementation at the terminal and the SM represents the actual implementation of the mentor functionality of the provider in the access networks (see also Section 2.4, where integration of the negotiation entities into existing architectures is discussed). Figure 7 below shows the location of the EQM and SM roles within a model for identifying the negotiated parameters between the peers and the mentoring entities, in consideration with the resources utilization within the terminals and the network. Furthermore, Figure 7 displays the role of the Network Management (NM) as the entity that defines constraints for the negotiation, which constraints are enforced by the SM at negotiations.

The presented considerations lead to the definition of a simplified architecture for end-to-end application interactions that takes into account only the negotiation parties and the negotiated information for the applications (Figure 7). In this end-to-end QoS negotiation and management architecture, the roles of the negotiation peers are the end-systems (i.e. the implementation of the EQMs) and the roles of the negotiation mentors are the application-management service proxies in the access network (indicated in Figure 7 as Access Provider Network and implementing the SM role).[7] Consequently, a set of conceptual models for the automated end-to-end QoS negotiations for the multimedia applications can be derived (see also [20] and [132]):

- A *Reference Model* presents relevant QoS parameter categories (Section 2.2.1).
- A *Description Model* (Section 2.2.2) provides a formal definition of system parameters and their associations according to the categories identified with the Reference Model along with additional management data for optimising the descriptions and the manageability of the model in terms of negotiation protocol.
- A *Negotiation Model* defines how the distributed multimedia systems are configured on an end-to-end basis (Section 2.3) and what scenarios shall be considered for advanced QoS adaptation, which may include combinations of multiple end-to-end negotiations (Chapter 3).

2.2.1 A Reference Model for End-to-End QoS Coordination

This section presents the Reference Model (see Figure 7). The model distinguishes between applied abstractions of distributed mobile multimedia services that are identified through the models presented in Section 2.1 (see Domain Model and Aspects of QoS Provisioning). These abstractions incorporate the complete process for achieving QoS on an end-to-end basis. The model helps to identify parameter classes in consideration with existing standalone QoS management technologies (see also beginning of Chapter 1 and Sections 1.1 and 2.1). These classes are later applied to specify an accurate description and management model for QoS (see Chapters 3 and 4).

In the following, several categories of QoS parameters are identified, considering the different aspects for QoS provisioning, the presented model about the User, Service and Transport Views on the application and the layered model for the QoS guarantees (see

---

[7] The terms end-device, end-system, end-point, terminal or peer are synonyms, used interchangeably throughout this work to indicate the negotiation parties that are interested in the results of the negotiation. Further synonym terms are provider, access provider, access network that are used to indicate the mentors in the end-to-end negotiation scenarios where provider network infrastructures are involved.

Section 2.1.2). Similar categories or parameter specifications are also described in works about management of QoS of multimedia applications [16][17] and in contributions about multimedia perception of humans [134][138]. The identified groups of parameters are:

- **End-to-End perceivable QoS parameters** – This set of parameters corresponds to the user's perception of the performance of the distributed application. The user may personalize the presentation features of the service in a *descriptive* style applying a user interface or profiles (e.g. "Video Conference with VHS quality" – Figure 7). Such parameters are typically utilized for guiding the user how to specify QoS in her/his natural understanding [134] and the translation of the perceivable QoS characteristics in more technical terms is implemented inside the application.

  The utilization of the end-to-end perceivable QoS as negotiable data is difficult considering the current state of the development of multimedia streaming applications. A precondition for the application of this class of parameters as negotiable data is that a standard definition and a framework for mapping between Application and Transport QoS (see below) is available, in addition to comparison statistics between all applicable codecs for presentation of a specific media type. However, such comparison statistics and mapping definitions are rather scarce and fragmentarily described in the literature (see [134][135][138][141]). Only the ITU standards (e.g. audio coding standard series G.72x [152] – [158] and video coding standards H.261 [159], H.263 [160] and H.264 [161]) provide application to transport performance mapping that can enable also cross comparison between codecs of the ITU series. Within the scope of MPEG-7 [109], XML descriptions (named *Classification Schemes*) on how codec profiles shall be defined are provided (see also Section 4.1.3.3). However, the current state of these MPEG-7 definitions represent requirements for how codec parameterisations shall be described rather then real statistics about the codec performance. Hence, the *End-to-End perceivable QoS parameters* are currently not appropriate for the definition of the negotiation language for QoS as the knowledge how to apply such parameters in consideration with different codecs and application-to-transport parameters mapping is not available. As a result, the application-to-transport performance mapping shall be applied explicitly in the negotiation language (see Section 2.2.2), especially for codecs that are in position to be flexibly re-configured to the respective transport conditions thus being able to utilize one and the same application configuration with different transports and vice versa (e.g. this is the case of dynamic video codecs, like WAVI [17][162], MPEG-2 [163] or MPEG-4 [164]). The *End-to-End perceivable QoS parameters* are considered in this work only as indication about what parameters are necessary for proper building of User Interfaces (see also Section 2.3.2.2.2) and in case that exact profiles for mapping between application and transport parameters are available (see also Section 4.1.3).

- **Application QoS parameters** are used to describe end-to-end application performance in accordance with software and hardware resources of the user terminals (e.g. in VoIP scenarios) and the end-systems that are automatic services (e.g. VoD). End-system-specific resource parameters are memory, CPU, battery power, etc. Such parameters are not negotiated, but may be applied to generate specific end-system policies for the terminal's resource-reservation management. Such policies can be used within the application to generate correct protocol payloads, when a multimedia session is opened or when an adaptation condition for the session occurs (see Section 2.3.2.2.2). The terminal application communicates to its partner only the results of an adaptation condition, the actual decision taking rules, techniques and algorithms do not need to be negotiated as each device and application in a heterogeneous environment takes its decisions on its own in consideration with the current environment conditions, which are in general different and specific for each device and its applications (the reasons for this individual behaviour of the negotiating devices are further detailed in Section 2.3.2.2). The *Application QoS parameters* class considers the media coding techniques (i.e. the audio or the video codecs, etc. – see Figure 7) and their parameterisation capabilities (e.g. Typical parameters for video are frame size, frame rate, visual quality) as common information to be negotiated. These QoS parameters are negotiated between the peers for coordinating QoS end-to-end in form of QoS contracts at application level using the language of the negotiation protocol (see Sections 2.2.2 and Chapter 4). The end-systems may specify and negotiate multiple alternative configurations of application QoS in order to efficiently adapt in cases of network congestion or handover between different access technologies and thus be prepared ahead of the congestion to cope with the emerging restrictions. Compared with multiple mechanisms presented in the literature [16][17][35][36] that react spontaneously to QoS changes without being in position to plan, the planning of devices reaction through the negotiation of future QoS variation possibilities before or at the very beginning of a multimedia session is a new concept presented here. Unplanned reactions to QoS changes during the execution of the service may lead to instability or disruption of the service as the spontaneously chosen reactions may contradict to user and/or provider application constraints. The access providers may participate in the negotiation process of the end-systems and could use the application QoS contracts for admission control (e.g. to restrict the usage of certain codecs or limit their performance, if the transport parameters of the codecs do not correspond to the user's profile – see also Section 2.3.2).

- **Transport/Network QoS parameters** are used to describe end-to-end requirements of the system in terms of network resources like bandwidth, delay, etc. (see Figure 7). The system must derive these values based on: the actual capabilities used (multimedia codecs) and their specific QoS configurations, the media characteristics (e.g. amount of video motion for variable bit-rate video), the available network access-technology (e.g. UMTS, WLAN, Ethernet, etc.) and the transportation optimisation technologies in use (e.g. FEC or ARQ, or local buffering at the terminals). End-systems can negotiate these parameters and apply them as an input for their network resource reservations. However, provider entities acting as negotiation mentors may also use these parameters for admission control or for negotiating and enforcing network QoS on behalf of those end-systems which cannot or are not allowed to explicitly perform resource reservations on the network.

- **Static configurations** – In addition to all the QoS parameters described above and in order to set up a valid end-to-end multimedia session, end-systems must specify connection points for communication (e.g. IP-addresses and ports) and other fixed presentation capabilities for the media (e.g. media packetization rules, transport optimisation technology in use, etc.)[8].

- **Configuration identifiers** – The negotiation protocol should be in position to identify uniquely the different levels of capabilities and QoS information in order that the negotiation peers are able to univocally address these levels in their QoS adaptation procedures and the negotiation mentors are able to recognize the adaptation levels in

---

[8] As long there is just a single well-known technology in use, the respective parameters of the application are not negotiable. For example, this is the case of RTP [1] for multiple multimedia-streaming services.

order to enforce restrictions for the adaptation process or to assist the adaptation process from within the network (see also Section 2.2.2). A definition similar to the subsequent description can be found in [17], however, the definition in [17] considers only the view of a single application and its management and the internal realisation of QoS management in just a single end-device. Another analogous concept about the configuration identifiers is available within the scope of the common usage of SIP and SDP [35][62][165], however an explicit definition like the following is not expressed in the literature. This work proposes as univocal identifier for the to-be-adapted multimedia streaming services and their negotiation information the tupel:

(*User*, *Application*, *Device*, *Session (Media and Negotiation Statement)*, *Association of Streams*, *Stream*, *Flow*)

The single identification levels have the following meaning for the negotiation procedure between the peers:

- **User** – The user has to be identified in the negotiated information as a peer or as a mentor of the negotiation process, i.e. the identity of a physical person or of a legal body (e.g. a telephone company) shall be explicitly recognizable. Different end-system users may apply for different service usage profiles that are associated with different communication-framework utilization rules, e.g. a user may be associated with different profiles that represent the utilization of different terminals (like PC or PDA) or of the same terminal for different purposes (e.g. business or private). The information about the user's identity allows thus the provider to recognize the user for A4C purposes. A provider mentoring the negotiation process shall be recognizable by the users with the provider's identity to avoid malicious involvements in the negotiation process (see Sections 2.3.2.2.2 and 3.2). Current realisations of the *User* identifier correspond to the username part of the address of the control-data carrier and such an address can be a combination of a user's name and a password (e.g. SIP address [62]).
- **Application** – Different types of applications have to be differentiated in order to apply different local (i.e. at the terminals) and network prioritisation rules according to the specific requirements of the services. For example, the delivery delays through the network can be used as a prioritisation criterion [137][138]. The sensitivity of the services towards interruptions of the network connections depends on the type of the service. The non-real time services are almost not sensitive to interruptions (e.g. at network congestions or handovers). The sensitivity of the real-time pre-recorded services (e.g. Video on Demand) depends on the service ability for buffering [166]. The most sensitive services are the real-time live services like VoIP (e.g. delays above 200ms are rated as annoying [138]). The identification of the application is also used to differentiate between QoS-aware and best-effort services. The identification of the service is usually applied through the used control-data protocol (e.g. HTTP [167] or SIP [62] User Agent fields).
- **Device** – The identity of the device is its IP address [119] – [121] (for IP-based networks) or the telephone number. Note that some technologies do not differentiate explicitly between the *User* and the *Device* identity, e.g. SIP [62] uses in general the *User* identity and only in its combination with SDP [86], device identities are revealed. The identification of the devices is necessary for providing completeness for the static configurations of the applications (see above). Furthermore, in mentoring scenarios, the device issuing the requests/responses for a negotiation might not be the finally affected device from

the results of the negotiation (see Section 3.3). For example, the negotiation and the application streaming might be performed under the same user identity from different devices in scenarios like session migration. Hence, the differentiation about who is negotiating and where the application actually would run is made upon the *Device* identity.

- **Session** – The identification of a session is made due to two reasons and it corresponds to two kinds of session identifiers. First, the information about the multimedia session have to be kept consistent throughout the whole session in order to perform proper interaction with the system's resource management at session establishment and adaptation (see Section 2.3.2). In this case, the identifiers are applied to recognise different negotiation statements and correspond to protocol PDUs (portable data units) of the End-to-End Negotiation Protocol (see also Sections 3.1 and 4.3.2). This identifier is the *Session Negotiation Statement.* Secondly, the sessions associated with the same application can be prioritised, e.g. a session with higher importance for the application may interrupt a session with lower importance as the user may set up prioritisation rules like "a call from my boss interrupts all other calls". The identification of the media session is also necessary in order to address the adaptation process for the complete session. This second identifier is the *Session Media.*

- **Association of streams** – The association of streams corresponding to different media types or media contents has to be identified in order to apply media interlacing rules, e.g. lip-synchronization between a video stream showing somebody talking and the corresponding audio output [165] or video and subtitles simultaneous play-out for hearing-impaired persons [72].

- **Stream** – The identification of a media stream is associated with the application of QoS restrictions for a single media type. It is necessary for identifying different configurations of the same media type as QoS-configuration alternatives of the single medium (see Section 2.2.2.2), e.g. video coded with different codecs or with different configurations of the same codec. Concurrent media types are also differentiated upon the stream identifier.

- **Flow** – The flow is a part of the stream. Depending on the type of the multimedia codec the stream can contain one or multiple flows (e.g. in case of using multi-layered codecs). The identification of a flow is necessary only in case of multiple flows when they are associated with different communication access points, e.g. in SDP it is possible to define different ports for the flows of a media [86].

- **Security Identifiers** – These identifiers are applied in cases that mentors interact with the end-to-end negotiation of the peers to detect the authenticity of the mentors when the mentors manipulate the data of the negotiating partners so that malicious third-party interactions can be avoided. The application of the security identifiers in the negotiation protocol is detailed in Sections 2.3.2.1, 2.4, 3.2 and 4.3.6.

Considering, the different classes of parameters and identifiers a conceptual model for the negotiation language is developed (see Section 2.2.2) that is later used for specifying the negotiation language of the End-to-End Negotiation Protocol for QoS (E2ENP – see Chapter 4).

```
E2ENP-URI        = "e2enp://" [ userinfo "@" ] hostport [path]
userinfo         = [ user ][ ":" password ]
user             = *( unreserved / escaped / user-unreserved)
password         = *( unreserved / escaped / "&" / "=" / "+" / "$" / "," /
                   "?" / "#" / "[" / "]") / "'" / "(" / ")"
user-unreserved  = "!" / "$" / "&" / "*" / "+" / "," / ";" / "="
escaped          = "%" HEXDIG HEXDIG
unreserved       = alphanum / mark
mark             = "-" / "_" / "." / "~" / "§" / "<" / ">" / "|" / "\" / "^" / "°"
alphanum         = ALPHA / DIGIT
hostport         = host / host ":" port
host             = hostname / hostaddress
hostname         = *( domainlabel "." ) toplabel [ "." ]
domainlabel      = alphanum / alphanum *( alphanum / "-" ) alphanum
toplabel         = ALPHA / ALPHA *( alphanum / "-" ) alphanum
hostaddress      = ipv4-addr / ipv6-addr
ipv4-addr        = digits"."digits"."digits"."digits
digits           = 1*3DIGIT
ipv6-addr        = hexpart [ ":" IPv4address ]
hexpart          = hexseq / hexseq "::" [ hexseq ] / "::" [ hexseq ]
hexseq           = hex4 *( ":" hex4)
hex4             = 1*4HEXDIG
port             = 1*DIGIT
path             = "/" application "/" sessionid [param]
application      = *( unreserved / escaped / user-unreserved)
sessionid        = DIGIT
param            = *("/" segment)
segment          = *( unreserved / escaped / user-unreserved)
reserved         = gen-delims / sub-delims
gen-delims       = ":" / "/" / "?" / "#" / "[" / "]" / "@"
sub-delims       = user-unreserved  / "'" / "(" / ")"
```

Figure 8: E2ENP URI syntax

2.2.2 Description Model

The basic idea behind E2ENP is the possibility to actively negotiate multiple system configurations (i.e. application- and transport-level QoS, and system capabilities) for multimedia session establishment and to flexibly classify this information for QoS adaptation purposes in a hierarchical structure. Furthermore, the E2ENP configurations can be repeatedly applied to sessions having similar context (e.g. session parameters can be saved in the form of an advanced address book and then reused). These two features correspond to the requirements for a successful negotiation that leads to stable results after it (see Section 1.2.1 and the respective considerations about *win-win* negotiations in [45] – [47]).

2.2.2.1 Protocol User Address Syntax

The E2ENP Unified Resource Identifier (URI) represents the identities of the *User*, *Application*, *Device* and *Session Media* as defined in the tupel of the *configuration identifiers* parameters class (see Section 2.2.1). The E2ENP protocol agent maps the URI to the specific syntax used by the given session protocol to piggyback E2ENP information, e.g. SIP [62], RMI [88] or CORBA-IIOP [74] (see Section 2.3). The E2ENP addresses are agnostic of the specific session protocol and the possible transport protocol used underneath by the E2ENP protocol agent (see Chapter 5). Furthermore, the E2ENP address is also applied for security reasons as identifier of mentors that enforce specific performance restrictions to the application (see Sections 3.2, 3.3 and 4.3.6). Thus, a new URI syntax for E2ENP based on the classical URI scheme [168] and in ABNF format [169] is proposed (see Figure 8)[9].

Example of the E2ENP URI:

e2enp://dave:vG$l809@acme.com/davesApplication/sessionID

Example of possible E2ENP URI mapping to SIP URI [62]:

From: sip:dave@acme.com

Call-ID: 12345-sessionID

Authentication-Info: pw="vG$l809"

User-Agent: IP phone (davesApplication)

Note that the E2ENP User Agent implementation may add some information to complete the identification requirements of the carrier protocol (see the E2ENP UA implementation in Section 5.2).

The respective full or partial usage of the E2ENP address components depends on the domain model (proxying) and the address resolution mechanisms, which are mainly part of the session layer protocol. The E2ENP protocol agent uses at least the host names to identify the communicating partners, if the session layer protocol is already offering some proxying and/or redirection mechanisms for identifying the network paths.

The E2ENP URI is only an identifier for QoS features at *User*, *Application*, *Device* and *Session Media* levels of the services. These levels symbolize the application of constraints for the overall performance of the service that restrict the general performance (expressed with the simple QoS contracts, see below) of the same service. E2ENP assumes that an adaptation of the description with application of specific environmental conditions has taken place before the final session description used in the negotiation is generated (see also Section 2.3.2 and Chapter 4) and that the corresponding parts of the description are properly associated to their respective QoS levels to build a hierarchical model (see also Section 2.2.2.2 below). The generated virtual environment for the session implemented through a hierarchical QoS specification is then the actual/real environment for the session performance and it is described using only named associations of the simpler, lower levels of the hierarchy represented by the configuration identifiers tupel (see Section 2.2.1 – *Session Negotiation Statement*, *Association of Streams*, *Stream* and *Flow*). Thus, the negotiation language of E2ENP applies predominantly descriptions of simple QoS configurations and it does not describe the methods and techniques how a session establishment or adaptation decision is taken, as this information is not relevant for the communication process of the negotiation (as suggested also for human negotiations, the revealing of personal positions at negotiations does not lead to a fruitful negotiation [45], see also the extended discussion on this problem in Section 2.3.2.2). Consequently, this E2ENP feature is especially important for the negotiations between heterogeneous devices as a mobile terminal with low processing power may utilize simpler decision algorithms while a complex conferencing server may use more sophisticated decision techniques, still they shall be able to understand each other at negotiation and be capable to establish/adapt multimedia sessions between each other.

---

[9] The initial E2ENP URI is published in co-authorship from the author of this work as European patent application [170].

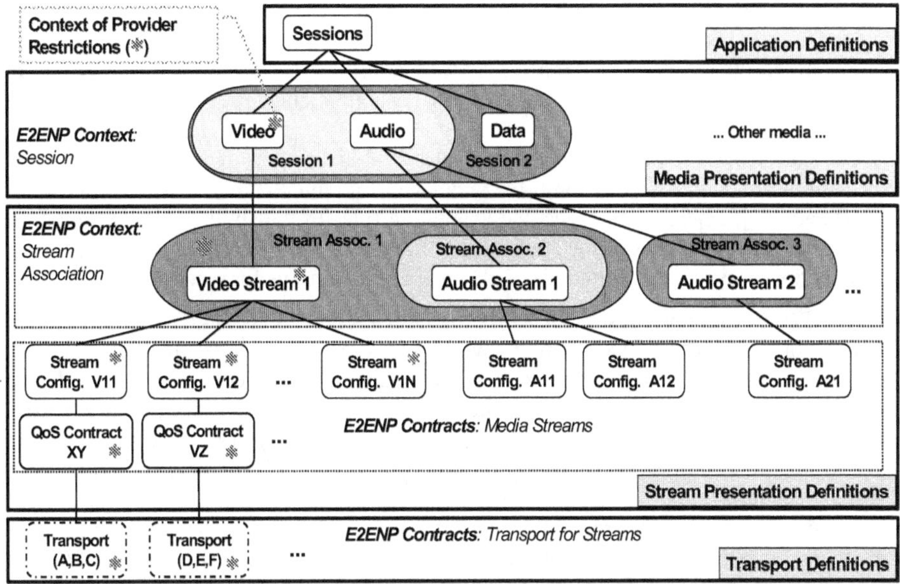

Figure 9: Example of hierarchical QoS specification

2.2.2.2 QoS Hierarchy

E2ENP defines its own description model using a hierarchical QoS specification that enables the management of multimedia systems at different abstraction levels (e.g. media stream, groups of streams, media session, media applications). The upper levels of the tupel representing the hierarchy were described in the previous section (Section 2.2.2.1). In the following, the identities of *Session Negotiation Statement*, *Association of Streams*, *Stream* and *Flow* together with their associated data are discussed. These identities of the E2ENP QoS hierarchy can represent interchangeable descriptions between applications, i.e. the same configurations of a multimedia session can be applied for different multimedia services as long as the context of the services is the same, e.g. different VoIP applications or upgrading versions of the same application may utilize the same QoS descriptions thus satisfying the requirement for heterogeneity support of the services both for configuration (e.g. upgrading or a new installation of an application may reuse already known configurations) and for negotiation optimisation proposes (e.g. a configuration applicable for multiple applications can be saved and only indicated as *in use* when appropriate – see also the E2ENP negotiation optimisation in Section 2.2.2.3).

Mobile multimedia applications can manage multiple media types (e.g. audio, video, data) independently of each other or simultaneously, synchronizing the information of different sources and with different contents. Each media type represents in the general case a media stream. The single streams can be logically grouped based on various criteria, thus building an association of streams. This bundling of streams allows assigning priorities that the services may then use to select the most appropriate QoS adaptation strategy (e.g. private conversations shall be downgraded to sustain business conversations, in case of limitations in the resource availability, or video streams shall be dropped and only audio shall continue in case of network congestions). The grouping of streams allows assigning priorities among different associations and defining appropriate adaptation strategies within the context of a single session. Hence, the QoS specification describing the parts of the session can be conveniently modelled in a hierarchical manner to cover features like: time synchronization, QoS correlation and resource distribution aspects among streams. A hierarchical model allows designing alternative QoS specifications, which the application needs for automatically adapting its resource usage with respect to the current resource availability and the rules and limitations representing the user's expectations and the provider's multi-user management policies. A similar model is shown in [17] and [70], where it represents the adaptation rules of the application. This model is partially reused and enhanced for the formal definition of the E2ENP description language to enable the specification of potential and actual configurations for services.

The E2ENP model consists of a tree of QoS specifications (Figure 9). These QoS specifications are defined at different levels of abstractions representing different *QoS contexts*. A *QoS context* contains predefined adaptation rules that steer the adaptive application to choose the most appropriate QoS specification upon given resource usage conditions for a group of media types. In terms of the E2ENP descriptions, a *QoS context* defines QoS correlation rules among media streams. For example, a QoS correlation can be a time-synchronization constraint on subordinate QoS specifications, i.e. time synchronization constraints are required in a movie play-out among the audio streams and the mouth-movements of the actors in the video. The *QoS context* corresponds to the *Association of Streams* identifier. If given a univocal name a *QoS context* can be differentiated from another *QoS context* in a negotiation process, thus allowing the recognition that different correlation rules may be applicable or shall be applied for the streams of a session (see the following example of a QoS configuration for a session below and Chapter 4). Each leaf of the tree is associated with a QoS specification named *QoS contract*. It represents the application and the transport parameters specification (see Section 2.2.1) for the given stream that determines a unique configuration of this stream. An application *QoS contract* is associated with a specific transport *QoS contract* to build one-to-one correspondence between application and transport QoS definitions. This is necessary on one side for the univocal local resource management of the application, on the other side for the accounting of the traffic through the provider infrastructure as different transport specifications are usually associated with different billing levels [68], [116] – [118]. The E2ENP model provides explicit application to transport parameters mapping, as a QoS configuration of a codec may correspond in some cases to different transport possibilities depending on the applied streaming and transport optimisation mechanisms of the service. The *QoS contract* corresponds to the *Stream* identifier. If given a univocal name, a *QoS contract* can be differentiated from another *QoS contract* in a negotiation process, thus allowing the differentiation between streams and between different configurations of the same stream (see the example below and Chapter 4). The *Flow* identifier corresponds to the association of a flow with a univocal port configuration in the static stream configuration (see Figure 9). The E2ENP model does not specify QoS for the flows as this specification is not available for some codecs and the management of the flows is specific for the performance algorithm of the respective codecs. However, if it is possible to define flows for the codec and these flows shall be streamed as separate sources, port specification of the flows is necessary and E2ENP provides such port identifiers for the flows (see Section 4.3.3).

The E2ENP incorporates also the context of an *adaptation path*. This path corresponds to any alternative configuration of a stream or of a stream association starting at the root of in the hierarchical tree and ending at a *QoS contract* or a *QoS context*. Thus, switches between different alternatives of adaptation can be described with the E2ENP language (see also Section 2.2.2.3).

Additional novel concept in the E2ENP description model is the association of provider restrictions with the *QoS contracts* and *QoS contexts*. This is especially important for the maintenance of the *QoS contracts* and *QoS contexts*, when the terminals are roaming within different provider domains (see Sections 2.3.2 and 3.2, and the example below).

Example of a QoS configuration for a session is (Figure 9):

Any given sub-tree originating from a specific branch node is associated with an adaptation-rule predicate. The resolving of this predicate selects a child node and hence instructs the system to enforce the QoS contract and/or QoS context associated with that child. For instance, "if Video Parameters V11 are no longer enforceable, switch to Video Parameters V12" or "if Stream Association 1" (Video and Audio) is not supportable (e.g. due to handover and thus lower data rate availability) switch to "Stream Association 2" (only Audio)". In the latter case, the adaptation process translates in a change of QoS context or, more simply, in a QoS context switch that is similar to the context switch concept as defined within the scope operating systems. The providers may mark (The mark is indicated in Figure 9 with an asterisk) the contexts and the contracts as not applicable in the respective providers' domain, if the proposals of the peers do not correspond to the providers network optimisation policies for the specific network segment of attachment of the terminals.

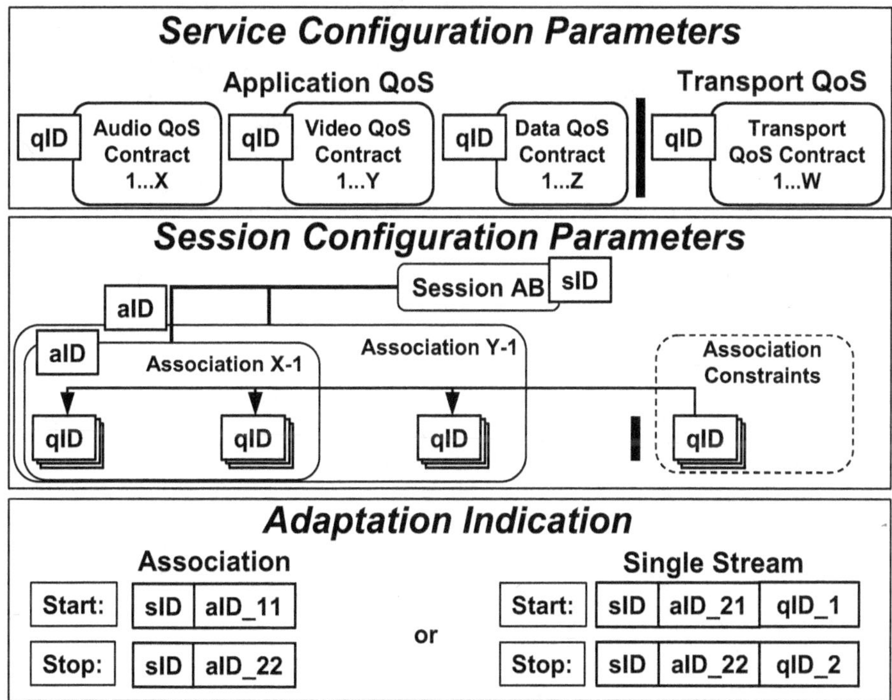

Figure 10: E2ENP referencing model

It is completely up to the adaptive application and its business logic to determine when to adapt (e.g. if packet loss ratio is between 0.1 and 0.5 % for 2 seconds), how to adapt (e.g. switch codec or remove video stream) and to what extent to adapt (reduce frame rate from 25 Fr/s to 10 Fr/s, where Fr/s=frame per second). E2ENP provides the negotiation mechanism so that peers can agree on such adaptation conditions and procedures and prepare resources and capabilities beforehand (as potential and actual configurations), based on the E2ENP tree model. A novel concept is the negotiation of the hierarchical QoS specification between the peers into multiple single negotiation statements. The statements correspond to the *Session Negotiation Statement* identifier in the E2ENP identifiers tupel (Section 2.2.1). The exact usage of the single negotiation statements is detailed in Section 3.1 (see also Section 2.2.2.3).

An example that combines adaptive applications and E2ENP to reconfigure the streaming process based on QoS feedback was presented in [65]. The concepts concerning the E2ENP hierarchical QoS specification and its application for QoS signalling optimisation (see Section 2.2.2.3) were published in co-authorship from the author of this work in [14], [131] – [133], [171] – [173]. The complete specification of the E2ENP description language in XML format is presented in Chapter 4.

2.2.2.3 Optimisation of the QoS Descriptions for Negotiation Purposes

In current systems, a complete set of session-configuration information is exchanged every time a session is negotiated or adapted (e.g. Offer/Answer Model with SDP [35], 3GPP IMS [21][22], ITU H.323 [107] or AEQUA [17]). This scenario is not efficient as the minimization of the signalling traffic is crucial especially at adaptation of running sessions. Furthermore, a session description containing advanced QoS specifications might cause additional processing impact for the negotiations at session establishment time, due to the increased complexity of the negotiated information when including the advanced QoS parameterisation. Thus, a novel concept for efficiently negotiating multimedia session configurations is introduced: *E2ENP defines an optional referencing mechanism to minimize data exchange and it gives the possibility to build a hierarchical QoS specification out of multiple, single negotiation statements of the peers. Each statement is associated with an identifier.* For this purpose, E2ENP distinguishes between service configuration, session configuration and adaptation indication (Figure 10) and associates configuration information with an identifier that corresponds to one of the levels of the E2ENP tupel (see Sections 2.2.1, 3.1 and 4.3).

Each of the contracts/contexts associated with the nodes of the QoS hierarchy (Figure 9) is labelled with a specific name (e.g. qID, aID, sID – Figure 10), which is used by the negotiating partners to address the correspondent contracts/contexts during some following QoS negotiation process. By exchanging only identifiers, every subsequent communication between the peers has minimized traffic and processing requirements.[10] When using the hierarchical QoS specification as a memory for possible adaptations, the adaptation process can be addressed at various levels of granularity. The application of a hierarchical QoS specification as a memory for possible QoS adaptations is a new concept here. In the literature [16][17][70][86], hierarchical QoS specifications are used predominantly for providing a snapshot of the current state of the complete system resource utilization for taking short-term management actions and for resources reorganisation between the system processes, but not as a memory for planning future negotiations and adaptations.

---

[10] This feature of the protocol is proven through tests (see Chapter 6).

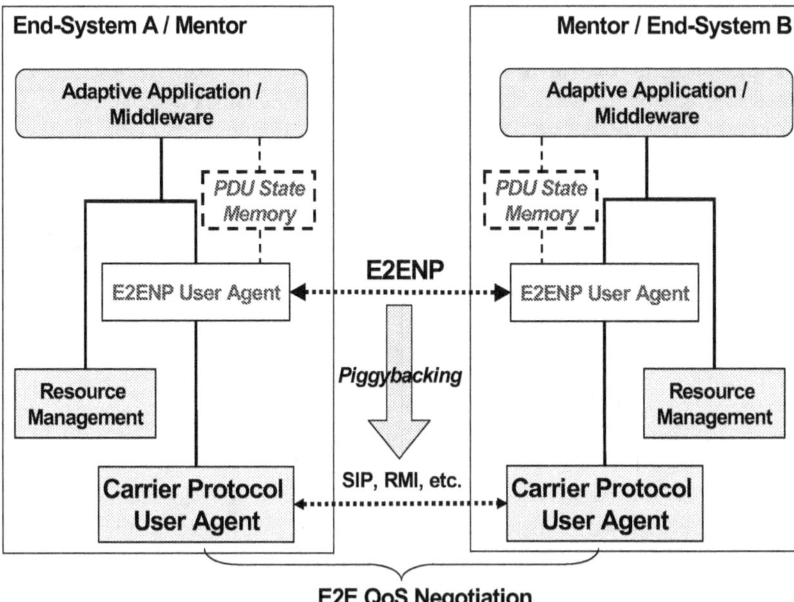

Figure 11: E2ENP basic negotiation architecture

The E2ENP *Referencing Model* defines specific description and management for the session-control data. E2ENP descriptions provided via this model are applied as configuration and control commands for managing multimedia sessions. Peers can exchange this information using any session management protocol (e.g. SIP [62], see Section 2.3 and Chapter 3). Examples on the E2ENP referencing mechanism are also provided together with the XML specification of the E2ENP language in Chapter 4 and Annex C.

2.3 Negotiation Models for System Configurations

This section details the *communication arrangement* principle (see beginning of Chapter 2) and it demonstrates the negotiation models utilized for implementing the End-to-End Negotiation Protocol concept. Existing and reused mechanisms that affect the E2ENP application are discussed, deriving requirements for the protocol and presenting the innovations necessary for the E2ENP implementation. Two major concepts of the E2ENP negotiation model are introduced: The organisation of the negotiation process between the end-systems (see Section 2.3.1) and the coordination of the resources reservations both in the terminals and on the network for providing consistency of the negotiated information (see Section 2.3.2).

Figure 11 shows the basic negotiation architecture of E2ENP. The system applying this architecture can be either a terminal or a mentor device. The logic of the negotiation decisions based on resources availability is a part of the adaptive application (or middleware). Therefore, it is up to the adaptive application to decide about and to execute the exact model being negotiated with the E2ENP (see also Sections 2.3.2.2 and 3.1). The protocol itself provides the negotiation language for creating hierarchical QoS specifications, for indicating changes in the resource availability, for performing negotiations at different modes and applying different phases (see Sections 2.2 and 3.1 and Chapter 4). Additionally, the E2ENP User Agents assure information synchronization between the parts of the adaptive application in order to insure consistency of the negotiated information between the peers and to provide deadlock avoidance (see Sections 2.3.2, 3.1 and 5.2). In cases that the application is in position to develop long-term memory about the exchanged configurations, the E2ENP UA provides partial memory support about which Portable Data Units (PDUs)[11] of E2ENP were already exchanged between the distributed parts of the application (see Section 3.1.4). However, the actual management decisions affecting this memory (e.g. PDU expiration, PDU re-validation, etc.) belong to the adaptive application/middleware and not to the E2ENP UA (see also Section 2.3.2.2.2).

The E2ENP User Agent is designed to piggyback the end-to-end negotiated information on protocols for session establishment and management or on other remote-call mechanisms as discussed in the next section (see also Sections 2.2.2.1 and 5.2).

2.3.1 Communication of the Negotiated Information

The following two sub-sections show the organization of the end-to-end negotiations with E2ENP and the applied existing technologies that are re-used to carry out the communication process between the negotiating peers.

2.3.1.1 Carriers of Capabilities Description

The idea about piggybacking device and session capabilities information on session level protocols is applied for several protocols developed within the scope of the IETF research and standardization efforts [174]. The session descriptions are provided as a separate piece of information and in a different format than the session control information. This is the case of the Session Description Protocol (SDP) [86] and its successor Session Description Protocol next generation (SDPng) [81] that can be piggybacked on protocols like SAP [82], SIP [62], RTSP [83], MGCP [84] or Megaco [85]. The IETF bearers of session descriptions are designed to fulfil different tasks, and consequently, they display different suitability as service and session negotiation protocols:

- Session Announcement Protocol (SAP) [82] – This protocol was designed to announce session configuration information to a potentially large audience, e.g. conferences or multicast services. The session configurations announced by SAP are described in SDP [86]. Initially SDP was designed for SAP only [69]. The combination of SDP and SAP provides only one fixed configuration per user and/or device accessing the service. This concept is similar to the *Yes/No Signalling* described in Section 1.2.2.2.1 and thus it is not suitable for negotiations. However, it is worth mentioning it to reveal the historical backgrounds of the SDP concept and to explain why SDP is developed predominantly for static configurations and why its extensibility model is so restricted. Because of SDP's inflexibility, a new more flexible and extensible description form is required for session configurations, and its development has been carried out under the name SDPng (see also [69][81] and the SDPng description in Section 1.2.2.2.3). The current work on SDPng is dedicated predominantly for covering scenarios that are more advanced compared to the mere session announcement and that can be implemented with the following three session protocols.

---

[11] Within the scope of the negotiation process with E2ENP, a PDU corresponds to a negotiation statement made by a single negotiating party.

- Real-Time Streaming Protocol (RTSP) [83] – RTSP is used for controlling non-interactive streaming sessions, unlike SIP that can support also other multimedia services (see below). Since RTSP is restricted to a specific type of sessions, it is not suitable as a general negotiation protocol for QoS. However, if the description format of the negotiation language is also applicable for RTSP, specific RTSP-only negotiations can also be performed. Nevertheless, it should be mentioned that RTSP follows the *Raw-Granularity-Signalling* concept (see Section 1.2.2.2.2) hence its extension as a negotiation protocol should be carried through the protocol piggybacked on it (see also SIP below).

- Media Gateway Control Protocol (MGCP) [84] and Gateway Control Protocol (Megaco) [85] are protocols applied to control multimedia gateways within the network. As the multimedia gateways are not typical service producers or consumers, they are also not typical negotiation peers but may take the role of mentors in the negotiation (see Sections 1.2.1 and 3.3.2). Since MGCP and Megaco are a special purpose protocols and since a media gateway is rather a mentoring entity in a negotiation, the description language for the case of MGCP and Megaco has to be able to carry such mentoring information as a support function of the negotiation (see Sections 3.3.2 and 4.3.6). Some approaches that are similar to MGCP and Megaco are shown also in [175][176].

- Session Initiation Protocol (SIP) [62] – SIP is used to establish and modify any kind of multimedia sessions. A description protocol (e.g. SDP, SDPng) is normally applied to deliver the parameters of these sessions. Applied with the Offer/Answer Model with SDP [35], SIP already provides a simple negotiation framework for system capabilities only (similar to the *Raw-Granularity-Signalling*" model shown in Section 1.2.2.2.2). As SIP is defined general enough to support multiplicity of media sessions and scenarios, it is the best candidate protocol for the realisation of E2ENP as general-purpose negotiation protocol, e.g. SIP is in position to perform not only end-to-end negotiations but it also provides communication possibilities for third-party call control [104] – [106]. To this extent, SIP is used in novel modes (see Sections 2.3.1.2 and Chapters 3 and 5) but its description structure and purposes remain substantially unchanged. E2ENP extends the basic SIP communication framework as a special purpose negotiation framework. E2ENP does not extend or modify the fields definition of the SIP protocol language, it only affects the SIP controls, respectively the SIP state machines (see Section 2.3.1.2 and Chapter 5), introducing several additional layers of control states between the SIP stack and the application. Consequently, the adaptive application is enabled to memorize previous control sessions and to plan future negotiation- and resource-management interactions ahead. The planning of interactions introduced through E2ENP is still an additional optimisation for the session signalling that can be combined with the short session descriptions (see Section 2.2.2.3 and Chapter 4) to minimize the volume of the negotiated information. These novel features of a piggybacking protocol not available in SDP or SDPng are introduced with the E2ENP concept and are presented in detail in Chapters 3, 4 and 5.

Further possibility for implementing and testing the E2ENP concept is the application of middleware for performing remote calls like RMI [88] and CORBA [74]. However, such general-purpose remote-call mechanisms are designed predominantly to provide abstraction of the connectivity mechanisms and to exchange data (e.g. streams, objects, etc.) between two points (i.e. Client and Server) [50][52][177]. Unlike the IETF protocols, the remote-call technologies do not provide particular mechanisms for performing special purpose communications like negotiations. Nevertheless, they can be applied for initial validation of the E2ENP concept and for testing of the E2ENP UA state machines as presented in Section 5.2 and Chapter 6. Remote-call mechanisms are used in the case of E2ENP as session protocol surrogates (e.g. RMI-based SIP surrogate – see Section 5.2 and Chapter 6) in order to easily emulate the E2ENP performance and to provide test possibilities in cases that the actual session protocol state machines display instabilities when integrated into the E2ENP User Agent (see Chapter 6).

Some additional reasons for giving preference to SIP as a carrier of the E2ENP messages are:

- SIP is a text protocol that can be interpreted using any programming language
- SIP provides its own security to the carried data (see also [62] and Sections 3.2 and 4.3.6).

2.3.1.2 Models for Organizing of the Communication

The E2ENP protocol agent reuses some of the negotiation features already available in the SIP protocol agent. In the following, the SIP protocol agent is explained and some initial discussion on how SIP is extended for the E2ENP negotiation purposes is given.

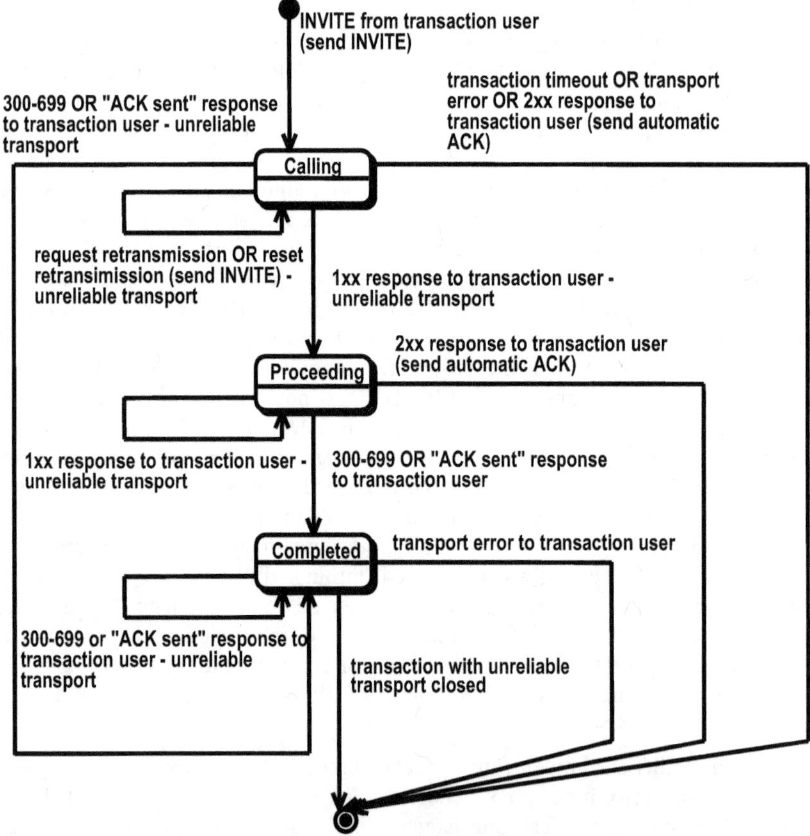

Figure 12: INVITE transaction – client (see also [62])

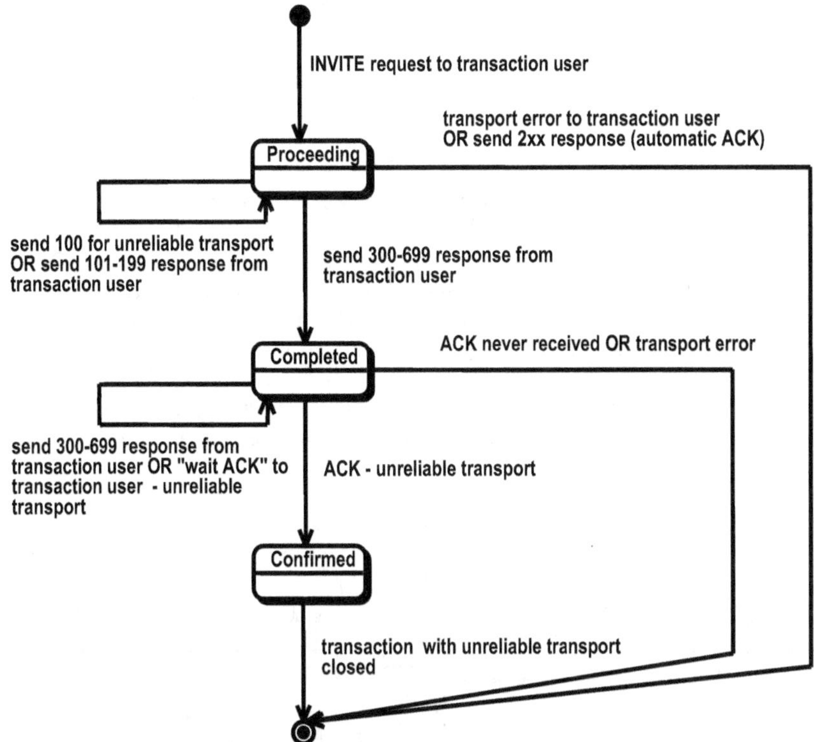

Figure 13: INVITE transaction – server (see also [62])

The SIP protocol agent consists of a protocol stack that is used to interpret the protocol language and of client- and server-side state machines (SM) [62][178]. The client SM corresponds to the party initiating a session management call and the server SM is the party processing the call. For E2ENP these SIP UA features build the basis for establishing the end-to-end communication. In a real implementation, the accesses to the SIP stack for protocol language interpretation and to the SM actions are represented through an application-programming interface (e.g. JAIN SIP API [178]). E2ENP reuses this API concept to also emulate the session protocol through a remote-call mechanism (e.g. RMI-based SIP surrogate). Furthermore, E2ENP UA applies partially the JAIN SIP API to link the E2ENP state machines to existing SIP protocol implementations (see Section 5.2 and Chapter 6).

The SIP specification defines the so-called methods (e.g. INVITE, OPTIONS, REGISTER, ACK, etc. [62]) that correspond to the required actions initiated by the client-side state machines. The responses to these actions are generated by the server-side state machines. These responses are classified into six groups that have respective numbered codes with 100 possible codes per group, i.e. the numbers between 100 and 199 are provisional responses, the numbers 200-299 indicate successful processing, responses 300-399 are used for redirection and the codes 400-699 represent different failures in reply to request failures or indicating server internal problems or network misbehaviour. A short designation of a specific response-code group is usually made using the major group number and the indication *xx*, e.g. *3xx* is the redirection group of responses. The major SIP specification [62] provides only a basic set of methods and responses. However, there are also extension specifications that present additional functionalities like resources-reservation coordination [36] or third-party call control management [101]. As SIP is a protocol under development, still further extensions are also possible in the future (see the respective IETF contributions done within the scope of the SIP management working groups named SIMPLE, SIP and SIPPING [179]).

The SIP specification differentiates between two kinds of transactions, i.e. dialogue building transactions (named INVITE transaction) and non-dialogue building (non-INVITE transactions).[12] The dialogue building transactions (Figure 12 and Figure 13) are applied for the case that the SIP UA shall remember what media sessions controlled through SIP are currently running, e.g. using the SIP Call-ID and CSeq header fields [62] the application indicates which session identifiers are already assigned to SIP and its controlled session/-s. The dialogue building transactions are applied for establishing and for intermediary management (e.g. adaptation) of the multimedia sessions established through SIP and the only method using this kind of transaction is INVITE [62].

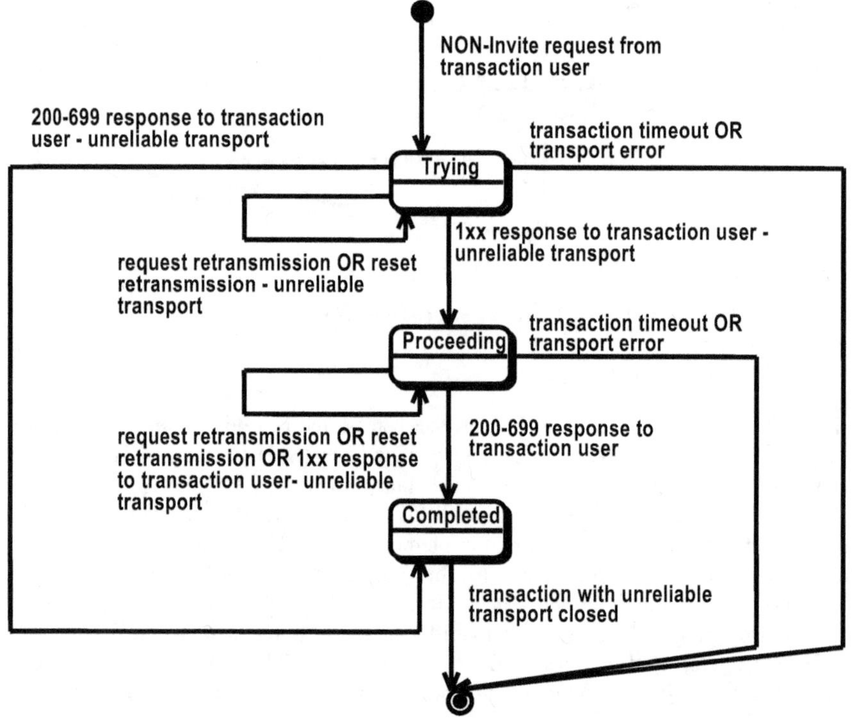

Figure 14: Non-INVITE transaction – client (see also [62])

---

[12]Although named transactions the SIP call management functions are not typical transactions as known within the scope of the OS [50][51]. This is because of the fact that the SIP transactions are not atomic, as they can be interrupted (e.g. using SIP CANCEL method [62]) before achieving final durable results. It is up to the SIP application then how to treat intermediary results.

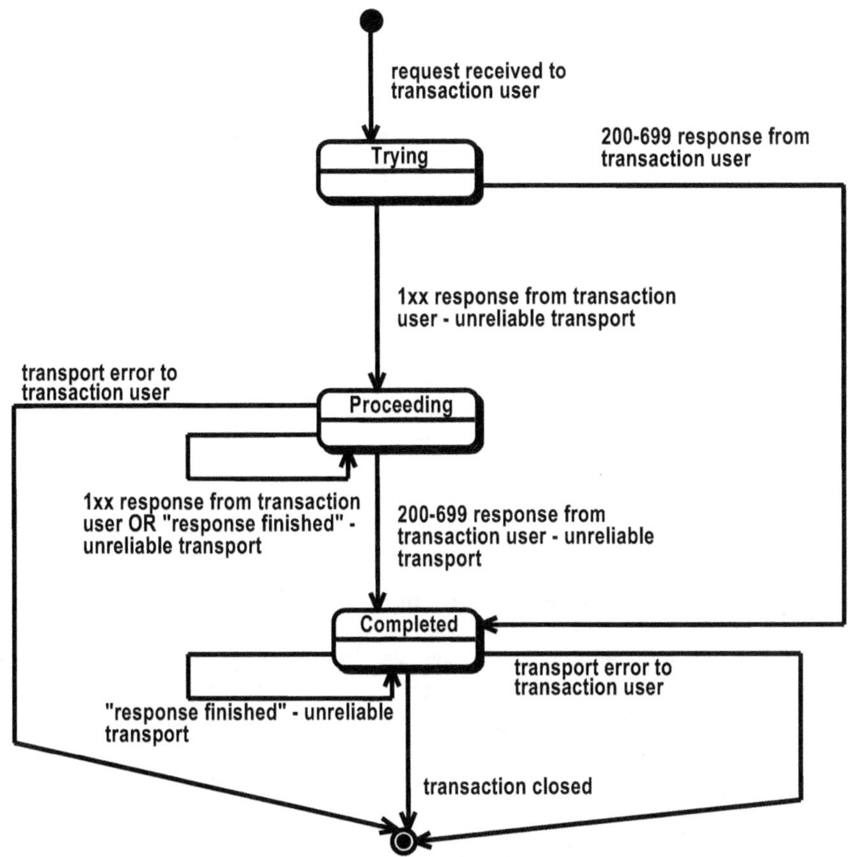

Figure 15: Non-INVITE transaction – server (see also [62])

The dialogue building transactions are the first step for providing negotiation management for the applications. However, SIP has no notion about the contents of the session descriptions carried as a protocol attachment. SIP merely controls the case if the multimedia session is already established or not, e.g. this is used for the scenario where a telephone application shall produce a ringing tone until the users' communication is established and shall not ring in case that the respective communication session is adapted. Hence, SIP provides just a differentiation between the management actions of negotiation and of re-negotiation directly associated with just one single multimedia session. The data management of the media session is done using other mechanisms, e.g. like the Offer/Answer Model (see below and [35]). The three states in the SIP client and server INVITE transaction correspond to the three major actions performed through the SIP UA (see Figure 12 and Figure 13), namely, starting of or waiting for an INVITE call, waiting for or processing/sending the answer to the INVITE call (i.e. a numbered SIP response) and confirming or waiting for confirmation that the answer is received (i.e. SIP ACK) [62]. The SIP transaction is thus a *three-way handshake* and the SIP ACK method is specific only for the INVITE transaction to accomplish this *three-way handshake* [62].

The non-dialogue building transactions (Figure 14 and Figure 15) are applied for exchanging of information or for other management actions that are not directly associated with a running media session controlled through SIP (e.g. terminal registration with REGISTER [62] is such a transaction). The non-dialogue transactions provide reliable transport for the messages through continuous sending of the provisional response *100 Trying* until the actual response to the call can be sent (see Figure 14 and Figure 15). This response serves as a guarantee for resetting the client's waiting timer for the response reception and for accomplishing the call with just two messages (i.e. request and response) [62].

The SIP design allows communication transactions that can apply reliable transport (like TCP [122]) or unreliable transport (like UDP [123]). The management of reliable or unreliable kinds of transport is indicated correspondingly in Figure 12 through Figure 15. Depending on the kind of transport, the SIP SM(s) take different management decision paths. The utilization of unreliable transport with SIP allows the less problematic application of the protocol for managing services that run over radio technologies or perform handovers, as TCP does not tolerate link interruptions and its resetting after a link interruption causes additional overhead to the transaction management (as shown on example of RMI in [180]). The application of unreliable transport with SIP and respectively with E2ENP is the means that enables the negotiation protocol performance within mobile network environment. The efficiency of the protocol in such environments is proven through tests (see Chapter 6).

Further details on the SIP transactions and methods can be found in [62][104][105] or are discussed later in this work with respect to their application for piggybacking of E2ENP information. Furthermore, negotiation scenarios with E2ENP based on SIP are provided (see Chapter 3) where details about the re-used SIP methods and transactions are given. E2ENP applies both kinds of SIP transactions to build additional modes of SIP, integrating the SIP SM(s) into a higher hierarchical level of SM(s) to provide also advanced features for the protocol like reservation coordination, deadlock avoidance and resource planning management (see Sections 2.3.2 and 3.1 and Chapter 5 for further details). The negotiations with E2ENP create higher-level dialogues that span over both kinds of SIP transactions and may include multiple associations of SIP methods. This novel feature of E2ENP gives the application the possibility for management of the exchanged configurations also outside currently opening or already opened multimedia sessions thus providing planning of interactions based on available previous knowledge (see Section 3.1) and allows the realisation of advanced negotiation and adaptation scenarios (see Section 3.3).

The Offer/Answer Model with SDP provides the basic interaction model for negotiations between two peers and defines the negotiation procedure the following way [35]:

> *Two entities can make use of the Session Description Protocol (SDP) to arrive at a common view of a multimedia session between them. In the model, one participant offers the other a description of the desired session from their perspective, and the other participant answers with the desired session from their perspective. This offer/answer model is most useful in unicast sessions where information from both participants is needed for the complete view of the session.*

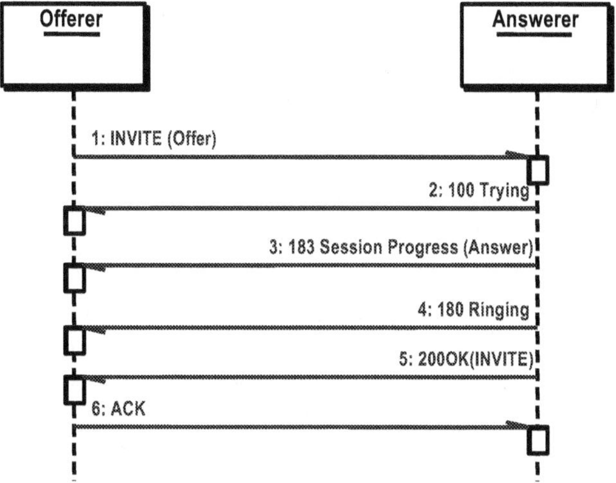

Figure 16: Basic Offer/Answer Model using SIP (see [35])

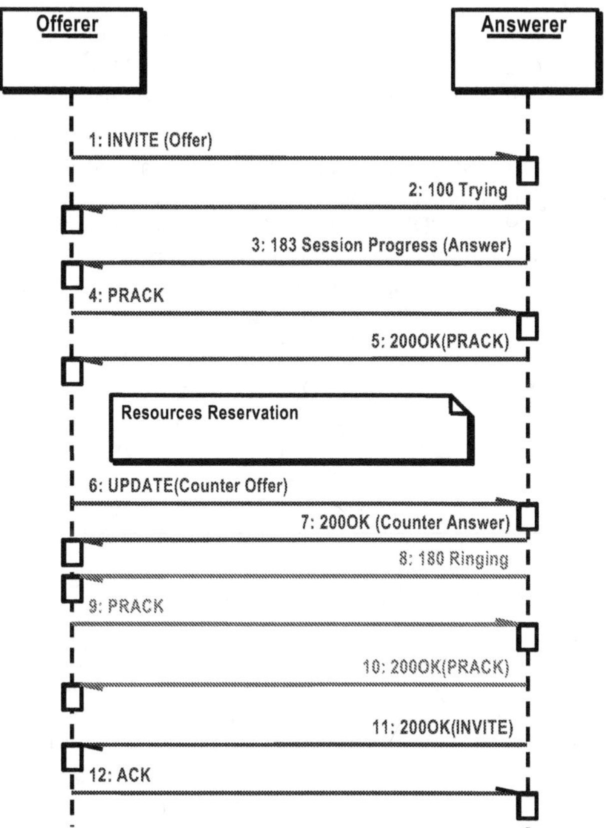

Figure 17: Offer/Answer Model with resource reservations (see [36])

A negotiation for session establishment with the Offer/Answer Model under application of SIP is shown in Figure 16. The session re-negotiation is similar but has one message less (i.e. Message Nr. 4 – 180 Ringing) as at automatic re-negotiations the user at the answering side does not need to be indicated of the re-negotiation, respectively the answerer-side application does not need to produce ringing or whatever signal to draw the attention of the user of the Offerer-device.

The Offer/Answer Model is being extended for resources-reservations management within the SIP scope in RFC 3312 [36]. The session negotiation/re-negotiation with this extended model is shown in Figure 17. As for the basic Offer/Answer Model the application does not need to draw the user's attention when automatic re-negotiation shall take place, hence the re-negotiation with the advanced model has three messages less then the negotiation (i.e. Messages Nr. 8, 9 and 10). The second pre-conditional acknowledgement (i.e. Messages Nr. 9 and 10) coordinates the finalization of the ringing process therefore it is not applied at re-negotiation. Furthermore, the advanced Offer/Answer Model (Figure 17) has the possibility to adapt the negotiated data after resource reservations have taken place (i.e. Messages Nr. 6 and 7).

E2ENP adopts for its peer-to-peer negotiation management the advanced Offer/Answer Model (RFC 3312 [36]). However, E2ENP introduces the following innovations to this model:

- E2ENP does not apply the Session Description Protocol, but a negotiation language designed in XML [73] and incorporating information and features stemming from SDPng [81] and MPEG-7 [109] that will enable negotiations of the *Fine-Granularity* type (see Section 1.2.2.2.3 and Chapter 4).
- Unlike SDP [86] and SDPng [81] applied with the Offer/Answer Model [35][36], E2ENP gives the possibility to associate different pieces of configuration information stemming from different PDUs (respectively different negotiation statements) thus achieving expressions of long-term memory for the application, which is used on one side for advanced resources planning (see Sections 3.1.2 and 3.1.4), on the other side it provides signalling optimisation through the PDU referencing mechanism (see Sections 2.2.2.3, 4.3.2 and 4.3.5 and Chapter 6).
- E2ENP introduces further extensions to the advanced Offer/Answer Model [36] considering the possibility of implicit reservations (e.g. QoS guarantees through DiffServ marking [5]). Note that RFC 3312 [36] considers only explicit reservations with RSVP [3][4]. Provider managed reservations, coordination of the terminal and the network reservations and provider or third-party negotiation mentoring are not considered in [36] (see Sections 2.3.2 and 2.4 and Chapters 3, 4 and 5).

2.3.2 Resource Management Aspects

This section details further the application of the advanced Offer/Answer Model [36] within the scope of E2ENP. The enhancements to this model proposed and used in E2ENP are explained considering the positions of the different partners and mentors involved in the peer-to-peer negotiation process. Furthermore, a validation algorithm for the E2ENP information is presented that determines the order of the application of resource management restrictions upon the negotiation and the reservation process of the adaptive application/middleware.

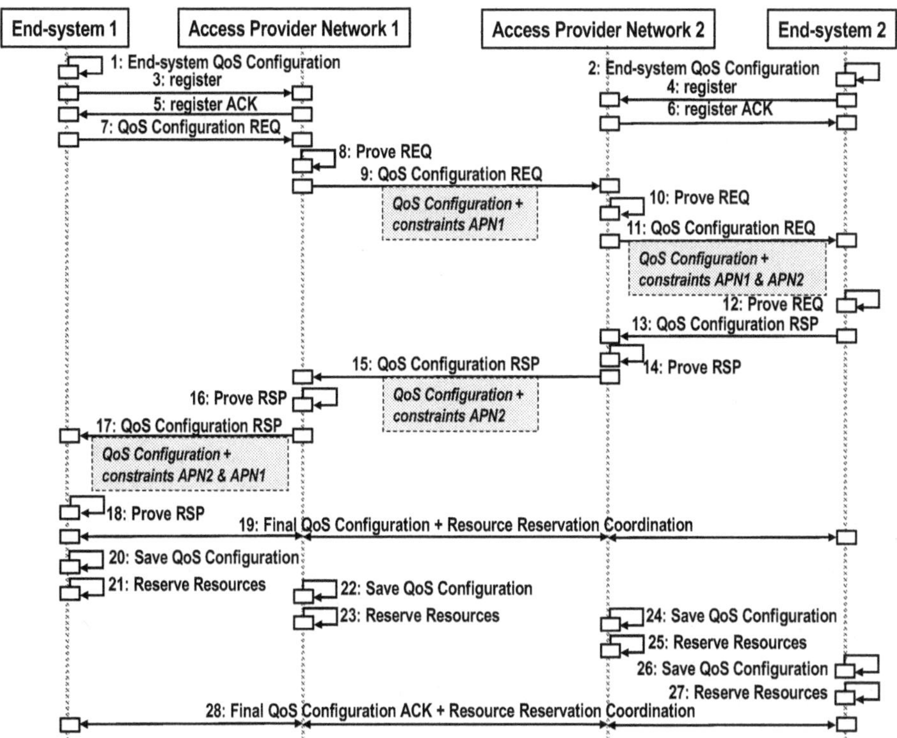

Figure 18: End-system and Access Provider Network interactions

2.3.2.1 Parties Involved in the Decisions Concerning the Negotiations

As stated at the beginning of Section 2.2 the major parties involved in the peer-to-peer negotiation process are the terminals (as negotiation partners) and some service provider entities (e.g. signalling proxies) on the network (as negotiation mentors). The assisted negotiation procedure is shown in Figure 18 and the initial concepts of it were published in co-authorship from the author of this work in [131]. The provider-controlled negotiation procedure is implemented as an extension to the Offer/Answer Model [35][36]. However, Figure 18 presents only the conceptual model of the negotiations with the provider as an overview of the resource-management aspects of the system, the specific scenarios with provider management are presented then later in Chapter 3. A negotiation assisted by the provider comprises the following steps:

- **Registration** – The End-systems first load their QoS configurations for the currently accessible network technology and register within the provider of reach (Figure 18 – Actions 1 – 6, the provider is indicated as Access Provider Network or APN). Note that the registration procedures between End-system 1 and APN1 and between End-system 2 and APN2 are two independent processes. Furthermore, an alternative scenario to the shown one is the operation of the involved end-systems within the same provider network, but this scenario is a simplification of the shown one. During its response to a registration request, the provider registering system instructs the end-system what validation policies, service/resource restrictions and reservation mechanisms are available within the access network of attachment of the end-system so that the end-system can validate QoS configurations on its own. Alternatively, the provider may also validate the negotiation statements of the peers (as in the example in Figure 18). These two validation alternatives cover two possible network reservation scenarios, namely, terminal-driven and provider-driven network reservations (see also the network resource-management interactions displayed in Section 2.1.2.1 and Figure 5, and the respective discussion provided in Section 2.4.1).

- **Negotiation** – When the *End-system 1* starts an E2ENP signalling session, it sends a QoS configuration proposal to its partner – *End-system 2* (Messages 7, 9 and 11). This proposal is validated by the local and/or the remote providers and eventually enhanced with Transport QoS and policy parameters (Actions 8 and 10). The validation at the provider side may take place explicitly, i.e. a provider proves the complete negotiation information in the E2ENP message, or implicitly, i.e. a provider relies on the validity statement of another provider without proving the complete E2ENP message itself. A validity statement of a provider applies both for end-systems and for other providers. Such statement can be expressed as a digital signature or a cookie inside the E2ENP message (see Sections 3.2 and 4.3.6 for further details on trust relationship establishment using E2ENP). The remote end-system generates an answer, corresponding to the supportable configurations at the remote end-system (Action 12) and responds to its partner (Messages 13, 15 and 17). The providers validate the response of *End-system 2* accordingly (Actions 14 and 16), as *End-system 2* might have inserted additional configuration information to the QoS Configuration document, which needs to be proven by the providers. When the response is received (Action 17), the initiator of the signalling phase can calculate the final QoS configuration (Action 18), and it generates a general system configuration or a particular session configuration. The negotiating end-systems might not immediately open a media session controlled through the negotiation session at the end of the negotiation, but could update some general configuration data in form of an advanced telephone-book entry for instance. This possibility gives the partners preliminary knowledge on how to quickly negotiate when a media session shall be opened or adapted at the end of a negotiation phase (see also Section 3.1.2). A particular session configuration is generated at session setup and/or at session adaptation signalled through E2ENP. During a session establishment or an adaptation, the notification of the final QoS Configuration (Message 19) can also serve as a trigger for a network resource-reservation (see the next item below).

- **Network Resource-Reservation** – End-systems need to explicitly coordinate their resource reservations as different underlying reservation mechanisms might be in use (e.g. RSVP [3][4] might be terminated at a network gateway within one of the APN(s) and/or one of the APN(s) might be pure DiffServ [5][61]). Therefore, E2ENP informs the peers about the state of network resource-reservation. The end-systems save the final QoS configuration within the QoS coordination process (Actions 20 and 26) and, if required, perform local resource reservations (Actions 21 and 27). During the resource coordination phase, network resources can be activated or authorized (Actions 19 – 28). The APN(s) save the final QoS configuration (Actions 22 and 24) in order to understand the reference keys to the QoS coordination document that contains the hierarchical QoS specification for the multimedia session of the two terminals (see also Sections 2.2.2.2 and 2.2.2.3). The end-systems may exchange such reference keys during the next E2ENP negotiation phases in order to switch between QoS levels or to enhance/restrict the hierarchical specification at changing environment conditions. Hence, the providers must be able to understand the keys and to assist the end-systems in their network reservations if necessary (Actions 23 and

25). When listening for the Resource-Reservation-Coordination message (Message 19) and interpreting this message, the APN(s) can reserve resources on the network on behalf of those end-systems, which cannot do this by themselves or are not allowed to do so.

- **Session Establishment** – When the answer to the Resource-Reservation-Coordination message is interpreted (Message 28), the APN(s) start the billing process for the reserved resources and respectively activate the metering devices within the network. Additionally, Message 28 details the currently selected QoS configuration, if multiple, alternative QoS configurations for the multimedia session are negotiated and one of them is chosen after the reservation process finishes. Finally, Message 28 is the trigger for the start of the media exchange between the terminals.

The so presented *Offer/Answer Model with Network Assistance* differs from the advanced Offer/Answer Model [36] on the following points:

- The model presented in [36] does not rely on reservation information exchanged during the end-system registration with the provider domain. However, as the resource reservations, the QoS guarantees and the provider billing systems are closely coupled systems within the provider user-management [20][92][181], it is clear that the service providers are those that will establish and try to enforce resource management policies for their end-users. Hence, exchanging such policies at end-systems registration procedures is an expectable step in the resource reservation management.
- The advanced Offer/Answer Model suggests that the terminals explicitly state the reservation mode in their control signalling. However, this mechanism does not tolerate systems that are not able to reserve network resources on their own. Furthermore, explicit reservation of resources on the radio links for mobile terminals cannot be guaranteed, as the performance of the radio links is unpredictable due to environmental conditions (e.g. weather, landscape – mountains or buildings, etc. may interfere with the faultless radio signal distribution [20][141], even if signal improvement techniques like the Software-Defined Radio systems [182][183] are in use). If the provider does not instruct the end-systems how to reserve resources on the network and how to manage the information flows in the radio networks (e.g. through FEC or ARQ mechanisms), the end-systems cannot expect that the providers will tolerate either the reservation mechanism of the terminals, or the QoS levels requested by the terminals. Additionally, explicit reservations (e.g. with RSVP) are also superfluous in ad-hoc networks due to the high structural fluctuations of such networks. Hence, the terminals in such network do not need to signal for network reservation but have to rely more on local adaptation mechanisms and on application redundancy mechanisms for the transportation of the media [65]. Eventually, the peers in an ad-hoc network may exchange information about network-resource and traffic management mechanisms ahead of an E2ENP sessions in order to guarantee E2ENP's network independence (see Section 2.3.2.1.2). Because of these considerations, E2ENP does not signal the reservation mode, but only the beginning and the end of the reservation procedure. E2ENP is designed to rely on preceding registration information in the provider-owned networks (see Section 3.2).
- Due to the restricted format of SDP used with Offer/Answer Model [35][36], it is not possible to express different alternative QoS configurations or different arts of restrictions the terminals or the provider may apply for the reservation. The hierarchical QoS specification (see Section 2.2.2.2) and the XML description model of E2ENP (see Chapter 4) enable planning and execution of network assisted negotiations and resource reservations that are not available with SDP (see also

Sections 2.3.2.1.2 and 2.3.2.2.2). E2ENP provides information about the necessary network reservations in a technology independent manner that is usable for any network-reservation technology (see Sections 4.3.3 and 4.3.4).

The resulting changes and enhancements to the Offer/Answer Model provided through the E2ENP concept are network-assisted negotiations and reservations, dependence on reservation information provided through the registration in provider-owned networks, signalling of the beginning and the end of the reservation without signalling the reservation mechanism and providing reservations and negotiations planning and restriction management through the application of the hierarchical QoS specification that implicitly guides the reservation process both on the terminal and on the network.

The following two sub-sections detail the peer resource planning mechanisms and the network utilization and coordination management applied at negotiations of opening or running sessions in cases of network congestions or terminal/service mobility.

2.3.2.1.1 Peer's Resource Management

In order to properly coordinate reservations at the partner terminals and on the network E2ENP applies the so-called *economy principle* [19][70]. This principle suggests that the actual reservation of the local resources at the end-systems shall be coordinated with the reservations at the network, in order to avoid waiting phases for the terminals and the network devices at some of the reservations steps. Furthermore, this principle dictates the order in which the reservations shall be done, i.e. end-system local resources should be reserved ahead of network resources. The reason for this order is that terminal's own resources are cheaper in terms of management overhead and economically. On the contrary, network resources are shared between multiple users, third party owned and associated with more complex access and management, inclusively accounting and billing [20][132][173]. Furthermore, special care must be taken while specifying and applying the corresponding coordination mechanism for the reservation, in order to provide consistency for and avoid deadlocks between applications utilizing the *economy principle*.

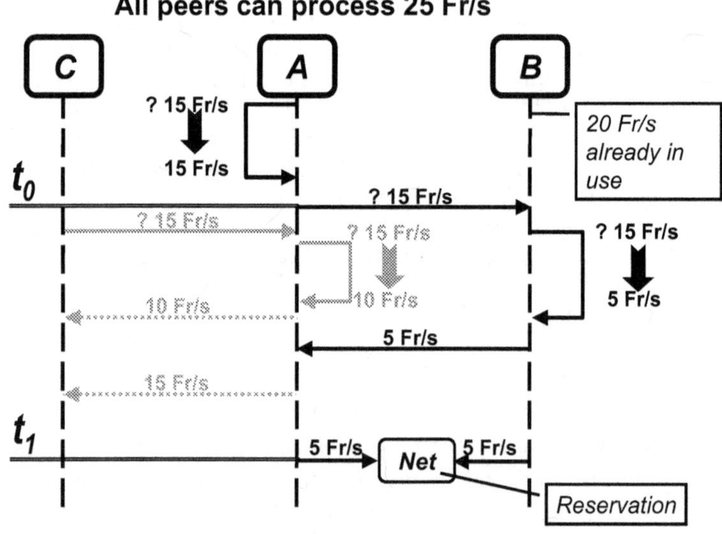

Figure 19: Consistency example

In the following, some example scenarios are shown that illustrate the application of the *economy principle*. They serve for specifying the E2ENP mechanisms for consistency of the negotiated information and for deadlock avoidance of colliding applications of the *economy principle*.

The first scenario shows why it is necessary to provide consistency (Figure 19). Assume that all the peers shown in the example can process maximum 25Fr/s. At time t_0 peer A initiates a negotiation with peer B for managing peer A's sending of a video stream to peer B at a frame rate of 15 Fr/s. Peer B has already an ongoing similar session that consumes 20 Fr/s with a different partner then A. Peer A successfully reserves local resources for processing 15 Fr/s and sends the negotiation request to peer B. Consequently, peer B reserves resources for 5 Fr/s for processing the incoming stream because that is all it can support. This information is propagated back to peer A and A releases the previously reserved resources for maintaining the negotiated frame rate value of 5 Fr/s and then starts reserving network resources equivalent to 5 Fr/s at time t_1. Taken that the network is not a limiting factor, finally peer A, peer B and the network are able to support 5 Fr/s for the given session between A and B. If at any point between t_0 and t_1 another peer C wants to create a session with peer A, peer A would only be able to allow peer C to admit 10 Fr/s (=25 Fr/s - 15 Fr/s) locally. However, if peer A receives the request from peer C at any point in time after t_1, peer A would be able to admit at least 20 Fr/s (=25 Fr/s - 5 Fr/s) for the new session with peer C, because the resource requirements for the first session are decreased due to constraints imposed on peer B, which are outside the control of peer C.

From this scenario, a requirement is derived that the protocol for managing local, remote and network resources between two peers shall not serve requests for resource reservations from another peer, until the protocol has succeeded in establishing an on-going communication session. This requirement influences the application of E2ENP in two ways:

- In case that two partners are involved in a negotiation process and an additional party wishes to join the negotiation as a peer sharing the resources of the first two parties, the applications of the *economy principle* have to either close their running negotiation and re-start the negotiation process including the additional partner, or have to ignore the additional partner until their own negotiation is finished with final stable results. The chosen carrier SIP of E2ENP can already support these two cases through its description language. However, the implementation logic of these cases is provided through the E2ENP UA implementation and the integration of the SIP User Agents into the E2ENP architecture (see Chapter 5). The first case (interrupting the negotiation) is executed using the SIP CANCEL method from the client side or SIP *488 Not Acceptable Here* [13] response from the servers side for interrupting an INVITE transactions between the negotiating peers under condition that the negotiating peers are not already in the final-result-exchange phase [62]. The second case (ignoring the additional partner) applies for all SIP non-INVITE transactions and for an incoming INVITE issued from the peer wishing to interfere with the negotiation. In that case, the answer of the already busy peers is SIP *486 Busy Here* response [62]. Since the issuing of all *3xx-6xx* SIP responses is a matter of coordination from the application side, it is

---

[13] *488 Not Acceptable Here* means that the server understands the session description but the required media or their configurations are not supportable at the server side. The client receiving such a message can repeat its request later using the same or different configuration, also in dependence with the fact if the server provides some reason for its answer or an alternative configuration within its response. [62]

up to the E2ENP coordination logic and up to the application decision how to exactly treat this coordination case. The current implementation of E2ENP recognises the cases of inconsistency of the negotiation upon SIP Call-ID, CSeq header fields and issues respectively *busy* responses (see also Section 5.2).

- In case that one peer coordinates the negotiation of multiple parties with it (e.g. a conference with a conferencing server), this peer has to assure that the single signalling sessions between it and the other communicating peers are coordinated respectively (see also Section 3.3.2). As the current SIP specification [62] and implementations (e.g. jSIP [184] or NIST SIP [185]) are not designed to support multiple applications simultaneously, it is a question of application/middleware logic to provide consistency for the reservations in that case. Respectively, the adaptive application/middleware can apply multiple instances of E2ENP UA (as described in the previous item and in Section 5.2) and coordinate them accordingly to assure consistency for all the applications from a single terminal point of view.

The second scenario involving deadlocks is depicted in Figure 20. This scenario explains why the E2ENP application must assure that there is no *Hold-and-Wait* condition or that such a condition is present only for a limited amount of time.

Assume that peer A wants to send a video stream at 25 Fr/s to peer B and vice versa. At time t_0, peer A reserves the local resources, and sends the negotiation request to peer B. Partner B receives this request respectively at time t_2. Meanwhile, peer B reserves the resources at time t_1 for sending a stream to peer A. Peer A receives this request of B at time t_3. Therefore, peer A waits for the response from peer B starting from time t_0, whereas peer B waits for the response from peer A starting from time t_1. Consequently, when both peers try to reserve their local resources at time t_2 (peer B) and time t_3 (peer A) for serving the remote requests they will both fail.

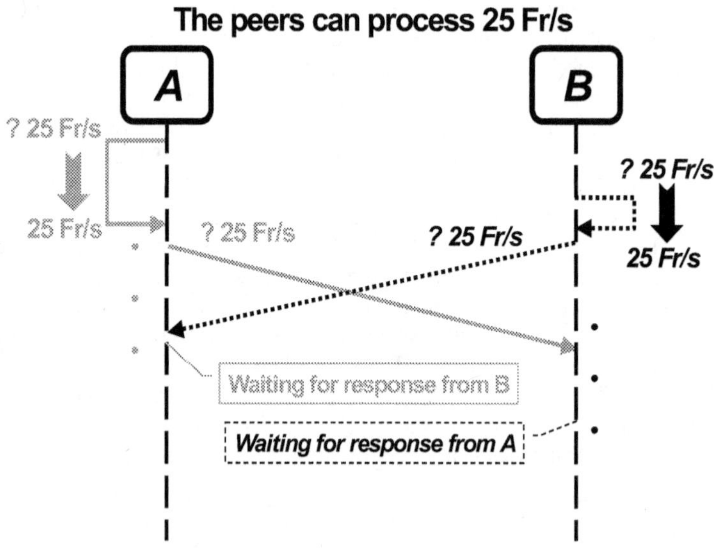

Figure 20: Deadlock example

From this scenario, the requirement follows that the protocol managing local, remote and network resources between two peers shall avoid deadlock situations or at least allow them to occur only for a limited amount of time after which the protocol shall be able to recover in any case.

E2ENP fulfils this requirement through implementing a *bully* process for resolving dual seizure between the colliding applications of the *economy principle* (see also *Election Algorithms* in [50][52]). The E2ENP UA (see Section 5.2) is able to store information about previous E2ENP sessions in order to provide consistency and completeness of the information used as a reference in successive E2ENP sessions (see also Section 3.1.4). Hence, the E2ENP UA caches also the identification of the peers involved in a previous E2ENP session, i.e. the couple of SIP *From* and *To* headers. If a dual seizure occurs, the E2ENP would detect it using such *From-To* couples. The deadlock resolution procedure used in the E2ENP UA applies as follows:

- If two peers are not already involved in a multimedia session, means that the E2ENP UA has no From-To-couple entry about the session in its local cache. At starting a session, the E2ENP UA generates the respective session IDs (e.g. the SIP Call-IDs, the SIP CSeq and the E2ENP PDU IDs (see Section 4.3.2)) and builds a hash out of them. E2ENP UA may also apply IDs generated from the application using E2ENP. The E2ENP UA caches the generated hash information to identify already successfully performed E2ENP sessions. At receiving an invitation for a session, each peer compares as first the From-To couple it generated with the one received from its partner to identify seizures. Furthermore, the hashes (i.e. the rest of the session IDs) are compared and the peer with the biggest hash value is elected automatically as an Offerer whereas the *loser* peer automatically moves to an Answerer mode. If the session is then successful, the E2ENP UA caches the respective hash information for providing memory to the application of the already performed sessions, which have served for exchanging negotiation information. Note that not all of the E2ENP sessions serve for exchanging negotiation information, hence, not all of the session hashes need to be saved (see also Section 3.1.4).
- If the two peers are already involved in a multimedia sessions, this means a From-To-couple entry is available in the E2ENP UA cache to provide dialogue management as prescribed for SIP [62]. In this case, the *bully* process described in the previous item is applied as well. However, the peer whose request is rejected could apply an additional and unplanned re-negotiation after the winner of the *bully* process has completed its negotiation. An example for this case is the simultaneous detection and re-negotiation reaction of both involved peers to a QoS violation. In this case, the *loser* peer may repeat the re-negotiation process with its partner if this is still necessary for further adaptation of the session, e.g. in a situation when the *loser* needs to further decrease QoS if the first re-negotiation session with its *winning* partner did not manage to do so.
- Since the generated identification hashes contain also randomly generated numbers together with the IPs, ports, and other information (e.g. the SIP Call-IDs are usually random), it is not to be expected that some of the peers would always be *winners* and some always *losers*. Hence, the application of the *bully* process is neither an advantage, nor a disadvantage for the peers, for instance, from the point of view of the billing process of the peers. This means that a constant *winner* of the *bully* process would be a *loser* from the point of view of the billing, as usually the *winner* has to pay for initiating the communication session. However, the application of random information in the *bully* process prevents such situations.

The benefit of the *bully* process of E2ENP compared for example with the current public switched telephone networks (PSTN) [186] is that the human users do not need to re-dial the connection (i.e. to resolve manually the deadlock) as this is managed automatically through the E2ENP UA. Furthermore, the deadlock resolution procedure of E2ENP provides an additional aspect to the Offer/Answer Model, namely, the establishment of an order when a peer can or shall apply Offerer respectively Answerer mode in the session coordination process.

The consistency provisioning and deadlock avoidance procedures with E2ENP represent the two major characteristics of the peer's own resource coordination management. These features of E2ENP were published in co-authorship from the author of this work in two patent applications (EP1414211 [170] and EP1331785 [173]).

2.3.2.1.2 Network and System Aspects at Application of E2ENP

The application of the QoS negotiation process has to be independent of the specific system and network management issues like reservation, network QoS guarantees, mobility or security in order to assure that the protocol is applicable in heterogeneous environments without being influenced from these environments. With this extent, E2ENP is coupled with some requirements to the application on how the application shall:

- Efficiently react to changing environment conditions corresponding to QoS changes resulting from network congestions or mobility, and
- Effectively prepare the results of its decisions for the negotiation process.

These requirements are specifically related to the protocol proxies on the network, to other negotiation mentoring entities and to the local application's decision logic with respect to when to apply explicit negotiations with E2ENP at adaptation and when to apply quick adaptations with other then E2ENP mechanisms.

The *first requirement* concerns the independence of E2ENP from specific other management mechanisms:

> *E2ENP shall not be a carrier of information about specific network reservation technologies, traffic management mechanisms and other non-QoS information that is not used in the preparation of the terminals' QoS contracts but may be applied to guarantee successful reservations and faultless transfer of the E2ENP payloads.*

For instance, non-QoS information is profile or authentication information that is not negotiable and represents static configurations, however such data can be used from the application to restrict its decisions when generating negotiation proposals to be presented via E2ENP (see Section 2.3.2.2.2 and Annex B).

The first requirement is covered through the application of an obligatory registration mechanism of the peers in the provider domains as specified for E2ENP's *Offer/Answer Model with Network Assistance* (see beginning of Sections 2.3.2.1 and 3.2).

The *second requirement* covers the protocol interaction with provider entities on the network. The end-peers are usually connected over one or a multiplicity of interconnected networks, including also provider entities. In terms of SIP, the SIP proxies are such entities. In the general case, the provider entities are in position to inspect and change the passing through them protocol payloads thus interfering with protocols like E2ENP. For instance, such proxy behaviour is described in [87] and [187], in the first but already obsolete version of SIP [62] and in the early versions of the 3GPP specifications for SIP application within the 3GPP architecture [22]. In the current version of SIP, the proxy modification of message bodies is forbidden due to some security reasons [62]. However, such requirement is too restrictive if the negotiation process shall be mentored from external entities that are not in the role of negotiating peers but are intermediaries on the communication path between the negotiating partners. As E2ENP shall not be applicable only for SIP (see beginning of Section 2.3 and Figure 11) and since E2ENP considers mentoring entities in the negotiation process, E2ENP provides its own rules for proxy interactions with the E2ENP payloads. In order to satisfy both the user's QoS expectations and the network provider QoS policies, the following mechanisms are suggested for E2ENP's interaction with provider entities (see also Section 2.4.1):

- The end-systems do not always need to apply the mentoring of the provider entities. However, the end-systems might need to be supported from the provider side for the respective network reservations. Hence, E2ENP should be able to be used in combination with (but independently from) provider entities. In this case, the end-systems make the negotiation decisions alone and the provider entities only listen in the negotiation process without actively participating in it in order to prepare the network and to guarantee QoS for the contracts agreed by the end-peers.
- Cases when the involvement of the provider is inevitable should also be considered, e.g. when the terminals are not in position to take the negotiation decision on their own or are not allowed to share provider policy information with the provider entities. Consequently, the providers should be allowed to interfere with and modify the E2ENP payloads, but the providers have to explicitly notify the end-peers about occurring problems and shall interfere only when end-peers are not aware of network utilization policies. However, provider constraints in the different access-networks of reach may differ and removing from the end-to-end negotiated information those QoS contracts that are inappropriate for the currently accessible network would negatively influence the terminals' own planning of QoS adaptations at network congestions or handovers. Hence, E2ENP provides a mechanism for intercepting negotiation protocol payloads and explicitly complementing them with provider constraints to help the efficient enforcement of provider policies without changing the technical specifications exchanged by the terminals end-to-end (see Sections 3.2 and 4.3.6).

Because of these considerations, the *second requirement* for independence of E2ENP from network aspects is defined as follows:

> *Mentoring entities applied within the scope of the E2ENP protocol are allowed to listen in the terminals' negotiation process in order to support the terminals at reservations of the network to guarantee QoS. Furthermore, the mentoring entities are allowed to only enhance the E2ENP payloads (i.e. the negotiation statements of the peers) with policy information. Removing of technical specifications of the terminals from the E2ENP payloads from the side of mentoring entities is prohibited.*

A *third requirement* for E2ENP follows from the second one and considers the specific E2ENP security:

> *At mentoring interactions, the mentoring entities shall explicitly identify themselves as such in the payloads in order to secure the E2ENP payloads against undesired interactions with malicious third parties.*

The third requirement holds true also in case of mentoring entities that are not protocol proxies, e.g. in third-party call control scenarios (see Sections 3.3 and 4.3.6).

The network intermediaries take the role of *mediators* at terminals registration process to provider-owned networks. End-systems can also take this role in ad-hoc scenarios when the peers exchange system-specific support information (e.g. profiles, security and authentication keys, provider policies, etc.). In the mediator case, the end-systems request the mentoring directly from their mediator. The mentors prepare then the necessary information for the peers through actively taking decisions on behalf of the terminals and supplying the terminals with system-specific information that is valid only in the current network segment. The mediator case is a necessary assumption to enable network-independent negotiations for multimedia session management and is discussed in detail in Section 3.2. At QoS negotiation procedures, the specific E2ENP mentors (i.e. the access provider proxies) act as *conciliators* of the negotiation process as the results of the QoS negotiations of the terminals shall be valid also in other then the current network of attachment in order to enable efficient handovers planning at the peers (see the corresponding scenarios in Chapter 3). Hence, the conciliating entities provide only the constraints as enhancements to the negotiation process without interfering with the information exchanged between the partner terminals end-to-end (see also Section 1.2.1).

Since the negotiations with session-level protocols are relatively expensive considering the system resource utilization and the time constraints at adaptation, a *fourth requirement* for E2ENP is put in place that covers the case when the applications shall use E2ENP and when adaptations can be managed with other then E2ENP means (e.g. through direct media stream adaptations as described in [89] or through media buffering at the terminals [8][166]). This concern in consideration with the price of the session management protocols utilization is mentioned also in [66]. If the applications use QoS contracts corresponding to spaces of the desired QoS performance (see [16], [17], [70] and [139] – [141]), the applications are in position to decide if the QoS provisioning is within or outside the desired performance, i.e. to evaluate when a QoS change can be compensated with cheaper then E2ENP means and when a QoS change shall be considered a violation of the QoS contract and respectively a complete re-negotiation or an explicit switch to another QoS contract is necessary (see also Sections 2.1.2.2, 2.2.2.2, 2.2.2.3 and 3.1.2). Consequently the *forth requirement* for E2ENP is defined as follows:

> *E2ENP shall be applicable in combination with (but independently from) other then E2ENP adaptation mechanisms known to the application. The applications of E2ENP shall thus differentiate between QoS changes that do not lead to contract violations and QoS violations that require adaptations with E2ENP.*

To this extent, the end-systems can perform adaptation at two different levels, i.e. using a fast, in-band signalling process at non-critical QoS changes (e.g. by changing at runtime in the RTP packet header the RTP payload type like in [89]) and/or applying a more structured process, based on the End-to-End QoS Re-negotiation Phase (see Section 3.1.2), whenever the first process is not sufficient to cope with a given QoS violation.

The requirements for system and network independence of E2ENP were published in co-authorship from the author of this work in [14], [131], [133] and [173].

2.3.2.2 Decisions about Negotiations

An important part of the negotiation procedure is the decision on what actually to negotiate and how to efficiently prepare the negotiation information in order to reach effective results that satisfy the involved partners in the negotiation [45][59] (see also [46] – *Prisoner's Dilemma*). As stated in Section 2.2.2.2, it is up to the business logic of the adaptive application to determine the exact steps of the adaptation process and respectively to decide what to negotiate. However, considering on one side the heterogeneity of the adaptive applications that may wish to communicate and on the other side the parties (e.g. network providers) that may wish to apply restrictions to the adaptation process, it is necessary to define an efficient process for preparing the negotiation information in an appropriate way to satisfy both the adaptation requirements of the terminal applications (e.g. end-user devices, VoD servers, etc.) and the limitation requirements of the intermediaries (e.g. the provider proxies).

In the following, an overview on decision taking algorithms, which could be applied for multimedia adaptation process, is presented. Furthermore, a validation process for efficiently preparing the results of a decision for negotiation purposes is discussed.

2.3.2.2.1 Backgrounds of Decision Taking

There is a multiplicity of methods and algorithms applied for decision taking in cases that a specific device or system configuration shall be enforced. They range between simple ones applying just plain comparison between values of parameters and very advanced ones utilizing long-term statistics and operating under conditions of uncertainty.

The mechanism presented in [66], [188] – [190] is applied within the scope of the Session Description Protocol next generation [81] for defining a *Collapsing Algorithm* for identifying the applicable media features after a communication process for exchanging system capabilities between two terminal devices has taken place. The algorithm generates Boolean predicates that determine if the given media parameters belong to a required set of values or not. Furthermore, the predicates estimating applicable feature sets can be associated with weighting coefficients that determine the importance of the set according to specific user or system requirements. However, such requirements are not expressed explicitly in the predicates. Similar algorithms are also proposed in [13][17][191]. These algorithms belong to the simple methods for taking decisions about a system configuration, as they rely predominantly on single statements about the current situation of the system and apply static weighting of parameters and parameter sets for determining the importance of the applicable features.

A more advanced set of decision methods includes mechanisms that can operate under conditions of uncertainty and incompleteness of information, relying on known sets of statistics (e.g. data sets for estimating user perception of application or network performance [134][135][137][138][141]) or on dynamically accumulated statistical data, thus being in position to apply previous knowledge for the efficient planning of decisions. Some mechanisms belonging to the advanced decision taking are:

- The algorithms of Constraint Programming [192] – This set of algorithms is in position of take decision under conditions of uncertainty through defining restrictions under which a specific decision holds true.
- Algorithms for decision taking within the scope of artificial intelligence argumentation [47][53] – This set of mechanisms specifies conditions for decision taking agents that have a *particular character* [47] in terms of communicativeness, cooperativeness and knowledge/forgetfulness of *own resources* or *foreign resources* respectively *own statements* or *foreign statements*. This set of algorithms relies on long-term memory of the agents (i.e. applications).
- Game Theory algorithms [46][59][193] specify series of conflict resolution mechanism that operate using dynamically accumulated statistical data, restriction conditions and function weighting for taking appropriate strategic decisions.

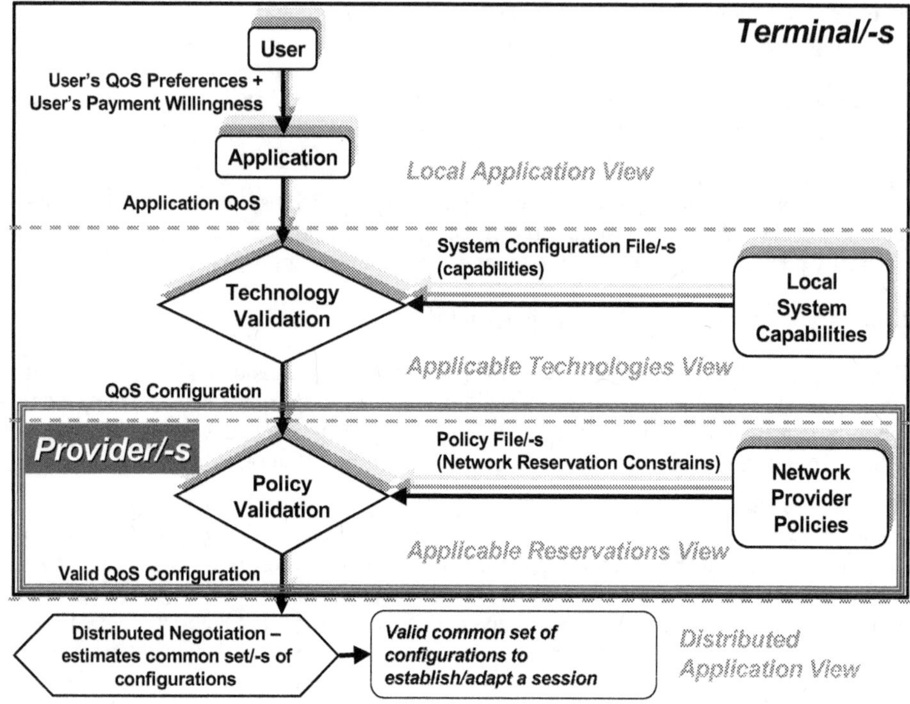

Figure 21: Validation procedure for QoS

Both in terms of negotiations as a social event [45] and in terms of game theory strategic problem resolution [59], it is stated that the *personal* decision of a peer with respect to what to negotiate is a peer's own problem. However, in order to meet a successful decision the peers shall exchange information containing commonly understandable and comparable values [59] and be prepared to create alternatives that will satisfy both parties [45]. Due to these reasons, the specification of E2ENP does not provide a specific decision algorithm for steering the adaptation process of the application, but only an optimisation procedure that shall help the negotiating peers and their mentors to prepare efficiently multiple alternatives for configuring an adaptive application through a negotiation process.

2.3.2.2.2 Validation Procedure for QoS

The specification of E2ENP includes a generation and validation procedure for QoS contracts and contexts as a recommendation for the adaptive applications (middleware) utilizing the protocol. This procedure gives the applications some possibility to optimise the preparation of QoS contracts for the negotiation process through restricting the application decisions, starting from more general decisions and ending with decisions that are applicable for just the current and specific adaptation case. The meta-algorithm provided through the E2ENP specification is named *Validation Procedure for QoS* and it is derived in consideration with some methods of the applied economics [194][195] (i.e. *Scientific Management Principles*) and of the computer science advanced design and analysis techniques [196] (i.e. *Greedy algorithms*). The concept behind the Validation Procedure (Figure 21) is the application of multiple steps corresponding to the views of the local application (at one terminal device), the applicable hardware technologies for running the local application (e.g. network technologies like LAN or UMTS) and the possibilities for network reservations and guarantees. The view of the distributed application resulting in applicable and valid QoS configurations is then visible after a negotiation procedure has taken place. Every *view* within the Validation Procedure (Figure 21) is associated with locally optimal and *view-specific* choices of the applicable QoS contracts/contexts. The specification of the different *views* is made in accordance with the model for QoS provisioning guarantees presented in Section 2.1.2.2.

Heterogeneous end-systems negotiate QoS in form of QoS configurations (QoS contracts and contexts – see Section 2.2.2.2) to agree upon common set/-s of supportable parameters for session establishment and adaptation. Any QoS configuration used in the end-to-end QoS coordination process should be proven against user's QoS preferences, end-system resources and provider-specific restrictions and recommendations (like user contracts or resource availability). E2ENP introduces a validation procedure (Figure 21) to guarantee the generation of correct QoS management information for the negotiation. Within this validation procedure, the application maps application QoS parameters to local-system capabilities, building one or several alternative QoS configurations. These configurations can also be associated with transport QoS parameters in accordance with the ability of the end-systems to support multiple network technologies and to reserve network resources on their own. This explicit association of QoS configurations with transport restrictions enables the adaptive applications to easily take adaptation decisions at changing application environment in cases that the terminals move to another network technology or provider network or the running sessions migrate to another terminal. Consequently, the applications can re-arrange in different associations of application and transport QoS specifications whenever such changes occur. The decision about building the QoS configurations through a *Technology Validation* process can be taken using the algorithms discussed in the previous Section 2.3.2.2.1. Depending on the complexity of the terminal device (e.g. mobile phone or PC) a simple or a more complex decision method can be utilized in this step of the algorithm, in accordance with the processing capabilities and the available memory for saving appropriate statistics and reference data in the terminal device. In the next step, the QoS configurations are validated against provider policies either at the terminal system (if provider policy information is downloaded during the terminal's registration within the provider system) or inside the network during the QoS negotiation process (Figure 21). This may lead to different results when generating the negotiation statements in accordance to the fact which entity (terminal or provider) has applied the restriction, e.g. the terminal may express provider restrictions as currently non-applicable transport QoS specifications or the provider may indicate that certain QoS specification is currently blocked (see Chapter 4). The art of performing *Policy Validation* and the results thereof depend on the trust models between end-systems and access network providers, and on the ability of end-devices to perform network reservations on their own [20] (see also Section 2.4.1), e.g. the provider may not trust the users, hence no provider policies would be downloaded at the terminal at registration and the provider would apply restrictions during the negotiation process between terminals. E2ENP specifies [14][173] that the *Policy Validation* done within the network during the negotiation process does not change the contents of the QoS configurations but explicitly includes descriptions of provider constraints associated with the QoS configurations (see also the principle of *communication arrangement* at the beginning of Chapter 2 and the requirements for network independence in Section 2.3.2.1.2). This enables the applications to associate their QoS contracts/contexts with different constraints whenever the provider network changes and the constraints of the previous provider are no longer valid. In this case, the already available QoS contracts/contexts do not need to be re-negotiated but are only associated with the new constraints. During the policy validation procedure, the terminals or the providers apply also a decision algorithm in accordance with their technological capabilities for processing the decisions and in relation with the application specific QoS configurations that shall be validated. The result of the policy validation procedure is a set of valid QoS configurations applicable for the terminal device within the specific provider network. The end-systems use the current provider constraints as a guideline for their resource reservations at session establishment or whenever an adaptation is necessary to guarantee system stability. After applying a negotiation process between the terminals, the resulting set of valid QoS configurations describes a complete technically supportable QoS range for the distributed application, thus the end-systems are enabled to change the access network technology and/or provider during an ongoing session by planning ahead appropriate management actions through using the alternative QoS configurations sets resulting from the negotiation [14][20][173]. All relevant QoS configurations exchanged between the applications build a QoS coordination document, used by the communicating partners to establish and manage multimedia sessions. The QoS coordination document describes the complete adaptation tree for the application (see Section 2.2.2.2 *QoS Hierarchy*) and provides identifiers to the different adaptation states and levels of the hierarchy. During ongoing multimedia sessions, applications can refer to the appropriate QoS state and/or level within the QoS coordination document exchanging only the negotiated ahead identifiers and links to the information in the QoS coordination document (see also Sections 2.2.2.2, 2.2.2.3, 3.1.2, 4.3.4 and 4.3.5).

The application of the validation algorithm and the QoS coordination document gives the applications the possibility to provide serialization of the complete session-adaptation state machine (see also [17][70]) with all its hierarchical levels, thus moving the costs of the expensive negotiation process before or at the very beginning of the multimedia session establishment where the time constraints for an adaptation are not that stringent. Furthermore, the configuration expenses at adaptation of a running multimedia session during the lifetime of the session are decreased through using knowledge and links to already pre-negotiated adaptation alternatives. The results approving this concept are demonstrated with measurements in Chapter 6.

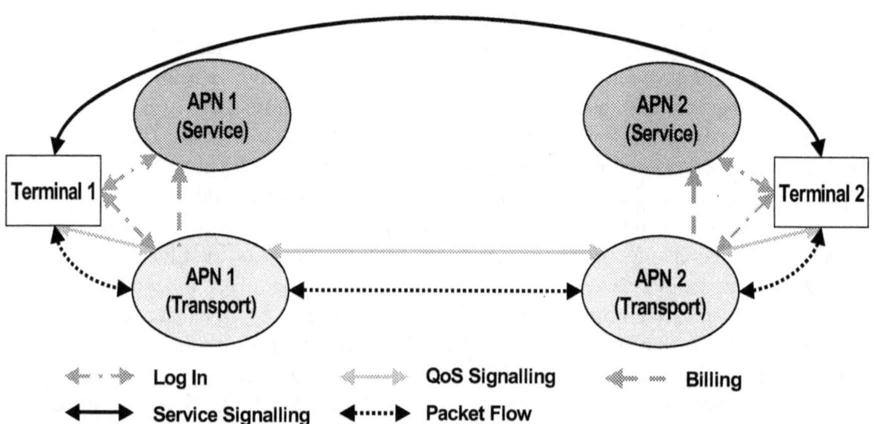

Figure 22: No Coupling interaction between the service and transport

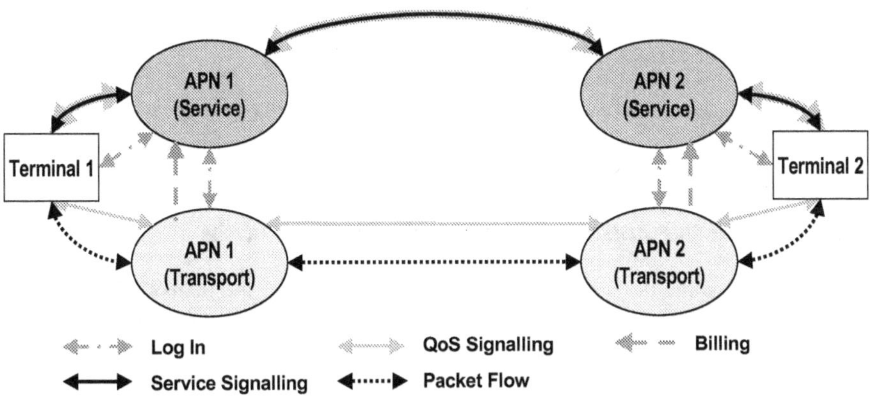

Figure 23: Loose Coupling interaction between the service and transport

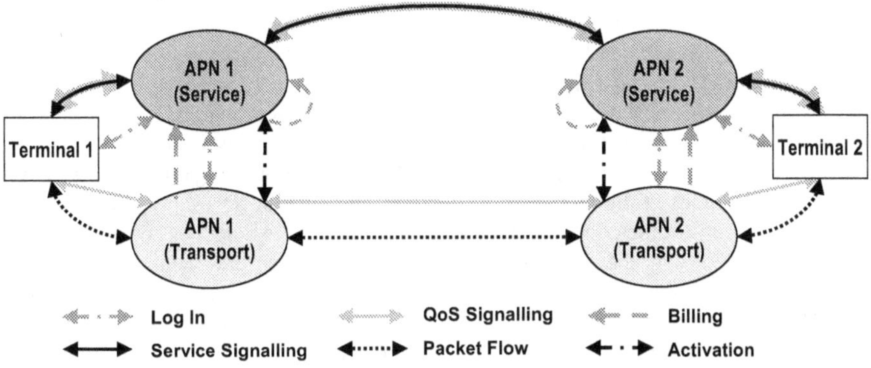

Figure 24: Tight Coupling interaction between the service and transport

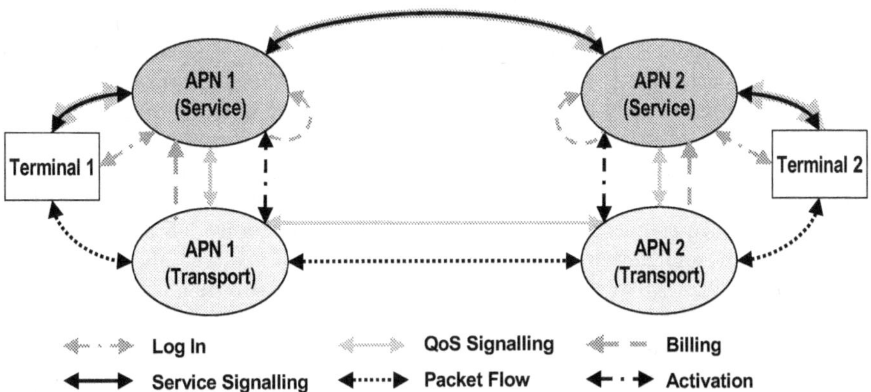

Figure 25: Integrated interaction between the service and transport

2.4 E2ENP Negotiations within Existing Architectures

This section presents some concepts on how the End-to-End Negotiation Protocol for QoS can be integrated into existing architectures that consider multimedia transport and provider management for QoS guarantees of such transport.

2.4.1 Service and Transport Interaction Scenarios

In the following, some basic scenarios are described depicting the inter-working between the session signalling for media transport delivery and the corresponding resource reservations within the network. The scenarios are referred to as *No Coupling*, *Loose Coupling*, *Tight Coupling* and *Integrated* in accordance with the amount of control and interaction between the service signalling and the resource reservations, resulting in different trust relationships between the provider and the terminals and in different billing options for the traffic. These scenarios are derived in consideration with currently existing architectures developed for provider-owned network infrastructures like 3GPP IMS, ITU H.323, MIND, MASA, DAIDALOS, etc. (see [20] – [22], [92], [107], [111], [112], [130] and [197] – [204]) and in accordance with the domain model presented before in Section 2.1. The E2ENP protocol can be applied only to some of the scenarios since in general the considered architectures take into account not only multimedia services but also legacy applications. The presented interactions in the following scenarios (see the respective figures – Figure 22 to Figure 25) have the following meaning: *Log In* is the exchange of login information at terminals' registration; *QoS Signalling* represents information used for reserving of resources at the terminals and within the network, *Service Signalling* are the management actions of the applications (e.g. for session establishment or adaptations of the service); *Packet Flow* is the actual transportation of signalling and media information in the network; *Billing* designates the information necessary for generating of bills for the users of the service and *Activation* is the explicit activation of the transport monitoring (e.g. for controlling the utilization of the network resources). The specification of the presented scenarios follows:

- **No Coupling** – The *No Coupling* scenario (Figure 22) depicts the situation where no session setup signalling is necessary (e.g. like the legacy FTP application [205]) or the session setup signalling is forwarded without further consideration when traversing entities responsible for the service management of the Access Provider Network (APN), e.g. the current Internet IP-telephones that use the basic SIP without attachments and apply stateless SIP proxies [62]. In this scenario, no interaction between the service and the transport entities takes place in terms of controls of the signalling data of the terminals. The end-devices care by themselves for separately registering for transportation services (e.g. per dial-in) and for high-level application services (e.g. SIP registration for classic IP-telephony without QoS support). Thus, provider service entities cannot support the terminals neither at QoS negotiations, nor for network resource reservations. The QoS signalling takes place only with low-level technologies like RSVP [3][4] or DiffServ [5] and is usually terminal initiated. The terminals can exchange capabilities information in their session signalling, however, the APN cannot guarantee the utilization and the support of these capabilities as it has no way to control and authorize them. The transport part of the network is responsible for delivering information about the network utilization of the services for user-billing purposes only. The *No Coupling* scenario describes the state of the current Internet and

PSTN session management and is suitable for legacy applications. Hence, E2ENP is not applicable in this scenario.

- **Loose Coupling** – This scenario is associated with coupling of the service and the transport signalling only at the terminal devices in cases a QoS-aware application shall be supported (Figure 23). The end-systems alone manage the service and the transport requests of the applications. Respectively, the communicating terminals couple their service signalling and their QoS signalling at application level using session descriptions to reserve resources locally (e.g. CPU or memory), but they may apply also QoS signalling at transport level to reserve network resources (e.g. with RVSP or DiffServ). The name of the scenario *Loose Coupling* is derived in accordance to the fact that the service part and the transport part of the APN operate almost autonomously of each other. The initial necessary configurations for this scenario are either static configurations of the terminals or are exchanged only once at network configuration and terminals' registration. The service part of APN propagates the terminals' registration information to the transport part of the network in form of a static profile, thus instructing the network how to treat specific application-requests of the terminals for traffic. However, the service part does not take care of actively supporting the terminals at negotiations for service configuration. A terminal in this scenario has a fixed configuration from its *home provider*[14] on how to utilize the network or the terminal may receive such information at binding-registration from the current provider of attachment (e.g. *roaming provider*[15]). The end-systems negotiate application QoS and reserve network resources on their own and the transport part of the network takes care only for the traffic billing. The transport part propagates only the final bill to the service part of the network. The *Loose Coupling* scenario is a typical example applied in the Offer/Answer Model with SDP [35][36]. It is a possible sub-scenario for transport resources management in the 3GPP IMS [21][22] and ITU H.323 [107] architectures. With respect to the E2ENP implementation, this scenario corresponds to the uploading of provider guidelines at the terminal at registration. The guidelines are used from the terminal alone to generate valid QoS configurations (see also Sections 2.3.1.2 and 2.3.2.2.2). These rules are not explicitly visible in the E2ENP payloads, as the provider does not prove them explicitly at the service level whenever application signalling occurs (see Section 4.3.3). This scenario assumes that the service part of the network trusts on one side the terminal that it would use properly the downloaded reservation guidelines, on the other side the service part trusts the transport part for the proper billing. Because of this trust relationship between the service and the transport parts of APN, the provider does not take care of explicitly activating the accounting meters at the transport part of the network, but they are activated as soon as traffic occurs. This may lead however to erroneous bills in cases that the transport part of the network misbehaves.

- **Tight Coupling** – The service and the transport parts of the APN are tightly coupled when the service part of the APN does not trust either the terminals, or the transport part of the network for performing respectively proper negotiations or proper billing. In this case, the controls upon the negotiations and upon the traffic management are concentrated in the service part of the APN. Consequently, the scenario is named *Tight*

[14] This is the provider where the initial identity of the terminal is registered (e.g. telephone number or e-mail address).

[15] A roaming provider is any other then the home provider and corresponds the network where a terminal is a visitor.

Coupling. However, the service part propagates terminals' login information to the transport part, as the terminals are allowed to reserve network resources on their own. The service part of APN listens in the service negotiation of the terminals and generates itself the bills upon the provided traffic monitoring from the transport part of the APN (Figure 24). The terminals couple their service signalling and their QoS signalling at application level to reserve resources at the terminals and may apply also QoS signalling at transport level to reserve network resources. The coupling occurs also because the service part comprehends the service signalling of the terminals and validates the negotiation payloads explicitly during the negotiations of the terminals (see Sections 2.3.2.2.2, 3.2 and 4.3.6 for the specific application with E2ENP). Additionally, the service part activates the transport resources and builds the final bills for the users upon frequently performed monitoring of the transport area. This scenario is also a possible sub-scenario in the 3GPP IMS [21][22] and ITU H.323 [107] architectures. The *Tight Coupling* scenario is more provider-centric, as the QoS provisioning in this case undergoes stricter management at the service part of APN compared to the *Loose Coupling* case (see above).

- **Integrated** – The service and the transport parts of the APN have integrated management as the name of this scenario suggests (Figure 25). The terminals operate using only service signalling, coupling it with application-QoS information to reserve local resources. The *Integrated* scenario is also a provider-centric scenario like the *Tight Coupling* (see also 3GPP IMS [21][22] and ITU H.323 [107]) and it is especially suitable for mobile terminals assistance. Such terminals have lesser processing power and battery capacity. Consequently, the mobile devices might not be in position to support all types of protocols and mechanisms for assuring of QoS (e.g. for signalling QoS reservations on the network). Hence, the providers have to support the network reservations when necessary for such terminals. As the service part of the network takes over the complete service and transport management, it does not propagate terminal login information to the transport part. The suitability of the *Integrated* scenario for assisting mobile devices results also from the fact that the behaviour of the radio links at mobility is not predictable. Therefore, it is not possible to assure QoS in such links explicitly[16], hence, the stable and predictable QoS guarantees are provided only in the fixed part of the network infrastructure where the service part of APN can reserve the resources explicitly. The application signalling in the integrated scenario is performed in the same way as in the *Tight Coupling* scenario (see above). The providers configure and assure QoS in the network through listening in and manipulating the service signalling (see Sections 2.3.2.1.2, 2.3.2.2.2, 3.2 and 4.3.6 for the specific application with E2ENP). Furthermore, the service part of the network builds itself the bills for the users like in the *Tight Coupling* scenario.

Since the above scenarios assume service and transport management with interactions between the application signalling and the network resource reservations only at the access parts of the network, combinations of the above scenarios are also possible, i.e. different access networks apply different scenarios. Such combinations are used at interoperation of networks with different capabilities, variable trust models between service, transport and user parts of the access network and different billing options for the end-users and in

[16] Implicit methods for optimising QoS at the receiver side exist both as hardware [206][207] and software [166] solutions. They assure that the receivers obtain optimised signal and lesser data loss in cases that unexpected environmental influences (e.g. weather, landscape – hills, tunnels) occur.

consideration with the fact that the terminals may have different capabilities for signalling transport reservations on the network. The management of the core network is for all the above scenarios provider-specific and out of scope for the end-to-end application signalling with E2ENP. It is a provider-specific issue how to manage the bulk traffic in the core of the network. Usually, the core network management is done with generalized methods for processing bulk traffic, which are not relevant for a single application (respectively for the specific signalling of a single application like done with E2ENP), but only for specific classes of applications (see also [142] – [145], [151] and the respective discussion at the beginning of Section 2.2).

2.4.2 Integration in Existing Architectures

This section discusses how the E2ENP concept can be integrated into existing middleware for QoS provisioning and into provider-based multimedia architectures and how these architectures profit from the application of E2ENP. The considered systems are:

- AEQUA (Adaptive End-system Architecture) and MASA (Mobility and Service Adaptation in Heterogeneous Mobile Environments) developed in [17][197][198] with participation of University of Ulm. These architectures treat the realisation of the device-internal QoS management (e.g. terminal or provider management system). The implementation of the architectures is maintained at University of Ulm (Distributed Systems Department) as a reference QoS Broker specification (see also Annex A)

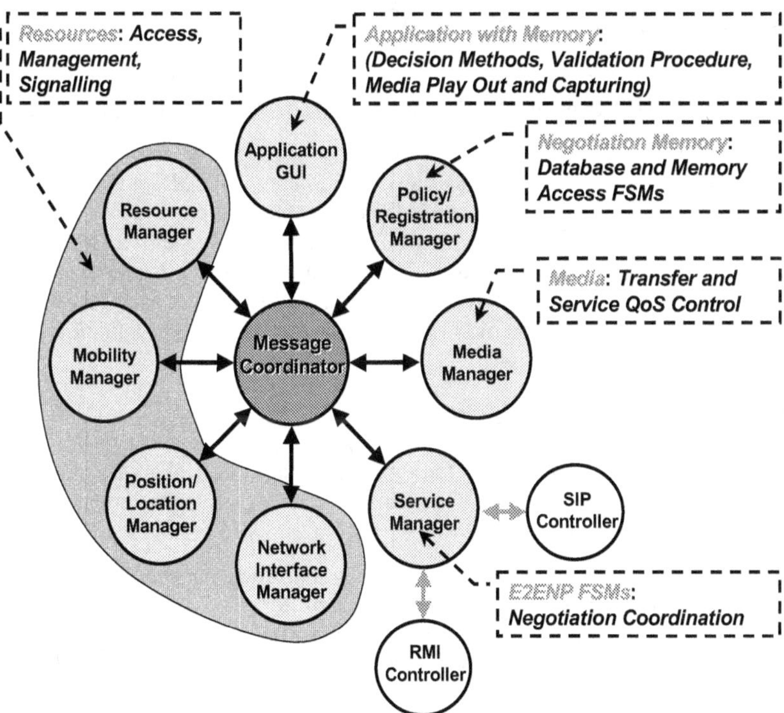

Figure 26: E2ENP view on AEQUA/MASA QoS architecture

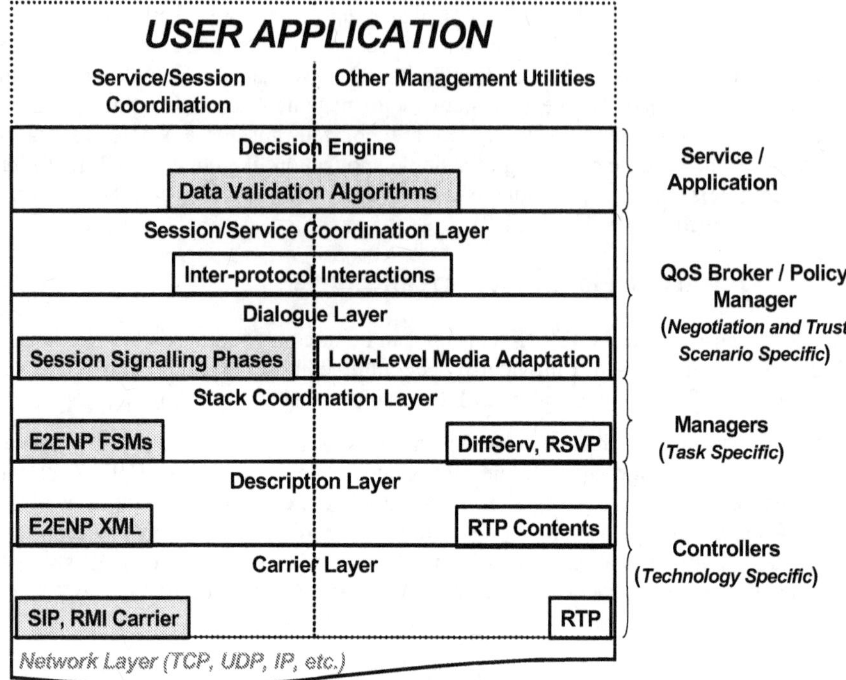

Figure 27: E2ENP contribution to AEQUA/MASA QoS architecture

- 3GPP IMS (Third Generation Partnership Project – IP Multimedia Sub-system) [21][22] and ITU H.323 (International Telecommunication Union – Packet-based multimedia communications systems) [107]. These architectures represent currently existing standards for realisation of IP-based multimedia services and treat the devices interoperation in provider-owned networks designed to deliver such services. Related work about the interoperation of these standards with the Session Initiation Protocol is shown in [103], [199] and [200].

2.4.2.1 AEQUA and MASA

The AEQUA and MASA architectures (designated also as QoS Broker) are internally build in a similar way (Figure 26). This QoS Broker consists of the following major sub-components (see also Figure 27):

- The *message coordinator* represents the communication between the different tasks (see below the *manager* definition) necessary for the entire QoS provisioning. Dependant on the negotiation types (see also Section 1.2.2.2), the trust models between the users and the providers (see Sections 2.3.2.2.2 and 2.4.1) and the utilization of different applications and services, the message coordinator can be configured differently in order to deliver the messages between the *managers* in different manner and order (see [197]). The coordinator arranges the specific communication, i.e. what message that is being sent from a manager corresponds to which of the other QoS-management tasks as receivers of the message. The message coordinator has the same interface to the different managers. This interface serves for bi-directional exchange of messages between the managers and enables the simple reconfiguration of the QoS broker through adding of more managers corresponding to additional QoS tasks necessary within the device (i.e. terminal or provider entity) that runs the QoS broker.

- Every *manager* fulfils a different specific task necessary for the delivering of the entire QoS at the respective device running the QoS Broker. The managers represent a generalized access to the specific technologies (see below the *controller* definition) used in the QoS provisioning. Note that the managers shown in Figure 26 are slightly different compared to those discussed in [17] and [197] as they represent the specific E2ENP view on the QoS Broker. The application of E2ENP with the AEQUA and MASA architecture for advanced QoS-management scenarios, including provider infrastructures and third-party call-control components, requires the internal rebuilding of some of the managers and the addition of some new managers (see the respective discussion below).
- Finally, the *controllers* represent the application of a specific local or distributed technology applied in the QoS provisioning. For instance, Figure 26 shows an example of the integration of the Session Initiation Protocol (SIP) [62] and Remote Method Invocation (RMI) [88] for service and session management. The controllers have specific interface to their manager as every specific technology may provide different functions, rules and expressiveness with respect to the QoS provisioning.

E2ENP contributes to the AEQUA and MASA QoS Broker in the following way:

- E2ENP introduces alternative adaptation scenarios with provider-controlled negotiations and with support of third-party call control negotiations (see Section 2.3.1 and Chapter 3) that are not available within the initial QoS Broker [17][197]. This feature of E2ENP leads to the definition of the following managers in the QoS Broker architecture (Figure 26):
    - The Resources Manager is controlling the local and the network resources and is used to inform the applications about changes in the resource availability
    - The Mobility Manager is controlling the access to different provider networks in order to optimise the roaming behaviour of the terminals
    - The Position/Location Manager is controlling the actual position of the terminals in order to be able to locate different and more suitable devices for transferring the services to them in cases that the current terminal running the application cannot manage the required QoS for the media and the application adapts then via migration to another device. Furthermore, the application can remain on the same device but not adapt on its own, it may locate adaptation services provided inside the network (see also the discussion on the ASPN role in Section 2.1.1 and the negotiation scenarios in Section 3.3)
    - Network Interface Manager is controlling the access to different network technologies in order to optimise the mobility behaviour of the terminals

  All these managers are seen from the E2ENP point of view as resources that provide corresponding information for the negotiation (see Section 2.3 and Figure 11). Additional sources of configuration information for E2ENP are the Media Manager, which provides the information about the specific media coding algorithms (i.e. the codecs), and the Policy Manager, which saves different usage policies and it represents the memory about the previous QoS negotiations.

- The E2ENP language introduces the possibility for fine-grained negotiations with the QoS Broker (see Section 1.2.2.2.3 and Chapter 4). Such negotiations are not available within the initial AEQUA/MASA concept [17][197], as the initial service and session coordination language of the QoS Broker can support only *Yes/No* negotiations (see Section 1.2.2.2.1). Within the AEQUA/MASA architecture the *Yes/No* negotiations are applied with the following meaning: when a peer receives a proposal for session establishment with QoS parameters, it accepts this proposal immediately (i.e. a *Yes* answer); the peer may request a modification (i.e. *No* answer) in cases that it cannot cope with the initial proposal or if during the session a QoS change occurs that leads to retrieval of a new QoS specification from the partner terminal (i.e. re-negotiation in terms of AEQUA/MASA). However, in terms of QoS management in heterogeneous terminal, network, services and provider environment such concept is much too optimistic, as reaching an agreement with just a single proposal and not giving the possibility to the partners to express their opinion under the conditions of high heterogeneity of the environment and of the positions hold by the user terminals on behalf of their owners is simply unrealistic (see also the discussions in Section 1.2.2.2). With this respect, E2ENP provides a sophisticated negotiation language (Chapter 4) and an implementation of a negotiation engine with memory for planning adaptations and future negotiations (Chapter 5). The negotiation engine (i.e. E2ENP User Agent) is designed as a manager for the QoS Broker. However, the application of the E2ENP User Agent is not restricted to the AEQUA/MASA Broker, as due to its modular structure, the whole agent or parts of it were successfully integrated into other architectures (e.g. MIND, for which E2ENP was initially designed, [20][130][132] or DAIDALOS [92][201][202] – see also Chapter 5 and Annex A).
- Last but not least, the Validation Procedure for QoS (see Section 2.3.2.2.2) and the implementation of consistency mechanisms (see Section 2.3.2.1.1) introduce the possibility to maintain proper and consistent QoS adaptation tree at all the partners negotiating with E2ENP, unlike the initial QoS Broker, which exchanges just a single QoS configuration at a negotiation round and does not take care of the coordination of simultaneous QoS negotiation requests coming from different partners. Thus, the implementation of E2ENP can guarantee stable adaptation behaviour of the applications due to the consistent common knowledge between the agents using E2ENP and due to the proper planning of adaptation actions ahead based on this knowledge.

The specific contributions of E2ENP to the QoS Broker architecture of AEQUA and MASA are presented in Figure 27 as grey boxes together with indications how the specific E2ENP functions are distributed among the different QoS Broker parts (Message Coordinator, Managers and Controllers). Examples of other then E2ENP technologies used in parallel to E2ENP in the Broker are also given, e.g. RTP is both a carrier and a description format for the streaming media data [1], RSVP [3][4] and DiffServ [5] are methods for fulfilling network reservations, other low-level adaptation mechanisms (e.g. FEC [10]) apply techniques for maintaining the dialogues between the distributed parts of the multimedia application. Furthermore, Figure 27 shows a layered model of E2ENP similar to the OSI layered architecture [54]. E2ENP is designed as a protocol applied at the session, presentation and application layers of the OSI model and it uses the abstraction of transport layer protocols like TCP [122] or UDP [123]. The Session Layer of OSI corresponds to the Carrier Layer of E2ENP and represents the ability of E2ENP to integrate different carrier protocols for managing its negotiation sessions (see Chapter 5). The Presentation Layer of OSI is associated with the Description Layer of E2ENP, i.e. the XML-based description language of E2ENP (Chapter 4). The rest, uppermost layers of

E2ENP correspond to the Application Layer of OSI and are dedicated for applications that are in position to perform negotiations using E2ENP's service and session signalling (see Chapters 3 and 5).

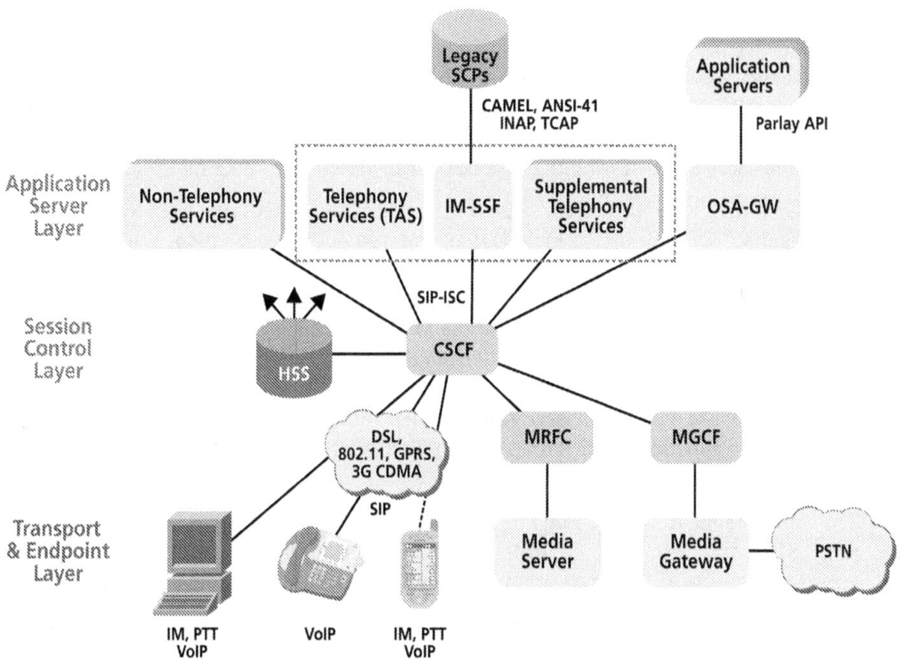

Figure 28: 3GPP IMS architecture [203]

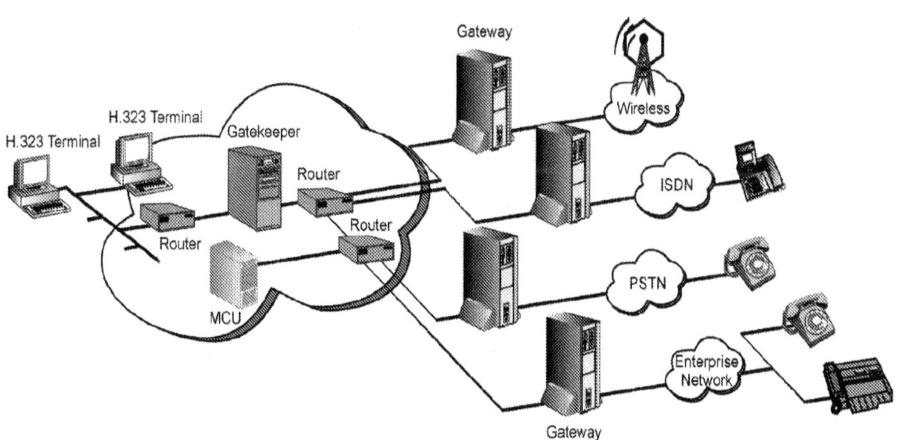

Figure 29: ITU H.323 architecture [204]

2.4.2.2 3GPP IMS and ITU H.323

3GPP IMS [21][22][203] and ITU H.323 [107][204] are the two standards that come into question when next generation networks and services shall be implemented (see also [111] [112][208][209]). This section gives a short overview of these architectures along with an initial study on how E2ENP can be integrated into such systems with provider management. As the work on E2ENP for provider-based interactions with the service and session management protocols was inspired by these provider architectures (see Sections 2.1, 2.3.1 and 2.4.1 and [131] – [133]), the following discussion serves as an initial validation of the E2ENP concept with respect to how providers can manage more efficiently the service and session management of IP-based streaming multimedia services.

The 3GPP IMS architecture consists of different kinds of terminal user-owned devices or of application servers and it includes several provider management components for establishing and supporting different arts of multimedia services (Figure 28). The central component implementing the specific provider management of the users is the Call Session Control Function (CSCF) and according to the 3GPP specification [21][22] this component is in position to *speak* SIP [62] with the terminal parts of the architecture for establishing and managing multimedia sessions. CSCF interacts with the Home Subscriber Server (HSS) that holds the unique subscription profiles of the users. The system implements multiple gateway functions with different complexity and purpose [21] (see also Section 2.1):

- The Multimedia Resource Function Controller (MRFC) provides signalling support (e.g. SIP) and resources controls for the media streams within the network. It corresponds to an enforcement point for provider restrictions on the network
- The Media Gateway Control Function (MGCF) provides an interaction point between the SIP-based session signalling and the signalling between multimedia-gateways on the borders of SIP-based and non-SIP-based network types
- Application Services (e.g. IP Multimedia Service Switching Function (IM-SSF), Open Service Access – Gateway (OSA-GW), Telephony Application Server (TAS), etc.) are used for interactions with different kinds of legacy and novel multimedia application servers

Of special interest for the application of E2ENP within the IMS architecture is the possibility to integrate E2ENP into the SIP transport (see beginning of Section 2.3 and Section 4.1.2). The 3GPP IMS specification has shown interest into interactions of the provider entities (i.e. CSCF) with the session establishment and management protocols [22], however, the currently used description format within SIP (i.e. SDP [86]) is very restrictive with respect to the implementation of such provider restrictions directly into the session descriptions (see also [69]). Hence, the 3GPP IMS can currently implement only terminal-driven resource reservation scenarios (see Section 2.4.1), which might be a drawback for the application of less powerful terminal devices, which cannot perform resources reservations on their own. Furthermore, in order not to disrupt the end-to-end negotiated information and to avoid undesired interactions with non-authorized third parties, the provider has to apply its restrictions and identification explicitly into the description protocol (using Security Identifiers – see Section 2.2.1), which is currently not possible with SDP [69][86]. The specification of E2ENP is in position to enable provider interactions with the session control protocols in a modular way so that different scenarios and combinations of scenarios (see Sections 2.3.1, 2.3.2.2.2 and 2.4.1) can be implemented within the provider management system (see Section 4.3.6). Hence, the E2ENP description format is a possible solution for the 3GPP IMS requirements for provider interactions with the session control protocol. Furthermore, 3GPP IMS can also profit from the hierarchical QoS specification of E2ENP (see Section 2.2.2.2 and Chapter 4) for optimally defining coordination rules for the media as such features are also not available with SDP [86].

The ITU H.323 architecture (Figure 29) is to great extent similar to the 3GPP IMS and displays similar problems (see above). However, this architecture relies predominantly on its own service control specifications [63][107][204] unlike 3GPP IMS where IETF mechanisms [21][22][174] are utilized. The major components of the ITU H.323 architecture are:

- The H.323 terminals that consume the multimedia services
- If available, the Gatekeeper may implement provider-side service management. This component is similar to the 3GPP IMS CSCF component, but its functionality is not fully specified. ITU H.323 [107] identifies only requirements for the Gatekeeper functionality, without going into details on how this component shall interact with the service and session management for the multimedia
- The Multipoint Control Unit (MCU) and the Gateways implement switching functions for enabling conference services (i.e. MCU) and network interoperability between H.323 networks and other systems

The E2ENP concepts and implementation propose a specific solution for provider interactions with the service and session control protocols applicable within the scope of ITU H.323. Hence, these E2ENP characteristics can be utilized for enhancing the Gatekeeper functionality. The protocol for service and session controls within the scope of ITU H.323 is ITU H.245 [63]. Although this protocol is specified in a different format compared to the E2ENP format (H.245 utilizes ASN.1 [210] and E2ENP uses XML [73][93][94]), interoperability can still be guaranteed as ITU specifies mapping rules between ASN.1 and XML [211][212]. Hence, the advanced features of E2ENP for providing hierarchical QoS specification and provider-mentoring interactions with the service control protocol are also applicable within the scope of H.245 and can be seen as its extensions.

The next Chapter 3 details further the interactions between the terminals and the provider entities for enabling advanced multimedia services with QoS support. Furthermore, it describes specific application scenarios of E2ENP for services similar to those recommended within the scope of 3GPP IMS and ITU H.323.

3 E2ENP Interactions and Scenarios

Many multimedia services can advantageously utilize the possibility to negotiate QoS end-to-end (e.g. telecommunication, conferencing, gaming, on-demand services, etc.). Each one of these applications has its own requirements regarding the way a QoS negotiation should be performed, i.e. which party starts the negotiation, which party sends first its proposal for system/session configuration, are there multiple configurations simultaneously possible (e.g. for conferences), etc. The services may profit from the possibility to remember certain QoS configurations and reuse them repeatedly, e.g. in a form of an enhanced telephone book that stores not only the addresses of the communication partners but also information about their service preferences. Thus, E2ENP introduces different modes and phases for performing the negotiations in order to satisfy the specific negotiation requirements and performance characteristics of different multimedia services. The application of different modes and phases of negotiations in the manner of E2ENP is an extension to the classical Offer/Answer Model [35][36] (see also Chapters 4 and 5) and it introduces new forms of application of SIP [62], without changing the initial header specification of SIP, but providing only additional service possibilities through enhancing SIP's state-machine model (see also Section 5.2.2.4).

Section 3.1 presents the E2ENP API of the extended Offer/Answer Model. The specific time-management of the E2ENP Portable Data Units (PDUs) is shown for the cases that the application is in position to utilize the E2ENP possibility to build a memory out of different PDUs and to refer to them in future negotiation rounds.

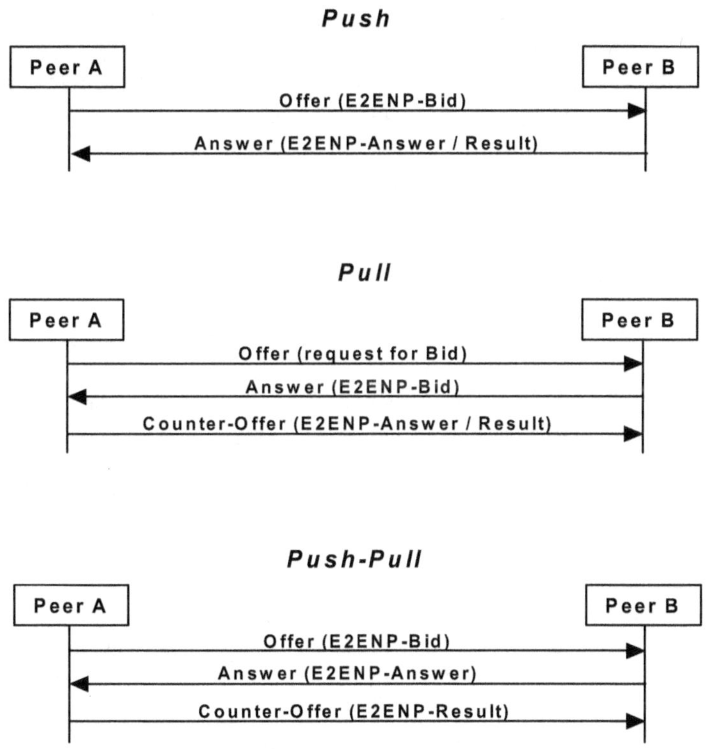

Figure 30: E2ENP modes

E2ENP depends on other then E2ENP's own interactions (see Chapter 2), e. g. in order to use E2ENP as a premium service in a provider-owned architecture the applications need to register with the providers and eventually to exchange with the provider infrastructure profile information that is used for example in the validation algorithm for the E2ENP PDUs (see Section 2.3.2.2.2). Furthermore, the usage of E2ENP with mentoring services introduces security and privacy problems for the applications in situations where the applications need to share their configurations with other entities (e. g. providers or third-party call control services). Considerations about the trust relationship between peers and mentoring entities are presented in Section 3.2 and necessary interactions that serve as preconditions for establishing of E2ENP negotiations are discussed.

Some specific scenarios with mentoring through third-party call control entities are presented in Section 3.3. These scenarios display the integration of the E2ENP Offerer/Answerer User Agents into higher-level adaptation engines that are in position to adapt services not only using server and/or receiver adaptation but also applying session mobility between two terminal devices or integration of intermediary multimedia adaptation services on the path between the actual senders and receivers. Sections 3.2 and 3.3 present thus innovations in the provider management under conditions that E2ENP's fragmentations of hierarchical QoS specifications and E2ENP's memory management for PDUs is applied.

3.1 Basic Protocol Structure and Application

This section introduces the basic end-to-end interactions with E2ENP. It presents the E2ENP User Agent API for negotiations towards SIP and towards the adaptive application/middleware. The E2ENP UA shares with the application a memory on the exchanged PDUs in order to provide consistency management at PDUs referencing and at system resource orchestration (see Section 2.3 and Figure 11). Hence, the management of this E2ENP UA memory is also discussed.

The basic E2ENP scenario requirements and interactions were initially published in co-authorship by the author of this book in [14][20][132][170][173].

3.1.1 Modes

E2ENP is in position to support three different modes for exchanging the negotiated data (Figure 30):

- **Push Mode** is the basic E2ENP interaction implemented through the E2ENP UA APIs (see Section 3.1.3). It is the shortest interaction of all the E2ENP modes. In this mode, the Offerer (Peer A) provides in its request the information about its QoS requirements and the Answerer (Peer B) replies with its requirements, providing in addition the final result for the service configuration (see also [20][35]). The Push Mode represents usually just a single E2ENP phase (see below Section 3.1.2) compared with the rest of the modes and their realisation in SIP, e.g. the push mode at Pre-negotiation uses just the SIP methods OPTIONS [62] or MESSAGE [213] for exchanging the negotiated information and just INVITE [36][62] at Negotiation/Re-negotiation (see also Section 3.1.3). This interaction represents the classical Offer/Answer Model [35][36] as defined within the IETF standardisation.
- **Pull Mode** is dedicated predominantly for automatic services where the call initiator (Peer A) does not send the initial QoS request, but asks its partner (Peer B) to deliver

first its QoS proposal which is used then from the initiator to compose its own QoS requirements and the final result for the QoS configuration of the service. This enables the initiator of the negotiation to adapt its QoS requirements to the specific configurations of the automatic service as such services (e.g. a VoD Server) usually are not in position to support any individually chosen codecs or QoS configurations. The realisation of the Pull Mode in SIP is performed, for instance, using a combination of two OPTIONS methods at E2ENP Pre-negotiation or with a combination of OPTIONS and INVITE methods at Negotiation/Re-negotiation. The actual implementation of the Pull Mode is accomplished not within the E2ENP UA API (see Section 3.1.3), but using the E2ENP language (see Section 4.3.2), thus the adaptive application/middleware is in position to use the same E2ENP UA API and SIP methods for realizing different E2ENP modes.

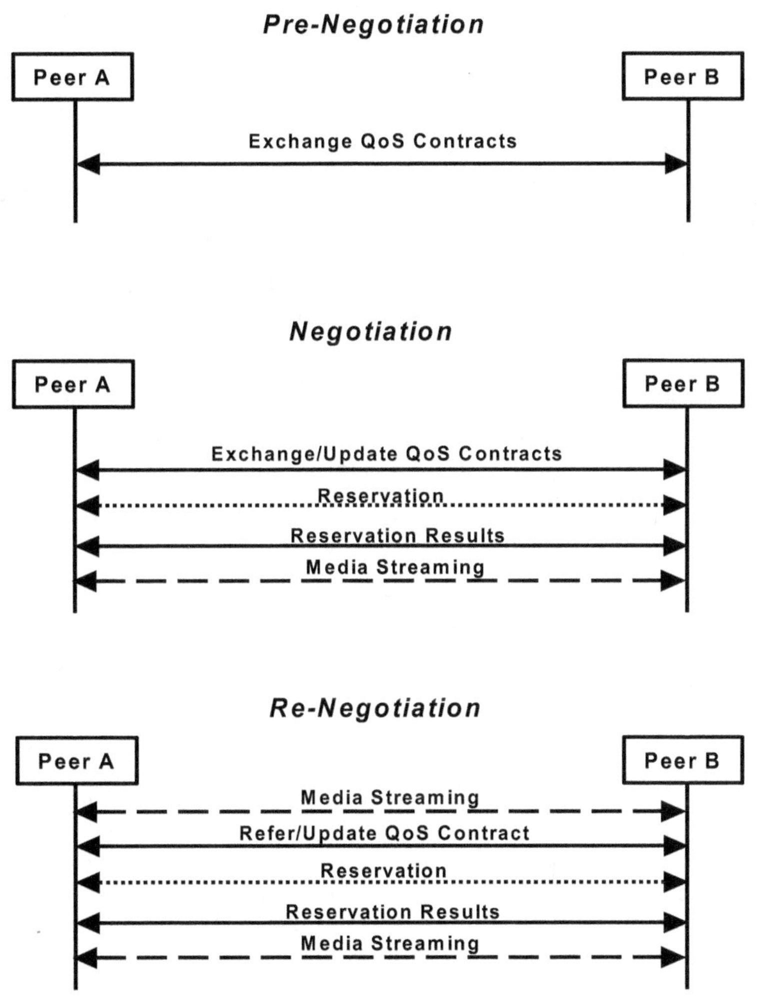

Figure 31: E2ENP phases

- **Push-Pull Mode** purpose is the support of interactive services, where the initiator (Peer A) and the responder (Peer B) have to exchange information about bi-directional communication and the final decision can be taken only after the Peer A has received the required configurations of Peer B. As in the case of the Pull Mode (see above), the Push-Pull Mode's implementation is a combination of multiple SIP methods and E2ENP phases.

The utilization of the three modes of E2ENP enables the applications to adapt to different negotiation situations in accordance with the specific negotiation needs of the services and in consideration with the flexibility of the application to react to negotiation requests coming from their communication peers.

3.1.2 Phases

The E2ENP consists of three major phases (Figure 31). The solid interaction lines in Figure 31 express the E2ENP signalling. The dotted and the dashed lines show the interactions of the peers for communication that is being orchestrated via E2ENP but is not a part of the negotiation protocol, e.g. *Reservation* with RSVP [3][4] (dotted lines) and *Media Streaming* with RTP [1] (dashed lines). Depending on the used services and scenarios, the different phases can be applied in different negotiation modes (see above – Section 3.1.1). The E2ENP phases are:

- **Pre-negotiation** – This phase deals with the negotiation of configuration information valid for more than one multimedia session, i.e. a service configuration. Such information remains unchanged over longer periods and it can be leased between the end-systems. The way the peers lease and remember configuration information is discussed in Sections 3.1.4, 4.3.2 and 5.2. The terminals exchange common basic configurations within Pre-negotiation phase, e.g. supported codecs, QoS contracts, etc. In some application scenarios the Pre-negotiation can be associated with the usage of advanced location services, which in addition to the location information (e.g. phone, host-numbers, etc.), deliver also codec descriptions and QoS contracts specified for the respective end-device entry in the location service (see Section 3.2.4). Having the Pre-negotiation information in advance, the end-systems can speed up the negotiation process for session establishment. The realisation of the Pre-negotiation phase with SIP methods is discussed in Section 3.1.3.

- **Negotiation** – The information exchanged during this phase deals with the configuration of, the establishment of and the enforcement of QoS conditions for a specific multimedia session. If the terminals cannot (e.g. due to limited memory) or are not allowed to perform Pre-negotiations (e.g. due to system policies or specific service features), the necessary general system configurations can be exchanged within the Negotiation phase together with the information about the specific session establishment (see Section 4.3). Thus, two types of Negotiation are defined – a *short* Negotiation, which uses references to the data exchanged in a former Pre-negotiation phase and a *full* Negotiation, which contains both service information and specific session configurations. The additional information for configuring a multimedia session within a Negotiation phase describes QoS contexts, i.e. how many media streams should be opened for the session; how these streams are associated with the applied codecs and basic QoS configurations; and what kinds of QoS-adaptation and time-synchronization constraints are applied for these streams. The QoS contracts and contexts are used for multimedia-session establishment and for any eventual QoS adaptation process that might take place during the multimedia session. The information exchanged in a Negotiation phase might also be leased between end-systems for easily applying QoS configurations to repeatable application scenarios, e.g. a multimedia conference concerning similar topics (see Sections 3.1.4). Details on the implementation of the Negotiation phase using SIP are shown in Section 3.1.3.

- **Re-negotiation** – This phase deals with the enforcement of QoS contracts and contexts within the system whenever *explicit* indication of adaptation conditions and global resources adaptation of the service including terminals and/or network reservations is necessary. Additionally and complementarily to the explicit E2ENP Re-negotiation process, fine-grained re-negotiations can take place transparently to E2ENP applying traditional means (e.g. changing the RTP payload-types within the media packets [2]). This process would be applied as long as the corresponding QoS change is not considered a QoS violation – this means that the streaming process stays within the limits of the predefined and currently enforced QoS contract/context (see also Section 2.3.2.1.2 – *fourth requirement for E2ENP*). If a QoS contract/context different from the currently applied and active one should be enforced due to QoS violations, the E2ENP Re-negotiation indicates this change. The Re-negotiation can also rely on information exchanged in some previous E2ENP phases or alternatively can specify new configurations to be applied for the further duration of the existing and active multimedia session. Similar to the Negotiation *short* and *full* Re-negotiations are defined. A *short* Re-negotiation uses only references to previously exchanged data. The *short* Re-negotiation is thus an efficient process due to the referencing of previously negotiated data that leads to minimization of the signalling impact within the duration of a media session (see also the measurements in Chapter 6). A *full* Re-negotiation enhances or exchanges the system/session configurations whenever service upgrades/downgrades occur that cannot be managed with the already known (from previously performed E2ENP negotiations) configuration information. The Re-negotiation phase using SIP is shown in Section 3.1.3.

Both the Negotiation and the Re-negotiation phases serve also as means of resource-reservation coordination at media session establishment or for QoS adaptations of running multimedia sessions (see Section 3.1.3).

The E2ENP design allows different combinations of the QoS configuration information to fit the requirements of different end-devices and system policies. This means that each E2ENP PDU that carries configuration information no matter to which phase it belongs can be referenced from within other PDUs in any of the E2ENP phases. This mechanism is used to build a memory for the application for performing the most appropriate session establishments and adaptations corresponding to the current system environment (see also Sections 3.1.4 and 4.3.2). With this respect, the Pre-negotiation, the Negotiation and the Re-negotiation are only the names of the corresponding E2ENP phases which indicate if the negotiation process occurs independently of a session (Pre-negotiation), at the very beginning of a session (Negotiation) or during a running session (Re-negotiation). However, as an example, a Pre-negotiation or a Negotiation PDU can be re-negotiations from the point of view of the overall service if the information within these PDUs refers to PDUs exchanged in earlier E2ENP phases. This means that the E2ENP PDUs can freely reference each other as long as the previously exchanged configuration information is applicable for referencing within the scope of the performance logic of the service (see Section 3.1.4), thus giving the adaptive services the possibility to minimize the exchange of configuration information through fragmentising the negotiation data into smaller negotiation bids and through using only references to known data, e.g. at adaptation. This special E2ENP characteristic allows the minimization of the signalling traffic to reduce the application costs for a negotiation process (see Section 4.3, the respective measurements in Chapter 6 and [14], [20], [131] – [133] and [173]).

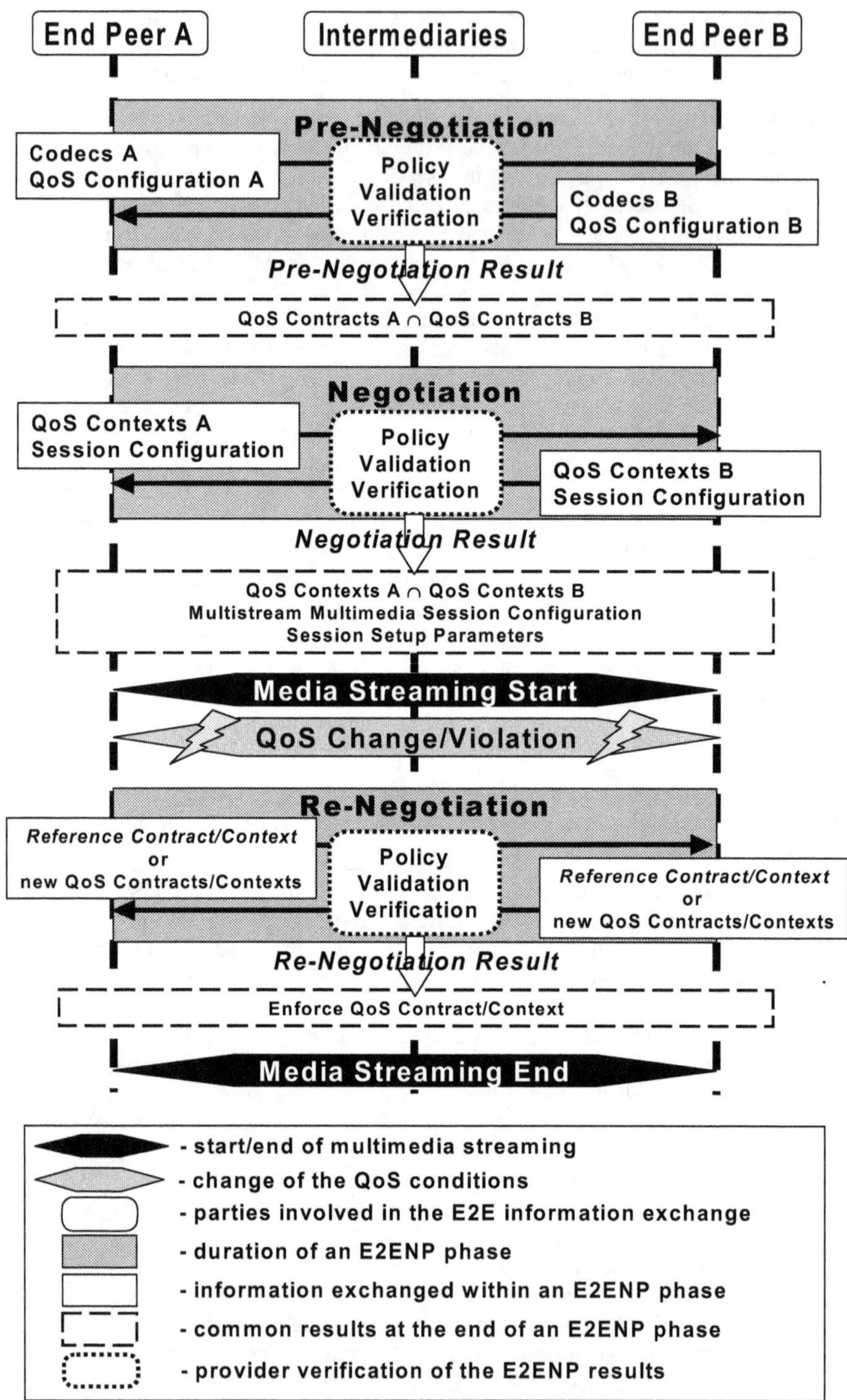

Figure 32: Application of the E2ENP phases with a single session

Current implementations of application signalling and negotiations [13], [16], [17], [19], [35] – [37], [81] that have features comparable to E2ENP are not in position to build such long-term memory, hence, they may appear inefficient in cases of managing complex multimedia applications as they have to exchange the complete service configurations at each QoS violation and respective re-negotiation of the service. Furthermore, some of the underlying network and transport mechanisms (e.g. the description carrying protocols like SIP [62] or connection management mechanisms like UDP [123], see also [214]) may apply restrictions to the sizes of the sent messages, which would make the application of the above mentioned state-of-the-art mechanisms impossible for complex multimedia services, as the descriptions of the services using such techniques would be too long for the successful transfer between the negotiating peers. The current session negotiation mechanisms cannot exchange configuration information piece-by-piece in form of multiple, referencing each other PDUs to limit the size of the exchanged data in the manner E2ENP does when applying different modes and phases. Of course, compression can be a solution for the state-of-the-art mechanisms. However, it introduces an additional processing step in the signalling that is avoided with E2ENP. Furthermore, using E2ENP and compression enables further reduction of the signalling traffic in comparison with the state of the art (see also Section 4.1.3.3). The specific E2ENP PDU building and referencing mechanisms characterizing the E2ENP description language are described in further detail in Sections 3.1.4 and 4.3.

Figure 32 shows an example of interactions of the E2ENP phases with a single multimedia session. This figure displays also the communication with the provider entities at validation of the E2ENP PDUs during the E2ENP negotiation phases (see also Section 2.3.2.2.2). The example shows the typical information exchange between the communicating partners in case they are in position to perform short negotiation phases, i.e. at Pre-negotiation the codecs and the respective QoS contracts for the service are exchanged, at Negotiation the QoS contexts of the multimedia session and the specific other information for establishing the session (e.g. ports for the media streams) are delivered and at Re-negotiation references to available contracts and contexts are communicated. Other combinations of the configuration information are also possible as discussed in the previous paragraphs (see short/full phases). Before starting of a session, the terminals (End-Peer A and End-Peer B) can perform one or multiple Pre-negotiations concerning the service to which the future session shall belong to and the session signalling proxies of the provider/-s (indicated with *Intermediaries*) prove the validity of the E2ENP PDUs. The Pre-negotiations may or may not reference each other. This depends on the logic of the service to build different session contexts or to enhance an existing session context with additional QoS contracts and contexts. Note that the E2ENP description language builds contexts up to session level (see Section 4.3) and if the service wishes to support contexts at higher level (e.g. of the kind "my service at home" and "my service for business") the service has to perform contextually different and independent Pre-negotiations that are however associated internally with one and the same application. Each group of mutually dependant Pre-negotiations would then correspond to an independent sub-tree of the hierarchical QoS specification that matches the context of a single session or of a single service configuration (see Section 2.2.2.2 and Figure 9 and Sections 3.1.4 and 4.3). The session start is indicated with a Negotiation phase and there is just *one* Negotiation per establishment of a single session. Multiple subsequent Negotiations for one session are possible only in case that the first Negotiation phase ends with a failure and no session is established or in case of multiparty services (Section 3.3.2). In case of a Negotiation, the providers also prove the Negotiation PDUs to validate the QoS context of the session. A Negotiation PDU may or may not refer to Pre-negotiation PDUs. The actual start of the session is indicated in Figure 32 with *Media Streaming Start*.

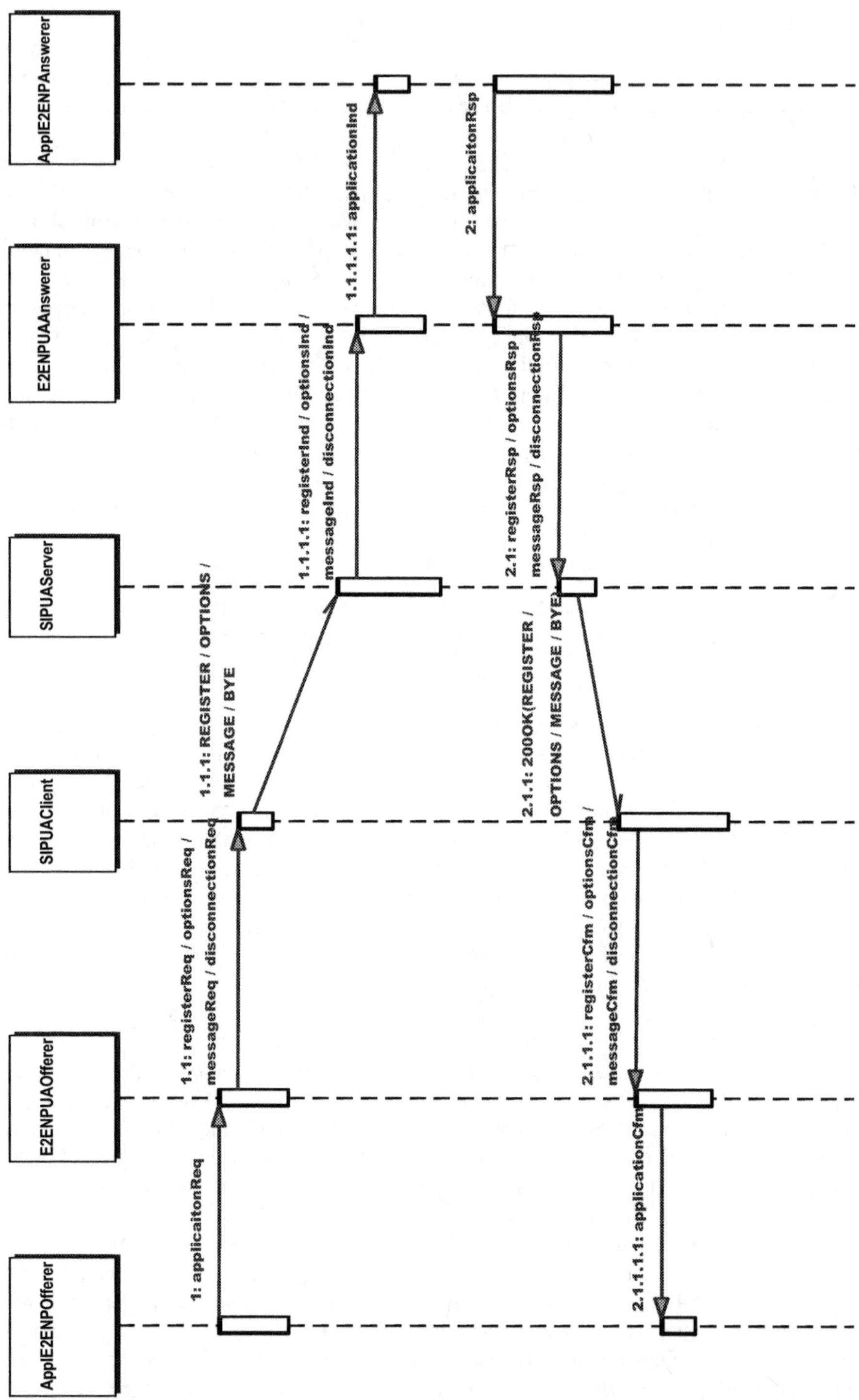

Figure 33: Pre-negotiation, registration and session-end using SIP and E2ENP

Whenever *QoS Change/Violation* occurs, the adaptive application may consider appropriate to perform an explicit Re-negotiation with E2ENP. As adaptations may occur multiple times within the duration of a session, multiple E2ENP Re-negotiations associated with a single session are possible. The providers prove also the Re-negotiations as the may enhance the context of a session with new QoS contracts/contexts. E2ENP contains also a commando for closing of multimedia session (*Media Steaming End*) performed in SIP with the BYE method [62] (see Section 3.1.3).

3.1.3 E2ENP Phases Implementation Using SIP Methods

This section presents the E2ENP signalling at User Agent (UA) registration procedures and the implementation of the E2ENP phases concept for realisation of similar communication to the Offer/Answer Model [35][36] using the Session Initiation Protocol [62] (see also Section 2.3.1). The E2ENP UA implements a low-level API for interacting with the carrier protocol and a high-level API for interacting with the application (see Figure 33 and Figure 34AB[17]). The design of the two APIs is general enough to enable the application of different carrier protocols (see also Sections 5.2 and 5.2.2.3) and to provide the applications of E2ENP with a generalized access for performing negotiations. Thus, the E2ENP UA represents the separation between the carrier-specific communications for realisation of the data transportation and the high-level application requests for negotiations, abstracting the application from the specific transport-management issues. The SIP specification [62] and implementations thereof (e.g. jSIP [184] or NIST SIP [185]) are designed for single application usage, as the SIP specification was initially planned for VoIP applications only. However, the E2ENP UA provides a new possibility to use SIP within middleware architecture to support multiple (also different then VoIP) services. Details about the internal structure of the E2ENP UA and its implementation are presented in Chapter 5. This section discusses the Message Charts (MSCs)[18] that represent the external APIs of the E2ENP UA Offer/Answer Model implementation, i.e. the communication APIs with the SIP UA and with the application (see Figure 33 and Figure 34). All the E2ENP UA APIs are implemented according to the communication primitives specification of ISO OSI for connection-oriented network services (see also [215]). The respective primitives of the Offer/Answer Model API of E2ENP UA are designed as follows:

- *Request* – The Offerer application uses requests for indicating the E2ENP UA about the start of a registration, a negotiation or a session disconnection action. The E2ENP UA at the Offerer side applies requests towards the SIP UA for initiating registration, data transportation at negotiation and session disconnection actions.
- *Indication* – The Answerer application is indicated from the E2ENP UA about a started registration, negotiation or session disconnection action from the Offerer side of the service. Respectively, the E2ENP UA is indicated from the SIP UA about corresponding actions initiated from the Offerer side of the service.

---

[17]Note that the separate pages of the split message chart in Figure 34AB display some overlapping of the single pictures and the reader has to follow the respective repeated message numbers (here 8.1.1) and names of the massages in order to understand the sequence of the messages.

[18]Note that the MSCs shown in this document display slight differences in their appearance due to the used different versions of the respective modelling software (i.e. Together [216])

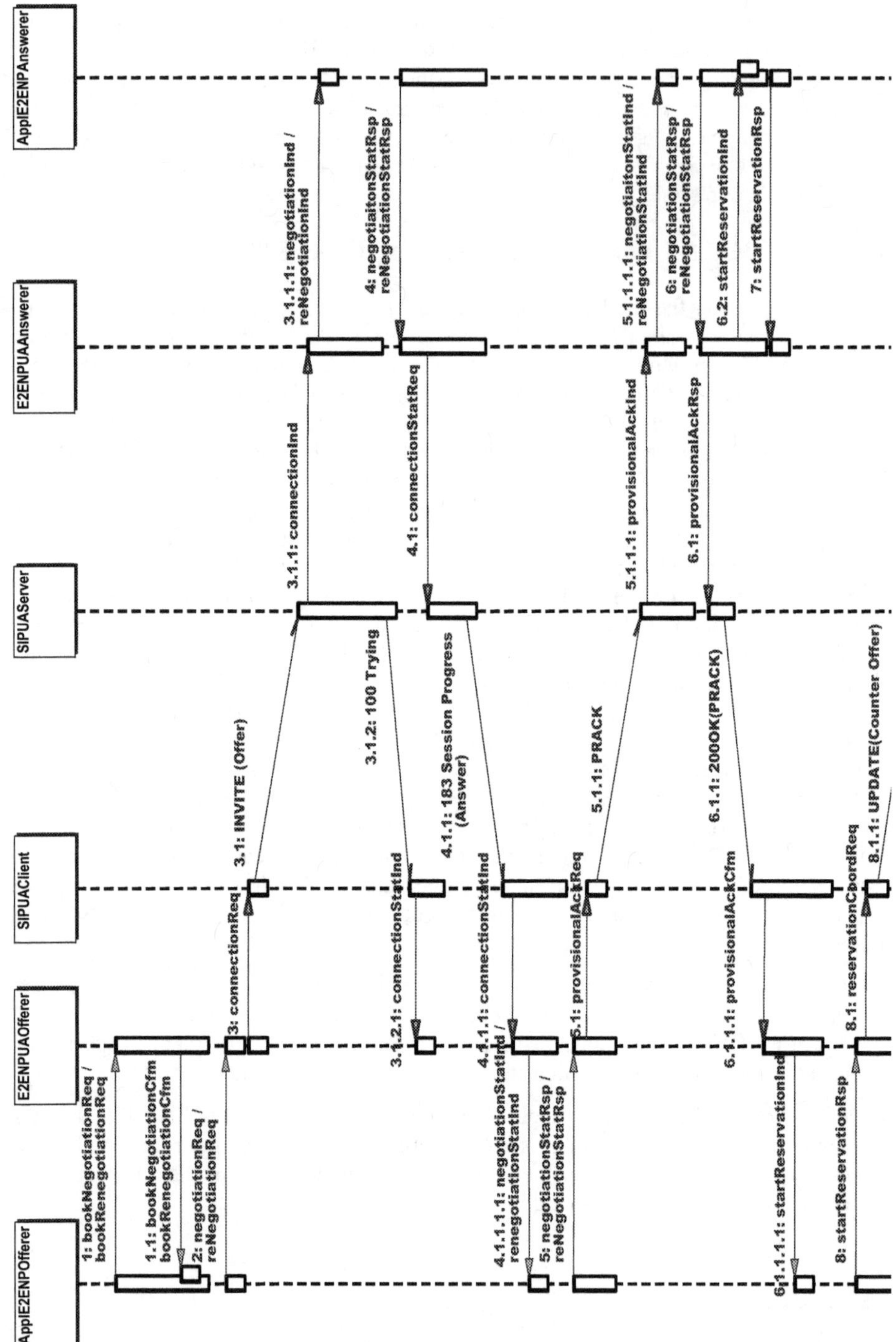

Figure 34: Negotiation and re-negotiation using SIP and E2ENP (A)

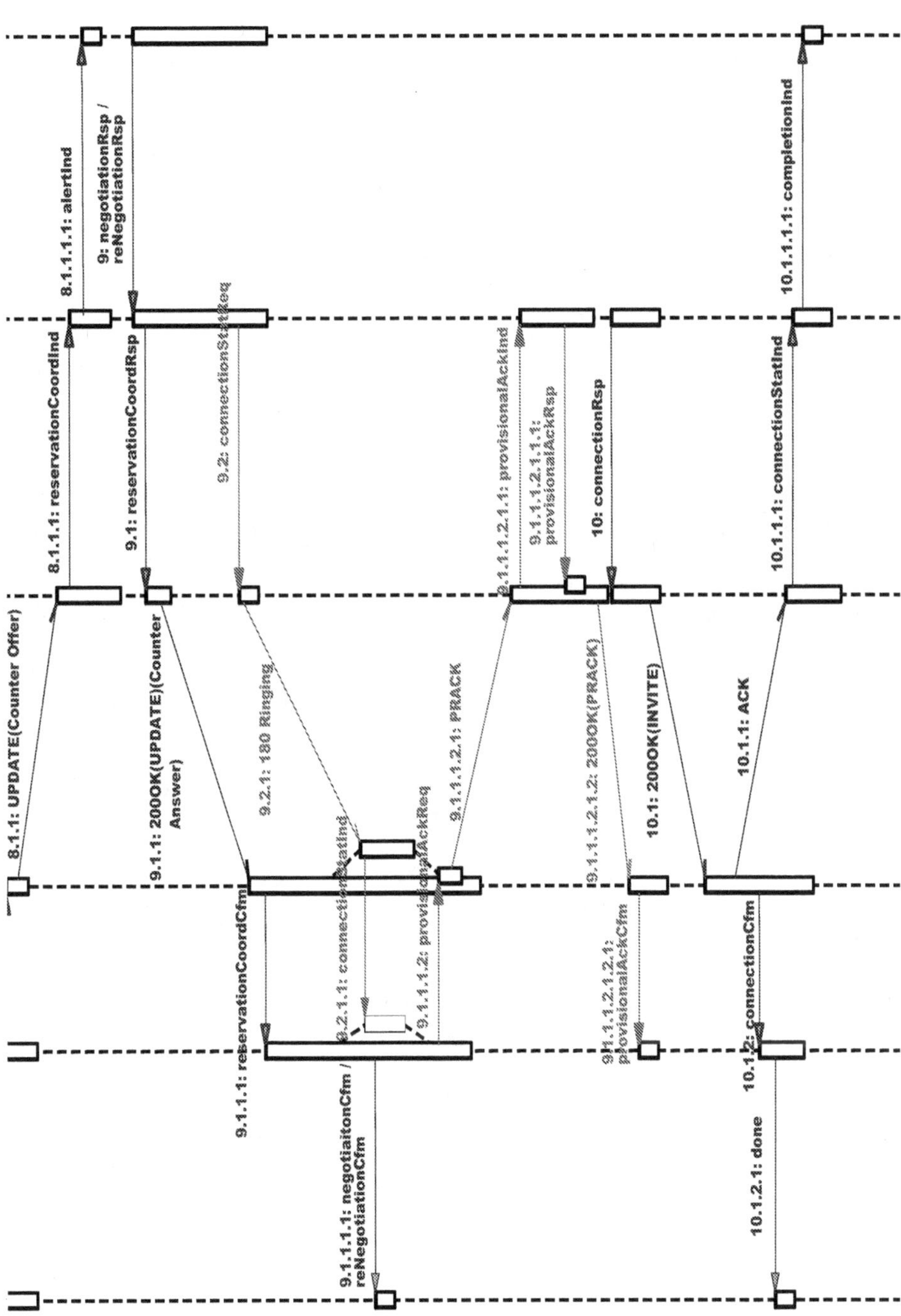

Figure 34: Negotiation and re-negotiation using SIP and E2ENP (B)

- *Response* – The Answerer application sends a corresponding response after processing the Offerer request to the E2ENP UA and the E2ENP UA delegates the respective answer information to the SIP UA.
- *Confirmation* – The responses are received at the Offerer side using confirmations. The SIP UA sends a confirmation to the E2ENP UA and the E2ENP UA delegates the appropriate information to the Offerer application using also a confirmation.

The E2ENP UA APIs expose also independent combinations of request and confirmation and of indication and response calls in cases that the adaptive application and the E2ENP UA are adjacent and co-located (see Figure 34). The application in such cases is the client side and the E2ENP UA the server side. The E2ENP UA and the adaptive application use this asynchronous client/server communication model (also producer-consumer pattern [146]) to exchange messages that serve the synchronisation and the coordination of only the service internal components at either the Offerer or the Answerer side (see also Chapter 5). Such actions are not associated with external communication between the distributed parts of the service.

Note that all E2ENP UA API methods are planned to carry additional descriptions that correspond to configuration and/or negotiation information, respectively to SIP header and/or message body information [170]. Although the current state of the SIP specification does not allow inserting message bodies into some of the methods (see [36]), the E2ENP UA API implementation disregards this rule in order to keep its negotiation framework open for other then SIP implementations. The E2ENP UA API uses where appropriate an abstraction from the SIP API in order to generalise the application-specific actions and to distinguish them from the actual protocol implementation. A simple action corresponding to a single SIP non-INVITE transaction (see Figure 33) is mapped directly to the application-logic API of E2ENP in cases that SIP already provides the corresponding univocal control feature. However, complex SIP transactions like the Offer/Answer Model with reservation (Figure 34) or the invocation of external actions (see Section 3.3.1.2) can be associated with further signalling abstractions, e.g. the Offer/Answer Model with reservation corresponds to two API functions, for session establishment/re-negotiation and for resource-reservation coordination. Since, the E2ENP UA takes over some coordination activities on behalf of the application, the generalised API corresponds to the invocation of these actions from the side of the application. Furthermore, the abstraction provided via the E2ENP UA enables the application to transparently use different then SIP carrier protocols.

The Figure 33 and Figure 34 display the E2ENP UA API for the E2ENP-based Offer/Answer Model. This model was initially specified in co-authorship by the author of this work in [170] and [173]. In the following part of this section, an extensive overview on the E2ENP-based Offer/Answer Model is given.

Figure 33 shows the E2ENP UA API in cases that SIP non-INVITE transactions shall be used within the scope of the E2ENP-based Offer/Answer Model (see also Section 2.3.1.2). Such transactions are applied to implement four different E2ENP actions that serve the configuration of the E2ENP UA, the pre-negotiations and session disconnection actions:

- Registration of the E2ENP UA with a SIP-based E2ENP Registrar (see also [62] and Section 3.2.2) – In this scenario, the Offerer side is a client E2ENP UA (e.g. a VoIP terminal) and the Answerer – a Provider Registrar (see also Sections 2.1, 2.3.2.1 and 2.4). This scenario is associated with the local configuration of the E2ENP UA (Section 5.2 and [170]) and with the binding of the terminal E2ENP UA with the provider network. The application call corresponds to the E2ENP API call sequence – *bindReq*, *bindInd*, *bindRsp*, *bindCfm*. In case that RMI [88] is used to simulate SIP behaviour (see Section 5.2.2.3), the binding corresponds also to the binding of the UA to the RMI registry. The respective SIP message used in this action is REGISTER [62]. The registration may or may not be associated with sending of description information between the terminal and the provider. If exchanging provider rules at registration, these rules are specified differently then the E2ENP language (see Chapter 4) as the provider rules are only the constraints to the negotiation that consider the network and the service specific characteristics and such characteristics do not represent complete negotiation proposals (see Section 3.2 and Annex B).

- Initial Pre-negotiation phase (see Section 3.1.2) is implemented using the SIP MESSAGE method [213]. Although planned for instant messaging between human consumers, the MESSAGE method can also be used to exchange E2ENP Pre-negotiation PDUs, due to the fact that the specification [213] does not limit the content-type of this SIP method to a specific Multipurpose Internet Mail Extension (MIME) [217] (see also Section 4.1.2) and secondly the E2ENP Pre-negotiation can be considered as a kind of instant messaging between the E2ENP UAs. An initial Pre-negotiation phase carries a new set of configurations (for initial building or for enhancing of QoS specifications) that are still unknown to the application, respectively to the UAs. The application call of a Pre-negotiation corresponds to the E2ENP API call sequence – *preNegotiationReq*, *preNegotiationInd*, *preNegotiationRsp*, *preNegotiationCfm* (see also Section 5.2.2.4.2). This call sequence is also used within E2ENP pull and push-pull modes at Negotiation/Re-negotiation (see Section 3.1.1). In such case, the messages carry Negotiation/Re-negotiation PDUs. The application knows that this message sequence shall be immediately followed by an INVITE sequence when the applied modes are push or push-pull (see below and the sequence in Figure 34).

- Renewal of E2ENP PDU information is associated with the management of the E2ENP PDU memory (see Sections 2.3, 3.1.2, 3.1.4 and 4.3.7) and is implemented using the SIP OPTIONS method [62]. Since the E2ENP PDUs can be timely limited and may expire or the configurations leasing between the Offerer and the Answerer may undergo premature termination, this method is used to extend the expiry time of the leased information and to keep the respective configurations in the application/E2ENP UA memory or to cancel non-expired PDUs and to remove them from the configuration memory (see Sections 3.1.4 and 4.3.7). The renewal of the E2ENP PDU leasing is a special kind of Pre-negotiation phase (Section 3.1.2) that can be performed before or during running sessions. However, the leasing is independent of the running sessions as it manages only the E2ENP PDU memory without affecting the state of the sessions' resources. The application call of the leasing renewal corresponds to the E2ENP API call sequence – *renewLeaseReq*, *renewLeaseInd*, *renewLeaseRsp*, *renewLeaseCfm* (see also Section 5.2.2.4.2).

- Termination of a running multimedia session is signalled using SIP BYE method [62]. This method finalizes also the SIP INVITE dialogue (see below and Section 2.3.1.2). The application call of the media session termination corresponds to the E2ENP API call sequence – *disconnectionReq*, *disconnectionInd*, *disconnectionRsp*, *disconnectionCfm* (see also Section 5.2.2.4.2). The application of additional information (i.e. message bodies) in the disconnection messages is currently not used, however it can be applied to provide reasoning information for closing the session, e.g. in cases that the session is interrupted due to erroneous behaviour of the application/network or due to pre-emption of a multimedia session by a higher-priority session or application (see also [170]).

Figure 34AB depicts the SIP INVITE transaction that is used within the scope of the E2ENP-based Offer/Answer Model for establishing of a multimedia session and for re-configurations of a multimedia session (see also the theoretical explanation of the Offer/Answer Model in Section 2.3.1.2). The sequence presented in Figure 34 describes the push mode of the E2ENP, respectively the second part of the push and push-pull modes (see above the description of Figure 33 and Section 3.1.1). Every initial SIP INVITE corresponds to an E2ENP Negotiation phase (see Section 3.1.2) and establishes the SIP dialogue (see also Section 2.3.1.2 and [62]). All the following INVITEs after this initial INVITE belong to the established SIP dialogue and correspond to E2ENP Re-negotiation phases (Section 3.1.2) under condition that no BYE is sent between the single INVITEs to terminate the media session and the SIP dialogue [62]. As already mentioned in Section 2.3.1.2, the Re-negotiation has fewer messages then the Negotiation due to the specific coordination of the ringing process used only at session establishment.

At Negotiation/Re-negotiation, the Offerer application has to book first this action within the E2ENP UA (i.e. API calls *bookNegotiationReq*/*bookNegotiationCfm*, respectively *bookRenegotiationReq*/*bookRenegotiationCfm*) to guarantee the atomic processing of the E2ENP transaction (see Figure 34A and the MUTEX explanation below). Unlike SIP [62] and the Offer/Answer Model [35][36], the E2ENP UA provides ACID[19] processing for the transactions of the E2ENP-based applications that have equal priority (see also Sections 2.3.2.1.1 and 5.2). However, the atomicity and the durability are not that strictly applied in the E2ENP UA compared to the classical sense of ACID [50][52]. An E2ENP transaction can be interrupted, for instance, due to application/network misbehaviour and it is up to the service logic how to treat the intermediary communication results (see Section 5.2.2.4.1). In order to keep the intermediary results consistent, the E2ENP UA implements a data-caching mechanisms for supporting the applications' decisions (see Sections 2.3.2.1.1 and 5.2.2.2). The E2ENP UA alone provides persistency for the negotiated data only within its life period, i.e. if due to termination of the E2ENP UA some PDU information contained in the UA is lost, the application of E2ENP should take its own measures to guarantee persistency for its PDUs or to re-invoke the E2ENP UA memory at new instantiation of the UA (see also Section 3.1.4 and Chapter 5).

The priority of the transaction is determined by the adaptive application/middleware on behalf of the user and her/his preferences and can be signalled between the peers using the SIP subject and/or priority headers (see [62][218][219]). The priorities are applied to provide transaction postponing of lower priority calls thus giving preferential treatment to higher priority calls at resources distribution within the terminals. The booking of the E2ENP UA produces a lock within the E2ENP UA MUTEX[20] (Section 5.2.2.4.1) to enable the atomic processing of a negotiation transaction (i.e. an E2ENP phase). This MUTEX is used to accomplish several actions, i.e. to guarantee the consistency and the deadlock avoidance at application of E2ENP (see Section 2.3.2.1.1 and the following explanation of the resource-reservation processing for network reservations) and to enable the enforcement of priorities through application of a queue for storing the low-priority calls while a high-priority call is processed (see Section 5.2.2.4.1). If the booking of the E2ENP

---

[19]ACID – atomicity, consistency, isolation and durability are the features of a distributed transaction [50][52].

[20]MUTEX – Mutual Exclusion is a management procedure for concurrent processes [50][52]. The E2ENP UA implements the MUTEX as a state machine (Section 5.2.2.4.2).

transaction is successful, the connection can be opened and the Offerer sends its proposal to the Answerer (i.e. call sequence *negotiationReq* or *reNegotiationReq*, *connectionReq*, INVITE, *connectionInd* and *negotiationInd* or *reNegotiationInd*). The booking of the MUTEX at the Answerer side is triggered through the *negotiationInd/reNegotiationInd.* Note that booking and releasing of the MUTEX at the Offerer and the Answerer applies as well for non-INVITE transactions when managing E2ENP calls with different priorities (see above and Figure 33). At a non-INVITE transaction, an application tries to lock the MUTEX once, as soon as a call is launched at the Offerer or the Answerer. In case of not being successful, the application may try again later, since the non-INVITE transactions are not that critical for the resource management of the system (see Sections 5.2.2.4.1 and 5.2.2.4.2). On the contrary, an INVITE transaction is critical for the resource management, consequently, the explicit booking of the MUTEX enables the Offerer/Answerer to try successively several times to get the MUTEX lock before ending with a failure indication. Furthermore, the implementation of automatic successive attempts to lock the MUTEX is a part of the E2ENP UA abstraction for the application, to prevent that the services actively wait for being scheduled for network communication (see Sections 5.2.2.4.1 and 5.2.2.4.2).

The SIPUAServer at the Answerer side replies with *100 Trying* whenever a new request from the Offerer side arrives at the Answerer. The *100 Trying* message serves for securing the connection between the Offerer and the Answerer, e.g. when using UDP (see also the SIP transaction state machines in Section 2.3.1.2). The SIPUAClient delegates *100 Trying* to the E2ENPUAOfferer to provide the status of the network connection to the E2ENPUAOfferer (*connectionStatInd*). This message is shown in Figure 34A only once.

The successful decision of the Answerer application on how to configure/re-configure the multimedia session leads to an *Answer* that is sent to the Offerer side (i.e. call sequence *negotiationStatRsp* or *reNegotiationStatRsp*, *connectionStatReq*, 183 Session Progress, *connectionStatInd* and *negotiationStatInd* or *reNegotiationStatInd* in Figure 34A). The Offerer replies with its status about the fulfilment of the resource preconditions, which action determines the start of the network reservation process that is coordinated through the E2ENP UAs (i.e. call sequence *negotiationStatRsp* or *reNegotiationStatRsp*, *provisionalAckReq*, PRACK, *provisionalAckInd* and *negotiationStatInd* or *reNegotiationStatInd*, see also [36]). The Answerer replies that it is also ready for the resources-reservation coordination (i.e. call sequence *negotiationStatRsp* or *reNegotiationStatRsp*, *provisionalAckRsp*, 200OK(PRACK), *provisionalAckCfm* and *startReservationInd*). Both the Offerer and the Answerer E2ENP UA achieve preferential treatment at network reservations, as the MUTEX guarantees their critical phase through artificially increasing the priority of the corresponding control session (see Section 5.2.2.4.1). The call *startReservationRsp* from the application side signals the end of the critical reservation section. In cases that E2ENP coordinates the network reservations, the reservation process is not interruptible and the continuity of this process is guaranteed with the critical phase realisation of the MUTEX in the E2ENP UA. However, the reservation phase is bound to a timer in order to guarantee fault tolerance to the application in case that the network reservation crashes in the critical phase of the MUTEX, resulting in blocking of the complete system. At timer expiration, the MUTEX is released with an error message and it is up to the application if it shall start without having network resources reserved or shall cancel the processing of the transaction (e.g. using SIP CANCEL method or some SIP failure response in accordance with the role, i.e. Offerer or Answerer, that reacts first to the failure condition, see also [62]). Note that the actual realisation of the critical section for the reservation (see Section 5.2.2.4.2) lasts until the completion of the transaction (see below) due to coordination reasons concerning the SIP recommendation [62], i.e. the subsequent UPDATE is not obligatory and the execution of the ringing and the media-session-opening confirmations are per definition [62] not interruptible. Hence, if an UPDATE is not present the transaction completes together with the completion of the reservation. However, an UPDATE can also be launched from the side of the Answerer [220] for the purpose of bi-directional resources coordination. Consequently, such UPDATE has to be considered within the Answerer's resources-coordination mechanism (see Section 5.2.2.4.2).

The Offerer side call of *startReservationRsp* (and the sequence *reservarionCoordReq*, UPDATE, *reservarionCoordInd* and *alertInd* – see the transition between Figure 34A and Figure 34B) can be associated with a counter offer included in the call that indicates the level of success of the network reservation, i.e. in case that the Offerer and the Answerer do not manage to reserve resources for the chosen nominal QoS contracts/contexts, they can use one or several UPDATE sequences [36][220] to switch to another configuration for the session with which the session shall start (at session establishment Negotiation) or shall continue (at session Re-negotiation). The respective reply of the Answerer side is implemented with the call sequence *negotiationRsp* or *reNegotiationRsp*, *reservationCoordRsp*, 200OK(UPDATE), *reservationCoordCfm* and *negotiationCfm* or *reNegotiationCfm* and it can carry a counter answer about the session configuration/re-configuration.

The following call sequence shown in Figure 34B is typical only for the Negotiation process: The E2ENPUAAnswerer sends *connectionStatReq*, 180 Ringing and *connectionStatInd* to inform that the Answerer application is alerted about the incoming call (e.g. it produces a ringing signal for the user); the Offerer replies that it *hears the ringing* and waits (call sequence *provisionalAckReq*, PRACK, *provisionalAckInd*) and the Answerer replies with its final precondition fulfilment status that completes the three-way handshake for the ringing procedure (call sequence *provisionalAckRsp*, 200OK(PRACK), *provisionalAckCfm*).

The completion of the session establishment/re-configuration is signalled either automatically from the E2ENP UA (like in Figure 34B) or the application instructs the E2ENPUAAnswerer to send the finalization message at user action (call sequence *connectionRsp*, 200OK(INVITE), *connectionCfm*, *done*). Both cases are considered in the E2ENP UA state machines (Section 5.2.2.4.2). The final acknowledgement for finalizing of the SIP (respectively E2ENP) transaction is sent from the SIPUAClient to the Answerer side (call sequence ACK, *connectionStatInd* and *completionInd*). The E2ENPUAOfferer applies the SIP UA's automatic ACK. The reception of the *done* message at the E2ENPUAOfferer and of *completionInd* at the E2ENPUAAnswerer releases the respective MUTEX. Consequently, the E2ENP UA can process further transactions.

3.1.4 Managing a Hierarchical QoS Specification with E2ENP

The End-to-End Negotiation Protocol gives the adaptive application/middleware the possibility to produce fragments out of their hierarchical QoS specifications for the multimedia session management and to remember which fragments were already exchanged during a negotiation process. These fragments are termed Portable Data Units (PDUs). Furthermore, some of the E2ENP PDUs are not parts of the hierarchical QoS specification but represent only references to such specifications. The *reference-only* PDUs serve the management of the application resources in cases that the application needs to adapt (e.g. short Re-negotiation) or are applied to manage the E2ENP UAs' and the applications' memory about the PDUs at PDU-leasing actions. The fragmentation of the hierarchical QoS specifications and the application of *reference-only* PDUs represent the E2ENP concept for signalling optimisation (see also Chapter 4).

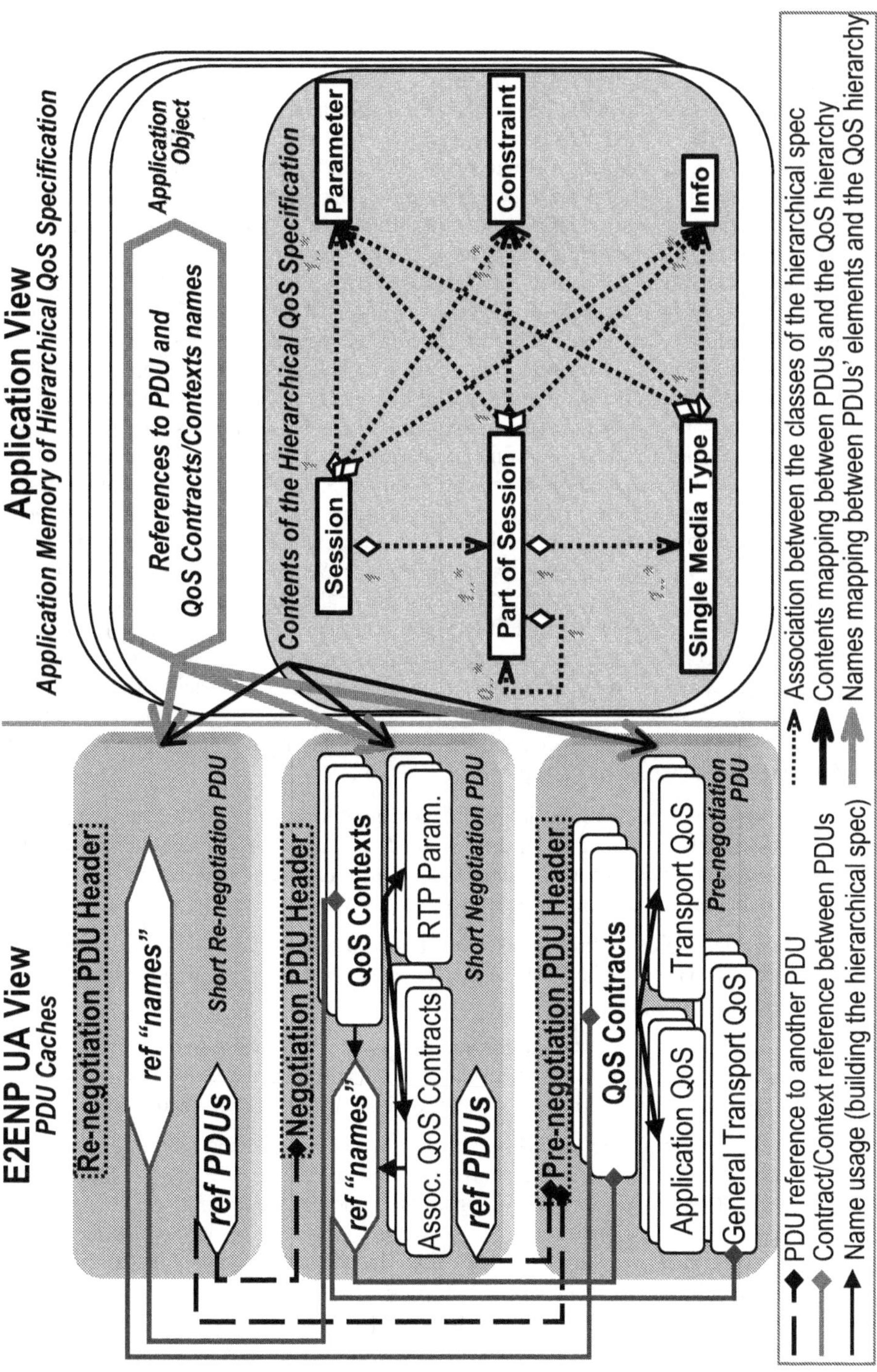

Figure 35: Association between the E2ENP PDUs and the session control object of the application

Figure 35 shows an example of the association of the E2ENP PDUs with a model of a hierarchical QoS specification of the adaptive application when short Negotiation and Re-negotiation phases are in use. The application's internal structure of the hierarchical QoS specification is different then the E2ENP XML specification. Due to this fact, the adaptive application and the E2ENP UA on a single terminal share a memory on the PDU-identifiers in order to uniquely distinguish the PDUs at negotiation and/or PDU-leasing actions (see Sections 2.3 and 5.2.2.2). Furthermore, the application has to associate the PDU identifiers with the identifiers of the fragments within the hierarchical QoS specification in order to be able to build the complete specification. The hierarchy for the PDU- and for the fragments-naming purposes was already discussed in Section 2.2. The E2ENP XML specification defines patterns for the PDUs' and the hierarchy-objects' names (see Section 4.3). The application's view on the hierarchical QoS specification can be described as a format-independent model that represents the natural hierarchy of an adaptive multimedia session (see Figure 35 and [202][221]). The hierarchical model of the session indicates what kinds of media and what relations between the single media types exist within a specific session. A multimedia session contains one or multiple parts, which correspond to the QoS contexts in the E2ENP language. The adaptive application has to uniquely name these parts in order to differentiate them at the hierarchical-QoS-specification fragmentation process and the following referencing upon the QoS contexts (see also Section 4.1.3.4). The association of *Part of Session* with itself represents the hierarchical building of QoS contexts. At the bottom of the model hierarchy are the single media types that correspond to the E2ENP QoS contracts per media type (e.g. audio or video). The QoS contracts are also uniquely named by the application in order to recognise them at fragmentation and referencing processes of the hierarchical specification. Each part of the application's QoS hierarchy is related to three kinds of descriptions that are associated with specific parameterisations available through the E2ENP description language (see Section 2.2.1 and Chapter 4):

- A *Parameter* indicates a negotiable and configurable description that may change its contents during a negotiation and an adaptation process (e.g. parameterisation of a video codec performance may change so that different frame play-out rates or picture sizes are in use, respectively different Application QoS applies).
- A *Constraint* indicates limitations upon used *Parameters*. For example, a network provider may wish to indicate specific transfer rates for the streaming media (e.g. max bandwidth as Transport QoS).
- An *Info* type describes static parameters that are not negotiable but are necessary for the correct session establishment, e.g. such parameters are the static payload types of RTP [2] or security identifiers of the peers and the mentors of the negotiation.

The PDU memory of the E2ENP UA is represented by a database build up from two caches [170]. The first cache is the short-term memory of the E2ENP UA that is used during a negotiation process to provide consistency information about the exchanged data between the negotiating applications to guarantee consistent resources access for the applications (see also Section 2.3.2.1.1). The second cache stands for the long-term memory of the UA, where only those PDUs are saved that represent fragments of the hierarchical QoS specification and that can be leased between the negotiating peers as a result of a successful negotiation procedure. The caches apply *java.util.TreeMap* class [222] for realisation of the E2ENP UA memory (i.e. the PDUs are ordered in a tree according to their cross referencing hierarchy, one PDU can reference one or multiple other PDUs. The PDUs are arranged according to the generated hash of their header contents and the *java.util.TreeMap* enables the quick access to the so-ordered PDUs). The caches give the application the possibility to manage this memory at negotiation, referencing and leasing actions associated with the PDUs (see also [170] and Sections 4.3.2 and 5.2). Note that the E2ENP UA understands only the PDU headers and the respective cross referencing and time-restrictions of the PDUs, but has no understanding of the hierarchical QoS specification and references upon it as this is a specific task of the application.

Figure 36 gives an example on how the E2ENP UA acts at management actions of the PDU memory. It consists of four pictures separated by vertical lines and each picture indicates one of the four PDU management cases. Note that the PDU memory contains the PDUs of different hierarchical QoS specifications. The PDUs relevant for the respective examples of PDU management shown below are indicated with shaded rectangles, any other PDU stored in the PDU memory is shown with a white rectangle. The management cases for the PDUs are:

- A *Normal* expiration occurs in the long-term memory of the PDU when a PDU that carries data expires, i.e. one or several fragments of a hierarchical QoS specification expire. The expiration time of a PDU is indicated in its header (see Section 4.3.2). The Normal expiration is shown in Figure 36 under the title *Normal*. The shaded rectangles with 'd' indicate the case where a non-expired PDU points at an expired PDU. In case of *Normal* expiration of data PDUs, the references associated with an already expired data PDU are also deleted from the E2ENP UA long-term memory. This activity of the E2ENP UA guarantees the consistency of the known/unknown hierarchical QoS specifications of the applications using the E2ENP UA. As an example, if a PDU that carries QoS contracts expires, it is deleted. Hence, all the QoS contexts pointing at such PDU turn useless and have to be deleted, as the knowledge they build upon is no more available. The *Normal* expiration within the short-term memory of the E2ENP UA is associated as well with either *reference-only* PDUs or with data PDUs that shall not be leased. This case is shown with a shaded rectangle with 'r' in Figure 36 under the *Normal* title. *Reference-only* PDUs and PDUs that are not leased are removed from the E2ENP UA memory as soon as a negotiation round finishes.

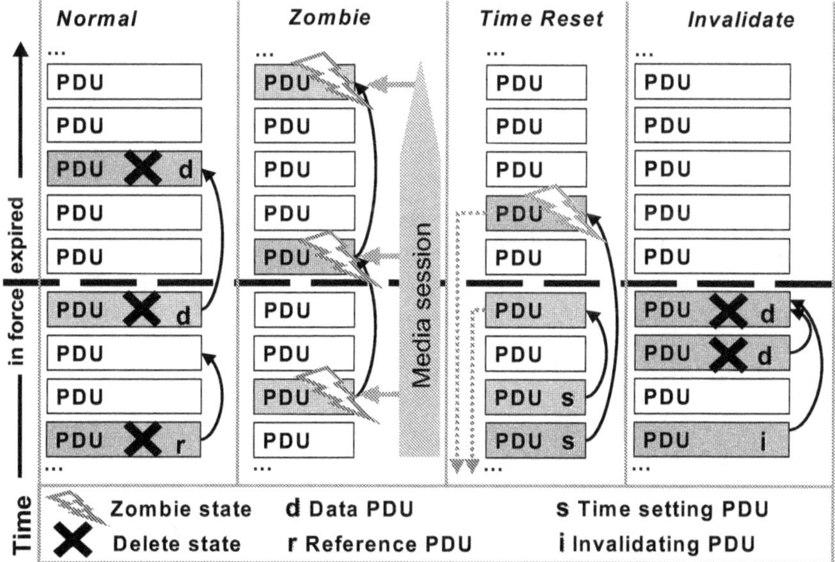

Figure 36: Time-management of the E2ENP PDUs

- The *Zombie* expiration of PDUs is associated with the condition if the PDUs that are expired are still in use by a multimedia session. The PDUs in a zombie state are shaded and indicated with a thunderbolt sign under the title *Zombie* in Figure 36. The E2ENP UA keeps in the long-term memory those PDUs that are expired or point at expired PDUs in cases that such PDUs are currently used by the application for its multimedia session management. During the time of a running session, the zombie PDUs can be re-activated by a corresponding time-reset PDU (see below).
- The E2ENP UA long-term memory is timely restricted for some of the PDUs. The applications using the E2ENP UA can request their corresponding peers to continue the leasing process for the PDUs by applying a *Time-Reset* PDU and a respective leasing Pre-negotiation (see Sections 3.1.3 and 4.3.7). This case is indicated in Figure 36 under the title *Time Reset*. The *Time-Reset* PDU (in Figure 36 an example of such PDUs is shown with shaded rectangles with 's') is either an *reference-only* PDU that can reset the leasing time for non-expired or zombie PDUs, or a data-carrying PDU that enhances the reset information with additional data.
- Some parts of the application's hierarchical QoS specification can be of special importance for the application and can be leased without time restrictions (e.g. telephone book entries). However, such entries may get also superfluous so that their removal from the E2ENP UA long-term memory becomes necessary. Furthermore, the applications may wish to remove from the long-term memory non-expired PDUs that are no longer useful for the application. In such cases, the applications are enabled to remove the required PDUs using *Invalidating* PDUs and respective leasing Pre-negotiations (see Sections 3.1.3 and 4.3.7). *Invalidating* PDUs are either an *reference-only* PDUs used to remove other PDUs from the E2ENP UA long-term memory, or data PDUs that overwrite the forcefully expired information with new data. The case of the *Invalidating* PDU is shown in Figure 36 under the title *Invalidate* and the PDU itself is indicated with a shaded rectangle with *'i'*. Similar to the *Normal* case above, the invalidating of a PDU terminates also all the other PDUs that point at the invalidated PDU.

The initial concepts of the E2ENP UA memory and its management were presented in [170] and [173].

3.2 Preconditions for Establishment of E2ENP Sessions

This section presents considerations and scenarios about the necessary security and trust relationships between the peers and the mentors as preconditions for the establishment and the accomplishment of E2ENP sessions. Furthermore, scenarios with provider interactions at terminal mobility and concepts of discovery of mentoring services using SIP are shown. These scenarios represent some specific E2ENP enhancements to the classical SIP and explain how the peers and the mentors of the negotiation process share the information on the fragments of the hierarchical QoS specification. The examples depicted in this section serve also as initial explanation for the extended functionality of E2ENP description language for supporting third-party call control and provider mentoring as a premium service for lightweight terminals that cannot perform all necessary management functions on their own (see Sections 2.3.2.1 and 2.3.2.2.2 and Chapter 4).

3.2.1 Peer, Provider and Third-Party Security at Mentoring

As a presentation layer protocol [54], E2ENP can profit from the security mechanisms provided by the underlying protocols. In cases SIP is used as E2ENP carrier, point-to-point (e.g. terminal-to-provider or provider-to-provider) and end-to-end (e.g. terminal-to-terminal) security for the communication of peers is guaranteed through the SIP-own comprehensive security mechanisms [62]. The SIP mechanisms allow securing the E2ENP information through the MIME signature and encryption mechanisms [223][224] for the SIP message bodies. Furthermore, the transport and the network security are available through mechanisms like TLS [225][226] or IPSec [227]. In order to apply E2ENP as a premium service for configuration of the multimedia applications, the E2ENP-enabled terminals have to register to the respective service providers. However, in order to verify the authenticity of provider entities and of other mentoring devices the E2ENP-enabled terminals have to receive at registration to the provider or at service discovery of mentoring services information about the authenticity of the respective services so that trust relationship between the terminals and the mentoring services can be established (e.g. through exchanging of certificates [224][228]). The authenticity of the terminals is also proven at registration from the provider and the mentoring entities (e.g. applying subscriber identification, using SIM [114] or certificates). The information exchanged between the negotiating peers or between a peer and a mentoring entity at third-party call control scenarios that should not be manipulated by entities on the path of the negotiation can be secured end-to-end between the negotiating peers using the standard S/MIME procedure for generating signatures (see [62][223][224]). However, if manipulation is desired by the peers in cases that a mentor acts on behalf of another peer (e.g. at session migration scenarios – see Section 3.3.1) or the provider wishes to enforce restrictions upon the negotiated data directly within the negotiation bids, this is no longer possible with the S/MIME alone as S/MIME provides just a single signature per single MIME. With this respect, E2ENP provides a placeholder for such provider and mentor authentication data in a similar way the XML signature syntax [229] does, so that the E2ENP mentors can add signature information within the E2ENP PDU and associate restrictions and utilization rules to the negotiation information in explicit way without interfering with the original bids of the peers (see Sections 4.1.3.4 and 4.3.6).

The security considerations presented here are necessary to enable provider mentoring and advanced negotiation scenarios with third-party call control, which scenarios are discussed in the following sections.

3.2.2 Terminal Registration

The simple end-system registration in the way an E2ENP-terminal registers within the domain of its home provider using the E2ENP API and SIP REGISTER method was already shortly presented in Section 3.1.3 (see Figure 33). Such registration is indicated as intra-domain registration (i.e. registration in the same domain of a single home provider) and it represents the association between a terminal and a provider domain where the initial identity of the terminal was created and where this identity is maintained by the home provider [20][92]. At intra-domain registration and in addition to the classical SIP registration information [62], the terminal may use the registration to upload at the provider side information about its user profile and user requirements for service support so that the provider can prepare for respective mentoring of the registered terminal, furthermore, the provider may send to the user information about the rules for applying services at the home domain (see Annex B). The sending of additional E2ENP-relevant information at registration between the terminal and the provider depends on the chosen service and transport domain interaction model (Section 2.4.1). In cases that the provider explicitly monitors and interacts with E2ENP PDUs, the provider has also to remember them in order that the provider keeps also track of the time validity of the E2ENP PDUs (see Section 3.1.4). At application of explicit interactions with the PDUs, the periodic re-registration procedure as typical for SIP [62] can be used to upload the current, relevant state of the peer's hierarchical QoS specification (see also Section 3.2.3). Re-registration occurs also in cases when the terminal has been switched off and is activated again within the same or a different domain (see also below inter-domain registration). The application of a re-registration is also appropriate in cases that a single provider supports technologically different networks where the relevant hierarchical QoS specifications for the service may differ. Hence, the change in the QoS specification at migration of the terminal between technologically different networks is also registered with the mentoring provider when the terminal moves in a new network. For the purpose of accomplishment of the uploading of multiple E2ENP PDUs at the mentoring devices (providers or third-party call control entities), the multipart MIME type [217][230] named *multipart/related* [231] is used to deliver the multiple fragments of a complex object in a single message in form of an aggregation (see Section 4.1.2). In the case of E2ENP, the fragments are the parts of a hierarchical QoS specification expressed as PDUs. The mentors using E2ENP has to know the PDUs of the peers in order to be able to build correctly the supervised hierarchical QoS specification and to understand the references the peers use in their short negotiation rounds (see Section 3.1.2 and 3.1.4 and Figure 35).

Advanced terminal registration occurs when a terminal roams foreign-provider domains and it is termed inter-domain registration, i.e. registration at mobility between different provider domains (see also [20][92]). This registration applies at handovers between different providers. Inter-domain scenarios are usually accomplished with combinations of SIP and other protocols like RADIUS [124] and DIAMETER [125] for authentication, authorisation and accounting and COPS [7] for profile management between providers. However, the SIP-only example shown here is relevant primarily for E2ENP and the E2ENP's ability to fragmentize QoS specifications and to reference parts of them.

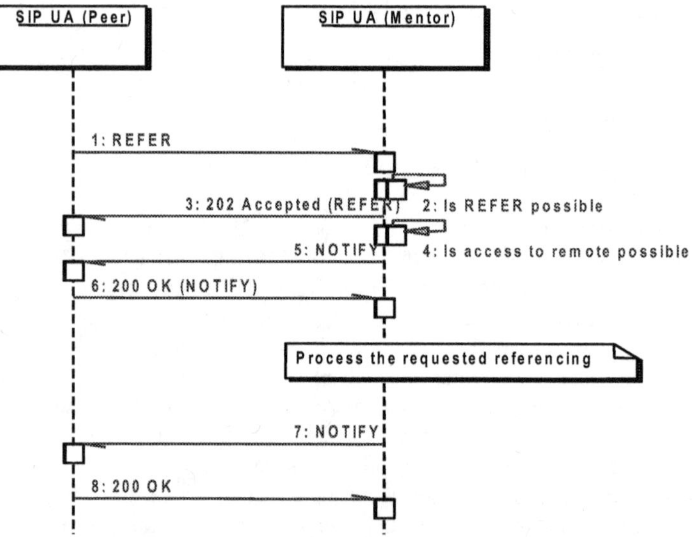

Figure 37: SIP REFER Method [101]

For the accomplishment of the inter-domain registration and of some of the following examples with mentoring, the SIP REFER method [101][102] is used that enables a device to request external actions from another device (see also Section 5.2.2.4.3), e.g. a peer may request a mentor to fulfil some action on its behalf using the REFER method (Figure 37). The requested party proves the possibilities for the external action and NOTIFY-s the requestor about them. Furthermore, the device requesting the action is enabled to receive results about the level of fulfilment of the external action, i.e. the results of the action are sent also with NOTIFY (the second NOTIFY in Figure 37).

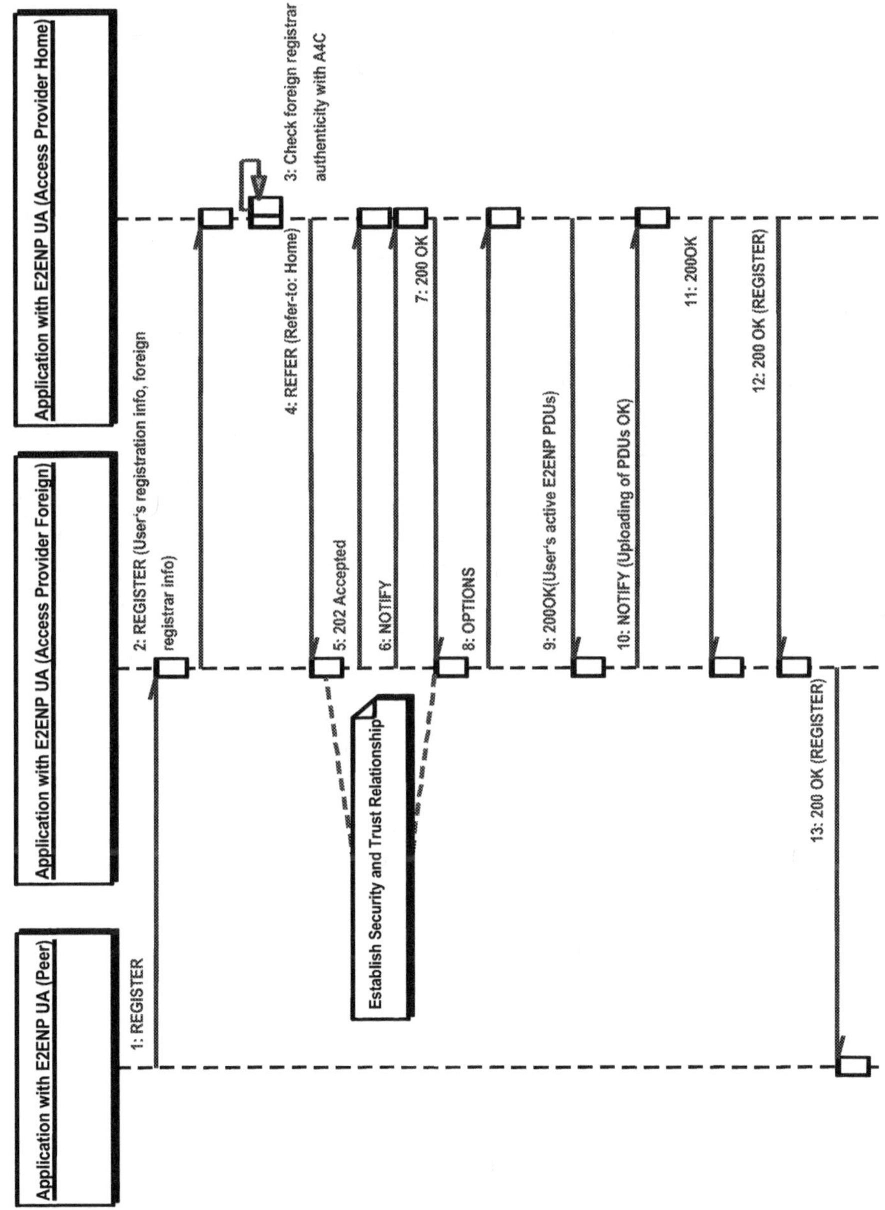

Figure 38: Inter-domain registration

If the foreign provider is not actively monitoring and manipulating the E2ENP PDUs at peers negotiation, the provider does not need to know the contents of every PDU but can use only the credentials given to the peer at registration for validating peer's PDUs (see scenarios *No Coupling* and *Loose Coupling* in Section 2.4.1 and the further explanation of the provider-management actions below). The registration in this case is similar to the intra-domain scenario above. The example shown in Figure 38 presents the case where the foreign access-provider has to remember the PDUs negotiated between the peers in order to provide adequate mentoring for the following negotiations between the peers, i.e. at every handover to another provider the new provider has to get acquainted with the PDUs exchanged in the past between the peers in order to learn to know the PDUs and to be able to interpret the respective cross referencing of PDUs that the peers may use in their future negotiation rounds (see also above the considerations about re-registration in home domain and Section 3.2.3). With this respect, it is suggested that the home provider of a terminal maintains a master copy of the terminal's active PDUs and this copy is uploaded at the foreign provider whenever a peer moves in a foreign domain. Arriving at the foreign domain, a terminal has to register before using the services in the new domain (Figure 38 – first REGISTER message). The foreign provider delivers to the home provider of the terminal information about itself (e.g. authentication information of the foreign provider) and about the terminal that wishes to register in the foreign domain in order to determine the levels for the service support for the terminal in accordance with the user profile that is maintained at the terminal's home (i.e. in Figure 38 the sending of the REGISTER message from foreign to home). The home provider proves the identity of the foreign (and of the remotely registering terminal at roaming) and if this information is acceptable, the home requests the foreign to access it and to obtain the active E2ENP PDUs of the terminal (i.e. the REFER request from home to foreign). The additional messages in the REFER transaction can be used by the home and the foreign to exchange further authorization and service support information. As a result, all or only a part of the terminals active PDUs are uploaded at the foreign. The foreign downloads the active E2ENP PDUs of the terminal from the home using OPTIONS method (the PDUs are received with the 200OK for OPTIONS, an alternative with using COPS transaction [7] instead of SIP OPTIONS is also possible on that place). The last NOTIFY is used to confirm to the home that the foreign successfully received the PDUs. The final answer of the foreign to the REGISTER request of the terminal includes information about which PDUs of the terminal are supportable in the new domain after the handover. The foreign can send for this purpose an E2ENP PDU that references all the relevant in the foreign domain PDUs using the PDU header referencing mechanism (see Section 4.3.2) and/or a *multipart/related* MIME as discussed above. If the new foreign provider does not support mentoring mechanisms it answers with a body that does not contain E2ENP information in the 200 OK to the terminals REGISTER request. However, the body may contain information about the supported services and their restrictions in the foreign domain so that the terminal validates its PDUs on its own (see Section 2.3.2.2.2). In case that the new provider refuses mentoring, the terminal has to re-negotiate its configurations with its corresponding peers if necessary, to adjust the configurations to the requirements for service and transport interactions in the new domain of the location of the terminal.

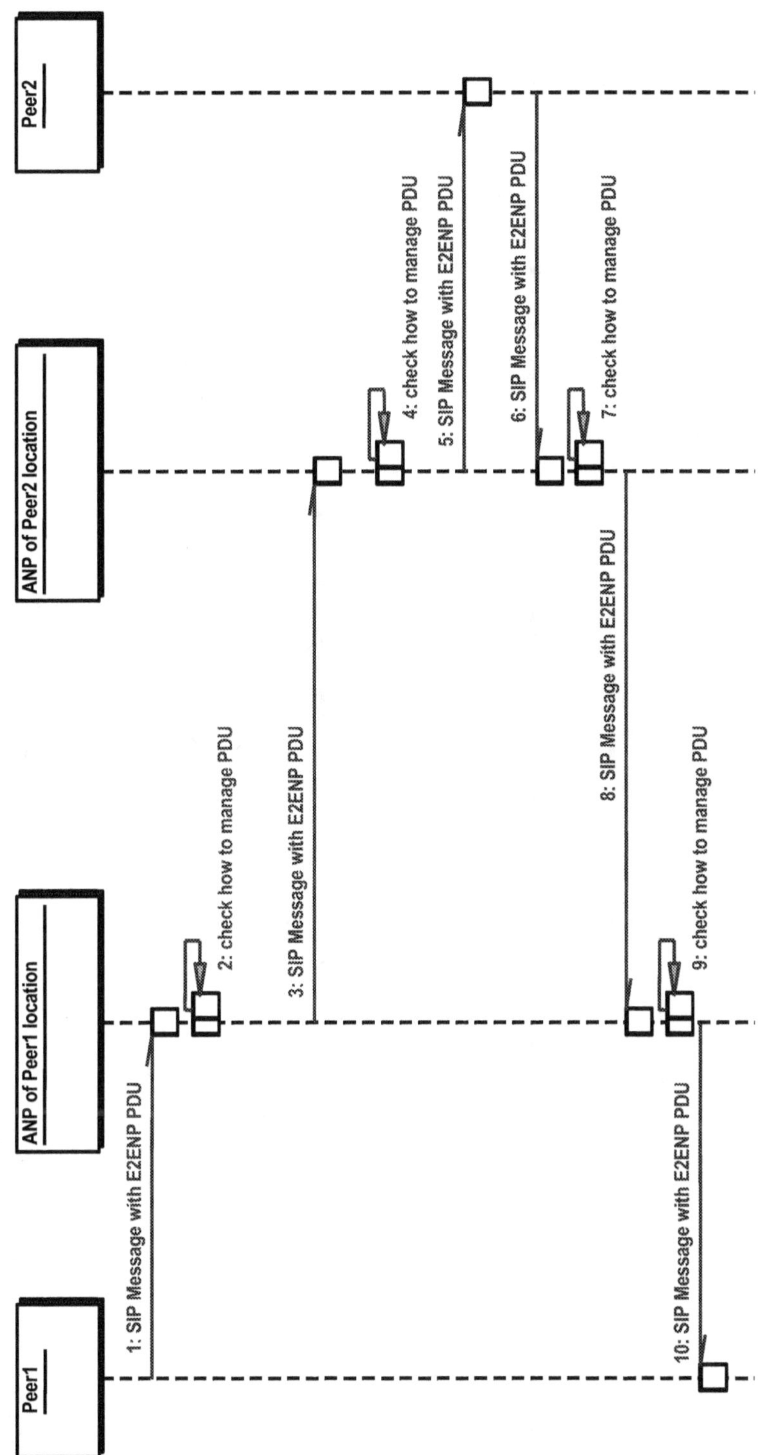

Figure 39: PDU management at E2ENP-based proxies

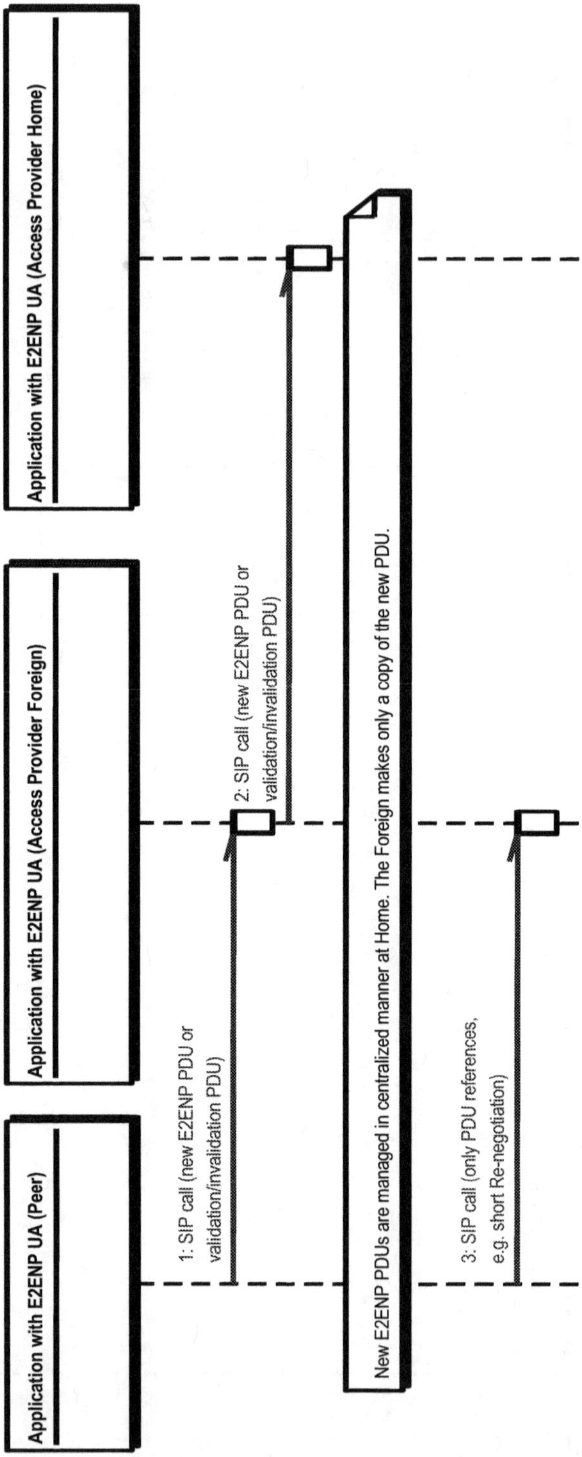

Figure 40: PDU registration in the foreign domain

3.2.3 PDU Registrations

The registration of PDUs at the provider is necessary only in cases when the provider mentors the terminals and interacts with their PDUs during E2ENP negotiations. Every time a PDU traverses the E2ENP-based proxies of the Access Network Provider (ANP), these proxies decide how to manage it (Figure 39, see also Section 2.3.2.1). In addition to the marking of PDUs with provider rules, the provider has also to prove whether to save the PDUs in order to enable the support of terminals migration between providers. As discussed in the previous section, the home provider offers centralized management for the registration of its terminals (e.g. as in the case of SIP). Hence, a precondition for supporting E2ENP scenarios is that the home provider shall support also the registration of the terminals' PDUs for terminals belonging to its domain (i.e. save the PDUs used in the mentoring process). At PDU registrations at the foreign domain, the foreign makes only a copy of these for its own management purposes and delegates the PDUs containing new information about the hierarchical QoS specification of the roaming peers to their home providers (Figure 40). This delegation of PDUs is applicable using the SIP re-registration and de-registration mechanism (e.g. if the peer moves to still newer domain the old domain uploads with the de-registration the still missing PDUs at the home). Alternatively and in non-mentoring scenarios, the terminals have to take care by themselves to upload and register their PDUs at their home provider in order to enable migrations between mentoring and non-mentoring providers. PDUs that contain only reference information that does not change in any way the hierarchical QoS specification (e.g. short Re-negotiation) do not need to be saved. Therefore, they are not delegated to the home provider (see last message in Figure 40). Note that invalidation PDUs can also be *reference-only* PDUs (Section 3.1.4), but since they change the hierarchical QoS specification, they have to be delegated to the home for QoS specification maintenance purposes.

3.2.4 Service Discovery of Terminals and Mentoring Entities

An additional precondition for enabling E2ENP scenarios, especially scenarios with third-party call control, is the ability of the system to perform service discovery of terminals and mentoring entities. The SIP framework containing registrars and location services provides already mechanisms for service registration and discovery [62][232]. The automatic services (e.g. E2ENP mentors in the role of third-party call control or conferencing services) register themselves in the SIP framework in a similar way as the user terminals do. Hence, they can be discovered using the location services in the framework [92][202][233][234]. The protocol for the discovery allows SIP-only solutions (e.g. combinations of OPTIONS method and registration methods like REGISTER [62] or SUBSCRIBE [235]) or non-SIP solutions like Service Location Protocol (SLP) [236]. Spontaneous discovery between terminals and mentoring services can be enabled through mechanisms including technologies like Bluetooth [237] (for short-range discovery) or Global Positioning System (GPS) (for wide-range discovery) [238] and enhancing these technologies with service-location information. Such mechanisms are applicable also in terms of ad-hoc scenarios with E2ENP [20][65] and scenarios with session redirection (Section 3.3.1). The relevance of discovered services for enabling of E2ENP scenarios has to be proven also with respect to the services' capabilities information. However, registration of capabilities is not widespread with the current discovery services [239]. The registration with service capabilities is enabled through additional description languages e.g. based on Resource Description Framework (RDF) [240] or on Web Ontology Language (OWL) [241]. Extensions of E2ENP for service discovery are also possible, in addition to the PDUs registration that is already a kind of capability registration.

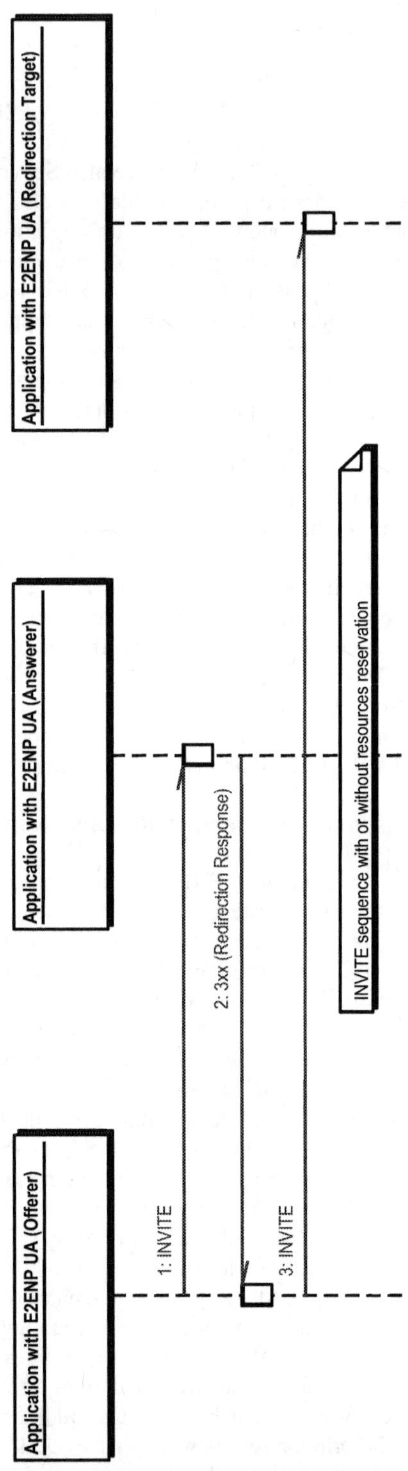

Figure 41: Session redirection

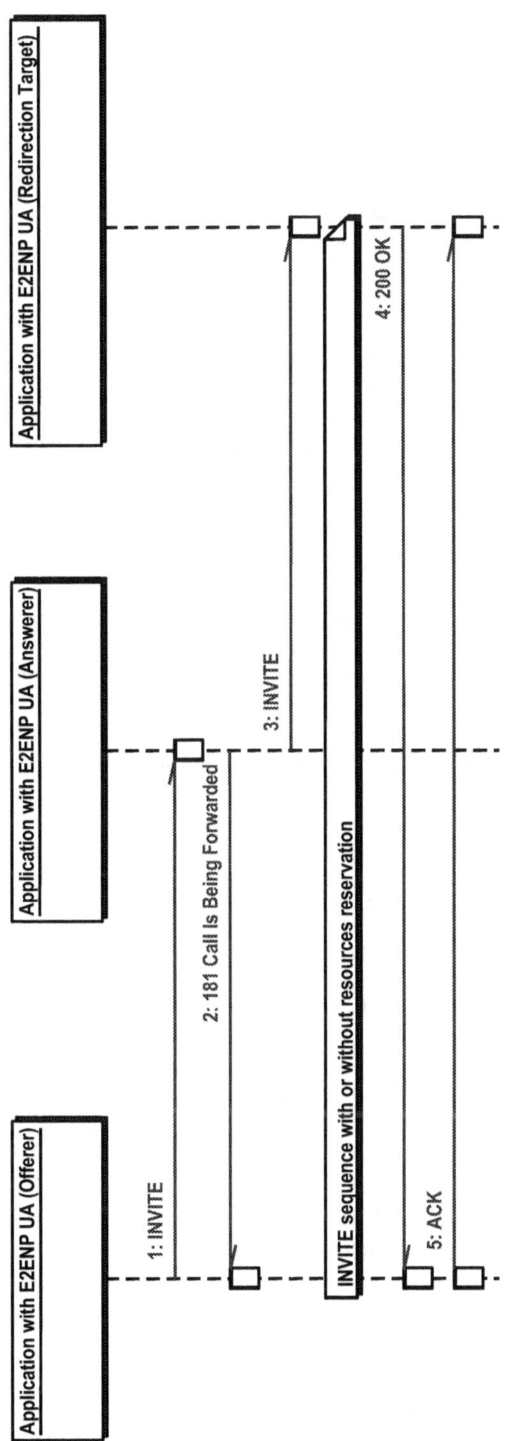

Figure 42: Session forwarding

3.3 Third-Party Call Control Scenarios

The currently available adaptation approaches for multimedia consider primarily server- and/or client-side adaptations based on control procedures similar to the Offer/Answer Model [35][36]. However, scenarios using advanced adaptation services can be an advantage for terminals and automatic services that are not in the position to adapt on their own due to limited capabilities (e.g. PDAs have limited battery and CPU power) or due to performance constraints (e.g. a single conference server cannot adapt to any individual receiver requirements) [20][92][202][221]. This section presents management scenarios where the peer-to-peer negotiations with E2ENP Offer/Answer Model alone are not enough to cope with the service requirements of the peers, so that the end-systems have to use external services for session redirection and/or media adaptation. Within the scope of the SIP and 3GPP IMS, similar scenarios are discussed [62][105][106][234][242] - [247], but they consider only the SIP call-control and not the session description management. In the following, an overview on these scenarios is presented discussing the impact of the advanced adaptation on E2ENP and on its mentoring functionalities for coordination and for management of the E2ENP fragmentations of hierarchical QoS specifications.

3.3.1 Session Mobility

The mobility of a session can occur either at the very beginning of a session before a session is being opened or during a running session. Respectively, a session establishment with an alternative service (Section 3.3.1.1) and a session migration scenario (Section 3.3.1.2) represent two different situations of session mobility.

3.3.1.1 Alternative Services at Session Establishment

The application of an alternative service represents the opening of a session on another device (alternative Answerer) then the initial Answerer device of the negotiation. The case where multiple alternative devices are involved in the session is considered a conference and it is treated later in Section 3.3.2.

An initial Answerer can decline explicitly a proposed session and it can point to the Offerer an alternative target for establishing the session (Figure 41), e.g. using the SIP *Contact* header [62]. In case that the initial Answerer device is configured by its user to be acquainted with the capabilities of the *Redirection Target*, this Answerer may send a reply to the Offerer as an E2ENP PDU. At receiving such an answer, the Offerer can adapt its request according to the requirements of the *Redirection Target* and send an adapted PDU to the new Answerer (i.e. the *Redirection Target*). With respect to E2ENP, the communication with the initial Answerer can be considered a Pre-negotiation that however does not have impact on the following negotiations, respectively the E2ENP UA Offerer has to forget the initial answer that came with *3xx* message of SIP as soon as its offer is adapted for the means of communication with the *Redirection Target*. The initial Answerer may point explicitly to the Offerer that it shall forget the respective mutual Pre-negotiation transaction using a *caching* directive of the E2ENP PDU header and setting the expiry time of all the corresponding PDUs to zero (see Section 4.3.2).

Alternatively, the initial Answerer may forward the session directly to the *Redirection Target* (Figure 42). In this case, the initial Answerer behaves in a similar way as a SIP Proxy. In terms of E2ENP, the initial Answerer may enhance the E2ENP PDU of the Offerer with constraints (as long as such constraints are known to the initial Answerer) or it forwards the respective message as it is to the *Redirection Target*.

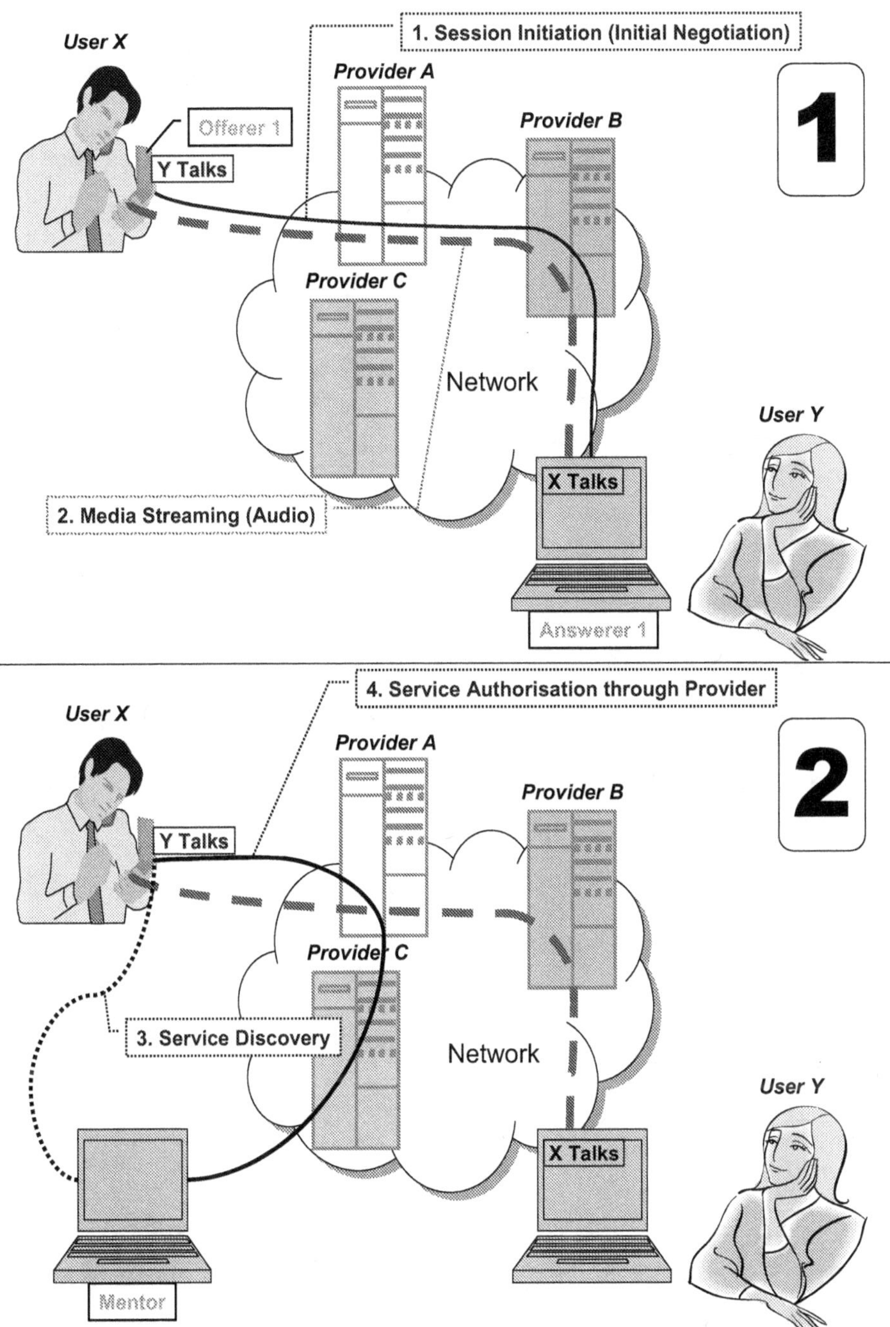

Figure 43: Session migration scenario
(A)

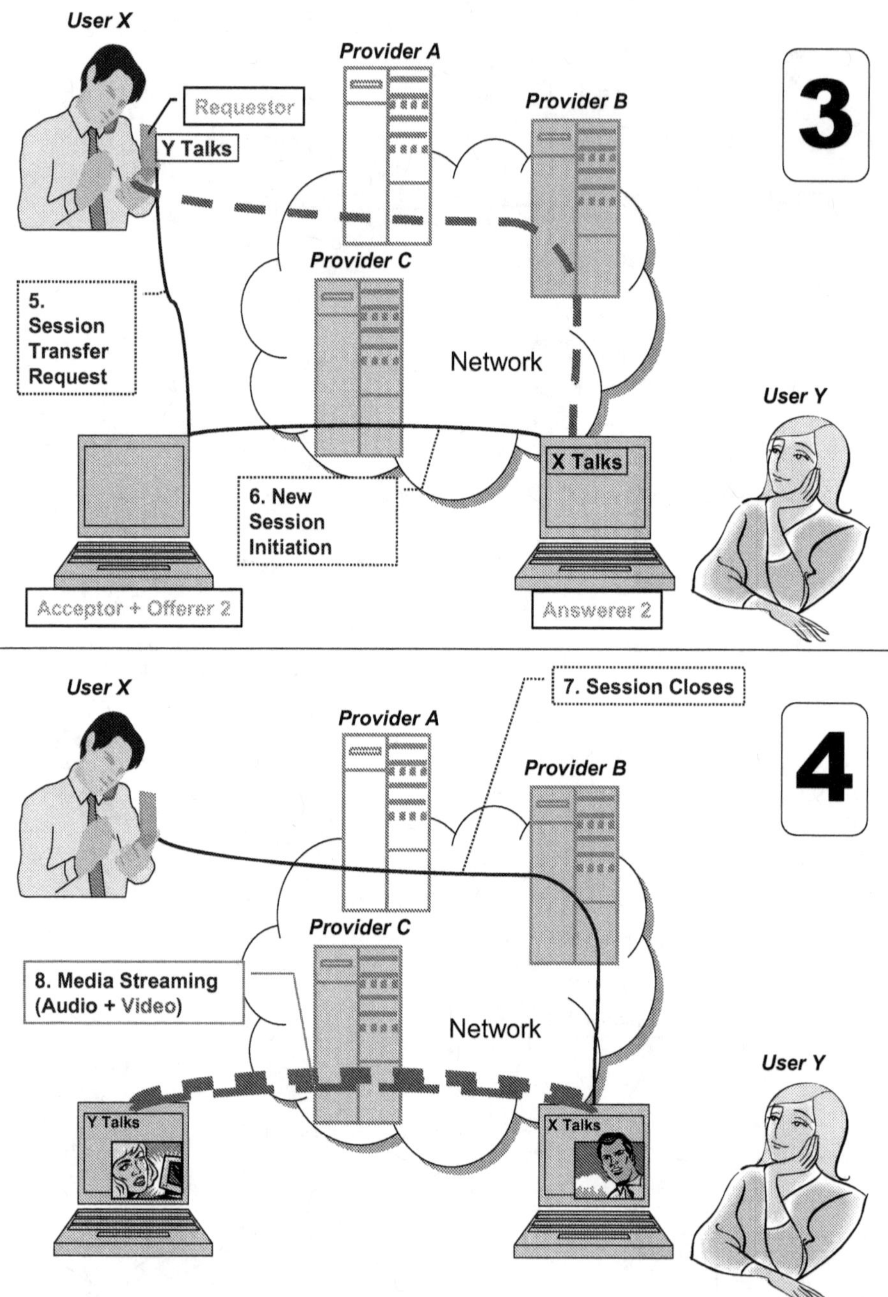

Figure 43: Session migration scenario (B)

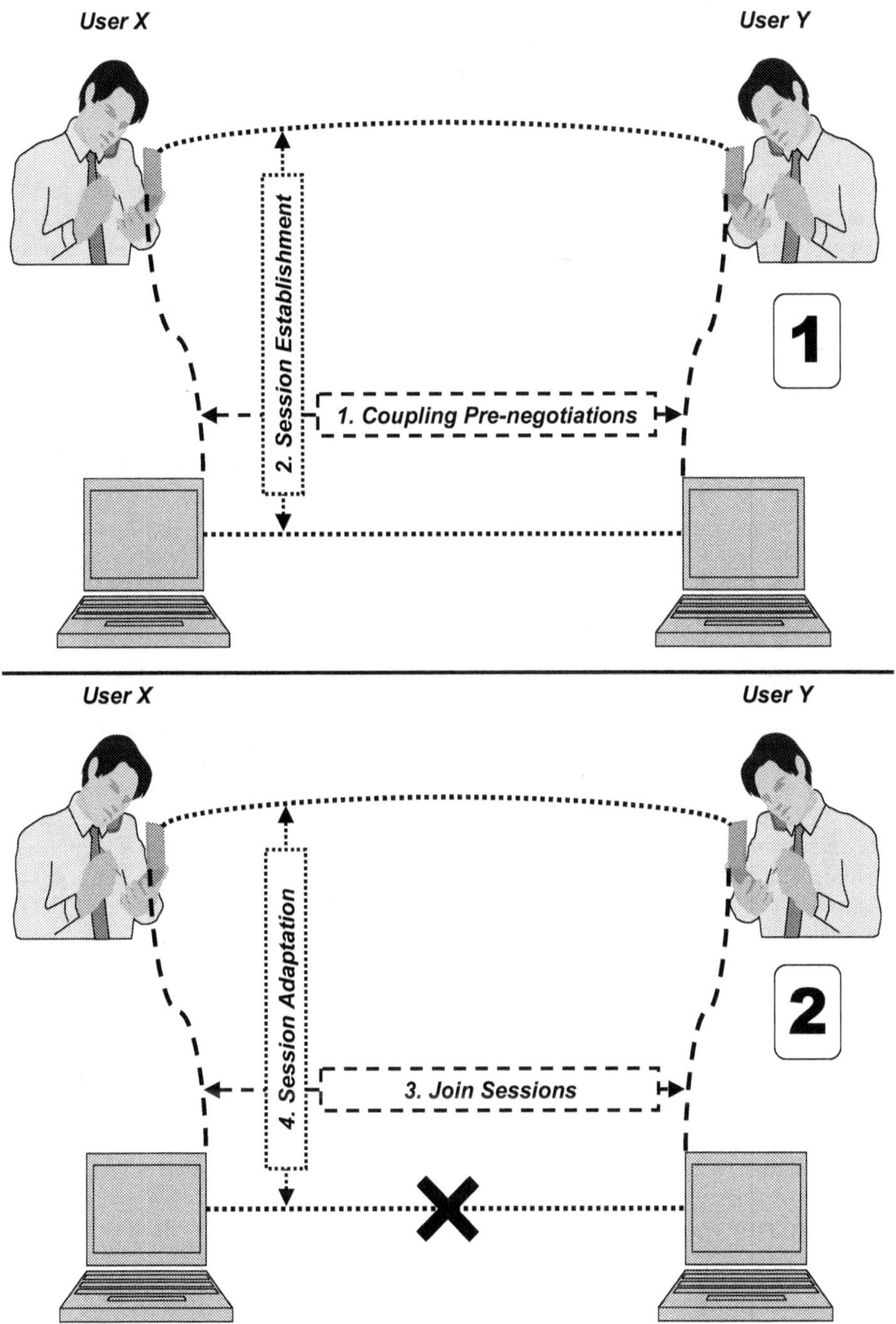

Figure 44: Coupling and joining of sessions

3.3.1.2 Multimedia Session Migration

The multimedia session migration describes the case when during a running session the device/-s used for presenting the session to a single user changes. There are three major sub-cases of this scenario, i.e. session migration from one device to another (Figure 43AB), session splitting in case that the initial session is split between several devices producing several corresponding SIP sessions (see also Section 3.3.2) and sessions joining when several dependant sessions join into a single session (Figure 44). Note that the scenarios shown in Figure 43AB and Figure 44 are presented in a comic-like manner, where the large numbers to the right of the single pictures (1 to 4 in Figure 43AB and 1 to 2 in Figure 44) indicate the sequence number of the activities within the scenarios.[21]

The session migration scenario is presented with an example in Figure 43AB. The example shows *User X* and *User Y* which have established a session with audio contents (Nr. 1 and upper part of Figure 43A). For the session establishment the terminals of these users play the roles of an Offerer and an Answerer, e.g. if *User X* makes the call his device is the Offerer, respectively *User Y's* device is an Answerer. Note that the initial order of the Offerer/Answerer roles does not matter for the establishment of the next session migration steps. If the *User X* moves around with his hand-held device and the device has the ability for service discovery, it may detect a more powerful device where a session with video can be established. The detection can be initiated from the hand-held device, from the stationary device or from the providers depending on the used protocols and mechanisms for the detection (Section 3.2.4). The hand-held and the new stationary device have to prove also their accessibility and the accessibility to the remote peer (i.e. the device of *User Y*) with their providers (Nr. 2 and lower part of Figure 43A). In case that the provider authorization for session migration is successful, the hand-held device issues a transfer request and the new device acts as a mentor of the session transfer with E2ENP. Note that mentors for the session transfer can also be the hand-held device, the terminal of *User Y* or the providers, however, in such cases the session transfer requires conference-like management, i.e. in case of SIP multiple and/or parallel INVITE transactions or multiple REFER transactions apply [105][234][246] (see also Section 3.3.2). The situation presented here considers for the migration just one REFER and just one INVITE transactions (see below and Figure 45AB). For the initiation of the transfer, the hand-held device takes the role of Requestor and the newly detected device the role of Acceptor. Simultaneously, the new device is also the new Offerer and the old Answerer is again an Answerer (Nr. 3 and upper part of Figure 43B). The result of the transfer is a new session with audio and video (Nr. 4 and lower part of Figure 43B). The old session closes. Furthermore, if the same situation is used to initiate only the video on the new device, the result is a session splitting and the closing message of the hand-held device is then a session adaptation with INVITE (see below and Figure 45). In addition to the discussed session splitting, a different situation is also possible, e.g. a mentor is used to initialise two new devices that were non-involved in the initial session [105][234]. However, such scenario is considered a conference with third-party call control (see also Section 3.3.2).

Since the session joining is a scenario that can be managed only having application knowledge of the hierarchical specification of the corresponding media types and their associations, such scenario is not considered in the SIP-only call control specifications [105][234][244] - [247], as SIP without the session description has no knowledge of the

---

[21] The multimedia session migration scenarios (Figure 43AB and Figure 44) partially apply Microsoft Office ClipArts.

possible associations between the media within different sessions. However, E2ENP has the possibility to deliver information on associations between sessions on different devices as E2ENP considers the user information as independent from the user's device (see also Section 2.2.1). Note that, in principle, SDP [86] could also describe scenarios where the different media are associated with different devices; however, SDP cannot associate the statements of multiple (more then two) negotiating parties as the corresponding Offer/Answer Model with SDP [35] works only for two parties. Consequently, associations of sessions through negotiations are not possible with SDP. E2ENP enables such information associations and consequently the application of the following session mixing and joining scenarios is possible.

For mixing of sessions and for sessions joining, the E2ENP PDU referencing mechanism is applied together with indications that the sessions (respectively their descriptions) belong together (see Section 4.3.2). For instance, two users can have a session that involves multiple devices and this session can be associated with multiple SIP sessions, e.g. between two pairs of devices. The closing of the session between one of the device pairs will cause adaptation that moves the session on the other pair of devices. A more practical example (Figure 44) of such session joining is the case with *User X* and *User Y* having a session between their hand-held devices for audio and PC-Monitors for video. In case that both users have to move, their video session migrate to the hand-held devices joining the audio session. Respectively and in case of SIP controlled sessions, the session between the PCs will close and the complete association will be adapted with E2ENP over the single SIP session between the hand-held devices, e.g. choosing more efficient audio and video coding to minimize the power consumption on the hand-held devices. The exchanging of E2ENP information for coupling of devices is a Pre-negotiation (i.e. in the example above the PC and the hand-held device of a single user are coupled). A Negotiation follows the coupling Pre-negotiation and it results in session establishment between the hand-held devices and between the PCs of the two users so that mixing of the sessions is achieved.

The communication for session mixing and joining implements the E2ENP API for negotiations as described in Section 3.1.3 (see also Figure 33 and Figure 34AB). The implementation of the adaptation scenarios that use session migration and session splitting requires the application of the REFER method [101][102]. Furthermore, REFER was chosen for implementation of the E2ENP state machines (Section 5.2.2.4.3) as this method gives the possibility for realisation of multiple scenarios with third-party controlled sessions and activities [234][243][245][246]. The E2ENP API implementation of REFER and the interaction for session migration with the basic E2ENP peer-to-peer negotiation API is shown in Figure 45AB below, indicating the respective negotiation and migration roles of the E2ENP UAs. The concept for the E2ENP API for REFER is similar to the E2ENP API for negotiations, namely, the APIs towards SIP and towards the application are separated in order to enable the integration of other then SIP protocols for initiation of externally controlled actions. Furthermore, the API is designed so that all messages can carry message bodies, e.g. for exchanging additional to the basic SIP status information and session descriptions.

Note that in order to understand the sequence of the messages in the two split parts of Figure 45AB the respective numbers and names of the messages in the MSCs should be regarded. The overlapping between Figure 45A and Figure 45B affects the following messages – *referInd*, *referRsp*, *notifyStatReq*, *notifyStatCfm*, *notifyReq*, *notifyCfm* and *BYE*. Furthermore, the E2ENP UA Acceptor appears in the both parts of the figure. The repeated messages and the position of the E2ENP UA Acceptor should be regarded when observing Figure 45AB in order to comprehend the correct message sequence.

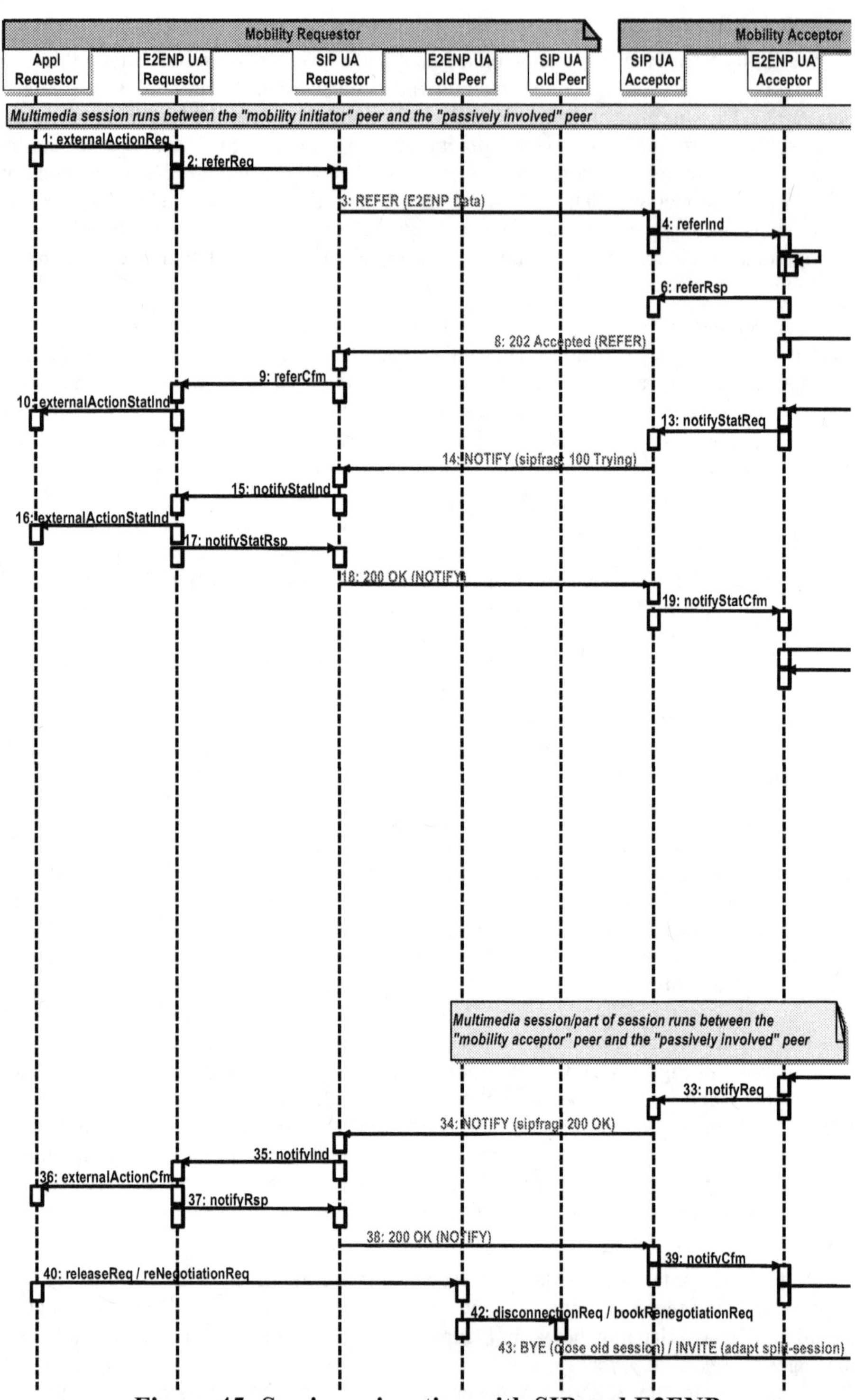

Figure 45: Session migration with SIP and E2ENP (A)

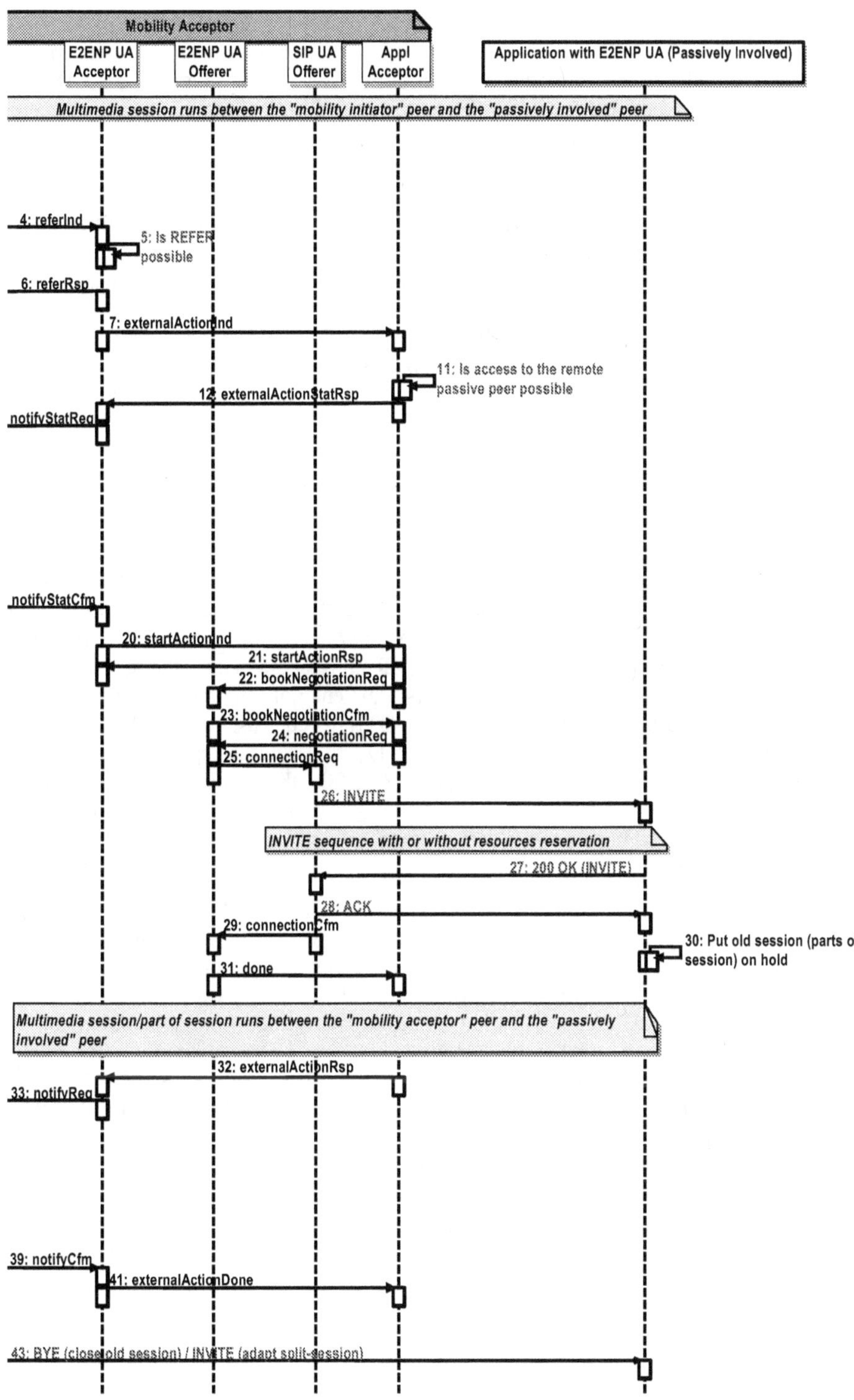

Figure 45: Session migration with SIP and E2ENP (B)

The migration scenario includes three parties Mobility Requestor, Mobility Acceptor and Passively Involved end-systems. The Mobility Requestor (Figure 45A) is the terminal that has the initial session with the Passively Involved device (Figure 45B) and the Mobility Acceptor is a device that is applied as a mentor for the migration and that accepts the whole or some parts of the initial session. As soon as the Mobility Requestor discovers the Mobility Acceptor and the access to the Acceptor is enabled through the provider (see Figure 43AB), the Requestor issues REFER request (*externalActionReq*, *referReq*, REFER, *referInd*). The Requestor may send E2ENP information (e.g. capabilities of the Passively Involved) about its initial session with the Passively Involved to the Acceptor, so that the Acceptor can be prepared to adapt its session information to the requirements of the Passively Involved. Since, the REFER message can be associated with multiple actions (i.e. other then session migration), the E2ENP UA has to prove first if the action described in the message is possible at the Acceptor, e.g. on this place the Acceptor may prove some of the SIP headers for identifying the action and its issuer like the fields *Subject*, *Organization*, *User-Agent*, *Content-Language*, etc. In case the action is possible, the E2ENP UA Acceptor replies with *202 Accepted* (*referRsp*, 202 Accepted, *referCfm*, *externalActionStatusInd*) and delivers the message to the application part of the Acceptor (*externalActionInd* - Figure 45B).

The Application Acceptor proves its possibilities for managing the requested external action and if the action can be fulfilled, it gives a positive answer to the Requestor, a NOTIFY message is send that contains *sipfrag*[22] [248] contents *100 Trying* (*externalActionStatRsp*, *notifyStatReq*, NOTIFY, *notifyStatInd*, *externalActionStatInd*). The application of *sipfrag* in the case of the REFER transaction enables the realisation of a request/response model using only SIP requests, thus enabling the applications of SIP to provide status and failure answers within the NOTIFY requests. In order to enable the proper SIP communication, the NOTIFY request has to receive a reply and this reply is issued by the E2ENP UA Requestor (*notifyStatRsp*, 200OK, *notifyStatCfm*). This reply serves also for securing the SIP communication on transport level if a non-reliable transport protocol like UDP carries the data packets.

At receiving the reply to the NOTIFY, the E2ENP UA Acceptor sends *startActionInd* to the Application Acceptor, which is replied with *startActionRsp* (Figure 45B). These two messages are used by the application to signal the handover of the communication thread to the external action (in this case the migration invocation) and to inform the E2ENP UA Acceptor that the action is successfully started so that it blocks and waits for the end of the external action. As next, the Application Acceptor starts the communication with the Passively Involved (Figure 45B – messages 22 to 31) in order to establish the session between the Acceptor and the Passively Involved devices and to realize the migration. The negotiation is done in the same way as shown in Figure 34AB. Although the E2ENP PDU referencing mechanism can indicate that the new session between the Acceptor and the Passively Involved is the same session as the session between the Requestor and the Passively Involved, an additional SIP-only mechanism is also available for exchanging one of the partners of the session with another device/-s (i.e. *Replaces* header [242]). This mechanism can be used in scenarios where the session description is not E2ENP or it can

[22] The *sipfrag* enables the SIP UAs to send a short form of replies within SIP requests. The classical specification of the SIP UA behaviour prescribes the sending of a single reply to a single request. However, some scenarios like REFER require variations from this behaviour, hence *sipfrag* is one of the possible solutions [248].

be applied in parallel and redundantly to E2ENP to enable backwards compatibility with SIP and its resolution for the migration scenarios only on SIP level.

The successful end of the session establishment between the Acceptor and the Passively Involved is signalled back to the Requestor (*externalActionRsp*, *notifyReq*, NOTIFY, *notifyInd*, *externalActionCfm*). The NOTIFY carries *sipfrag: 200 OK* to inform the user about the successful fulfilment status of the external action. If the external action was not successful, an error status is delivered back to the Requestor. The Requestor confirms again the reception of the NOTIFY (*notifyRsp*, 200OK, *notifyCfm*, *externalActionDone*), which message signals also the end of the REFER transaction.

The delivery of the final NOTIFY at the Mobility Requestor causes either the closing of the session (i.e. SIP BYE) if the complete session has migrated to the Acceptor or a re-negotiation of the initial session is done, resulting in session splitting. The final three massages (Figure 45AB – messages 40 to 43) can also go in the other direction, i.e. from Passively Involved to the Mobility Requestor, depending on the application logic reaction to session migrations of the Passively Involved. In case that both the Requestor and the Passively Involved try simultaneously to close or to adapt the migrated (respectively the split) session, the E2ENP deadlock resolution mechanisms determines which of the parties is the Offerer and which the Answerer (see Section 2.3.2.1.1 and 5.2.2.4.2).

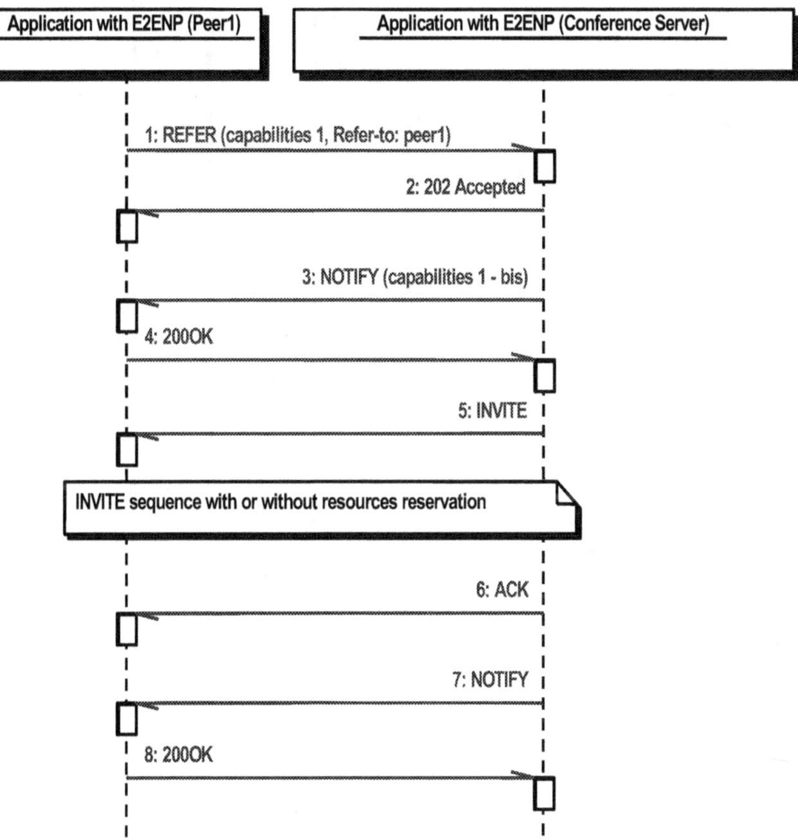

Figure 46: Conference with SIP and E2ENP

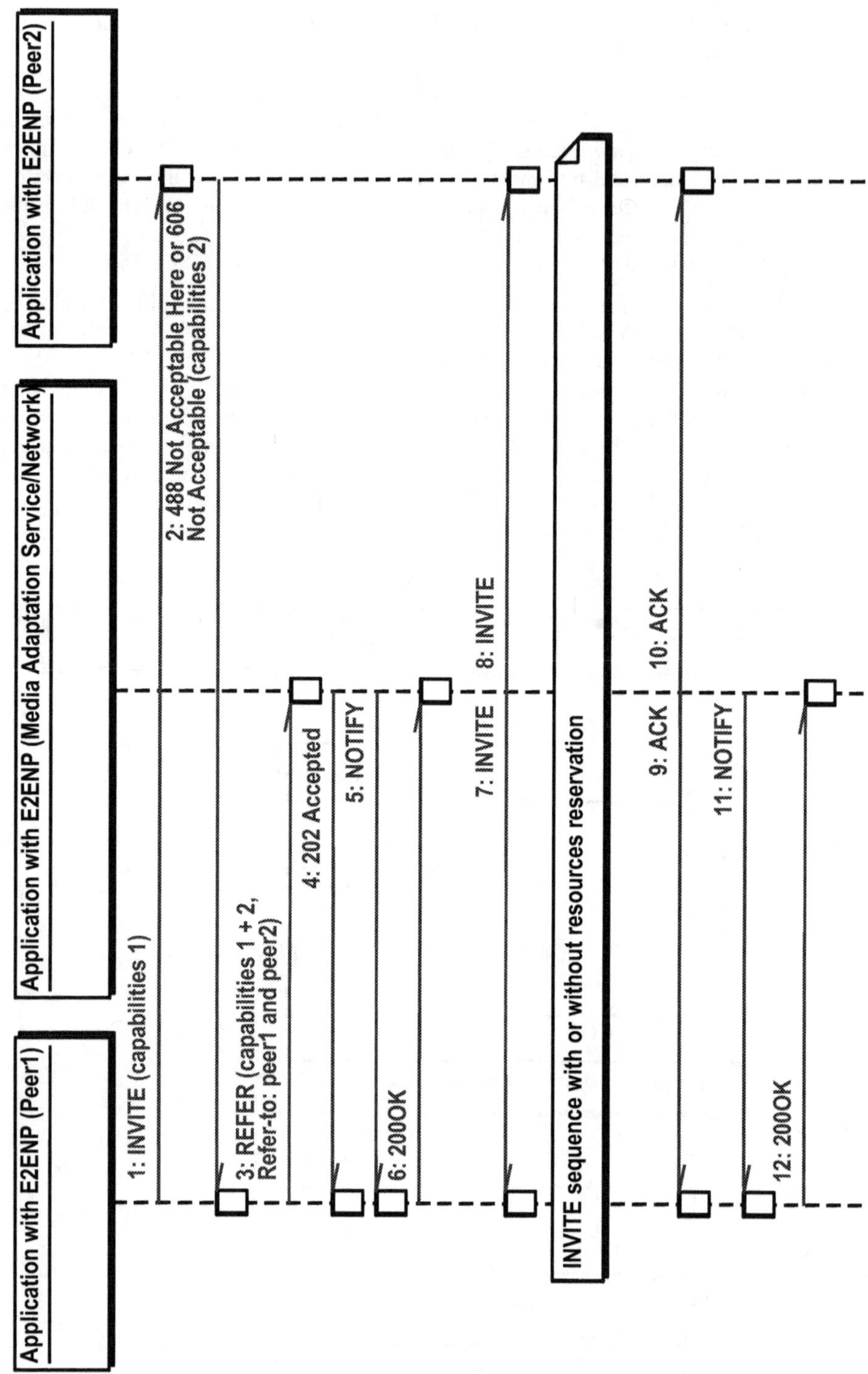

Figure 47: Media adaptation management with SIP and E2ENP

3.3.2 Multiparty Negotiations

Multiparty negotiations for peers that integrate and control multiple E2ENP UAs or multiple SIP UAs integrated in a single E2ENP UA (see also Section 5.2.2.4.1) on the same device are further scenarios for advanced adaptation, in addition to the E2ENP session description language optimisation (see Section 3.1 and Chapter 4) and the above described migration scenarios (Section 3.3.1). The following scenarios are not implemented in the E2ENP UA architecture. However, they are considered in order to provide completeness on the possibilities for performing negotiations and for executing advanced adaptations for the multimedia services (see also Section 4.3.6).

3.3.2.1 Conference

The example shown in Figure 46 presents session establishment with a conference server. The terminal *Peer1* and any other end-system that wishes to communicate over the conference server sends its capabilities to the server (Message 1 – REFER) and the server replies with its possibilities for supporting the session to the terminal (Message 3 – NOTIFY). In this case, the REFER transaction is used as a Pre-negotiation between the terminal and the server so that they get informed about the possible resources utilization for the following session. Since the server is the end-system that shares its resources between the multiple conferring terminals, it is the most appropriate device that can coordinate the shared resources and the respective communication produces the shortest signalling form with SIP (see also [105]). In terms of E2ENP, the server should be enabled to coordinate multiple UAs in parallel to orchestrate the reservations of network resources between it and the conference participants. For this purpose, the conference server issues several parallel INVITEs to the conference participants in centralized manner thus synchronizing the resources reservation coordination between the multiple peers. Similar scenario is also presented in [234] in terms of session migration and splitting when several new devices shall be involved in the session, e.g. that the session between peers A and B migrates from peer A to peers C and D. However, the aspect of resources coordination in centralized manner is not shown in [234].

3.3.2.2 Media Adaptation Services

Media adaptation services are usually applied in cases where the terminals that wish to communicate have no common capabilities and cannot transfer their sessions to other appropriate devices through migration. At invocation of media adaptation services, combinations of SIP and MEGACO protocols can be applied to manage the communication between the terminals and the media gateways (SIP) and between several media gateways (MEGACO) [92][202][221][250]. Two major scenarios describe the invocation of media gateways [202]:

- *Implicit invocation* means that the end-systems are not aware of the media adaptation. The application of the content adaptation services in this case is managed in similar way as the case where the providers reserve resources for their subscribers (see Sections 2.3.2.1 and Figure 18). The providers interact with the session descriptions in order to adapt them to the possibilities for media adaptation of the gateways. In terms of E2ENP, the provider adds the new capabilities in a modular way in the respective PDUs (see Section 4.3.6).
- *Explicit invocation* is the case where the terminals invoke explicitly the adaptation service when they discover that they do not have common capabilities. The scenario presented here (Figure 47) is still another possibility to communicate with adaptation

services using explicit invocation, compared to the cases treated in the literature [92][106][202][221][249][250] where no explicit coordination of resources is considered. Figure 47 shows an example where the adaptation service (or the edges of an adaptation service network, i.e. distributed adaptation service [250]) coordinates in centralized manner the resources reservations between the terminals and the adaptation service applying a combination of INVITE and REFER transactions. Alternative scenarios (with multiple SIP INVITE sequences) are also possible. The access network providers act then as mediators of the communication between the terminals and the adaptation service/-s [202] or one of the end-systems takes over the orchestration of the communication with the adaptation service [106]. According to the applicable scenarios, interactions with E2ENP PDUs for adapting the session descriptions to the requirements of the terminals and the adaptation services may also occur. Such interactions may also lead to modular addition of capabilities to the PDUs in order to enhance the basic service capabilities with the additional features of the adaptation services (see Section 4.3.6).

The session management scenarios described in this chapter serve as requirements and motivation for the development of the E2ENP negotiation language for optimising the session descriptions and their exchange between the communicating parties as presented in the next Chapter 4. Some of the major negotiation scenarios are implemented in the E2ENP UA state-machine logic shown in Chapter 5.

4 E2ENP Negotiation Language

The E2ENP protocol implements a description language defined in XML [73]. The specification of the language is an application of the XML-schema technology [93][94]. The choice of XML as a modelling tool for E2ENP is made because its schema is extensible and it allows the integration of XML descriptions belonging to different models and namespaces. The schema is also suitable for expressing object-based models (e.g. models like the hierarchical QoS specification of E2ENP) and for covering protocol-specific requirements when designing protocol languages (see also [94] and [251]).

The initial definition of the E2ENP language (as a concept for optimising the configuration of multimedia-enabled end-systems) was a partial contribution of the author of this work to the IST-MIND project [14][20][170][173]. However, the model presented in this contribution shows an additionally advanced version of the E2ENP language, considering the following issues:

- The initial version of the E2ENP XML schema was able to produce only well-formed XML-instance documents but not valid ones, due to mismatches in the namespace inheritance between the E2ENP and the SDPng definitions. The current version presented here produces both well-formed and valid XML-instance documents that were tested for correctness using XML Spy [252] and XMLmind XML Editor [253] (see Section 4.2 and the respective comments on the correctness of the namespace inheritance presented there).
- The initial E2ENP specification considered only theoretically the role of the provider and of other mentoring entities at interactions with the E2ENP PDUs. The new version of E2ENP implements such interactions in the description language (see Sections 4.1.3.4 and 4.3.6). Furthermore, descriptions for enhancing and optimising the PDU management issues are added to the E2ENP language (see Sections 3.1.4 and 4.3.7)
- The E2ENP advanced development is aligned with the latest tendencies and discussions within the scope of the session descriptions and service-control protocols [69][72][81][86] in terms of appropriate contents and descriptions for session management. Improvements in the definitions of QoS contracts (see Sections 4.2.4 and 4.3.3) are introduced to remove specific dependencies to reservation and resource-management technologies as an optimisation for the negotiations within heterogeneous environments (see the discussion in Section 2.3.2.1.2). Furthermore, the service and session definitions are enhanced in accordance with the latest session-management developments, e.g. concepts for integrating E2ENP with MPEG-7 and MPEG-21 frameworks and XML standards are considered (see Section 4.1).

The first section of this chapter (Section 4.1) introduces the concepts used for developing of the E2ENP language and it presents considerations that should be taken into account when using the E2ENP language as a negotiation protocol. Furthermore, Section 4.1 provides details on the reused external specifications within the E2ENP language and on enhancements that E2ENP provides to these specifications. The XML schemas of the E2ENP language are introduced in Section 4.2, followed by multiple, comprehensive XML examples in Section 4.3 that explain the application of the negotiation language. Furthermore, Sections 4.2 and 4.3 present the innovative features of E2ENP for multimedia-session and PDU management, compared with SDP/SDPng and with other similar session-control protocols.

4.1 Description Language

This section introduces some initial considerations used for development of the E2ENP negotiation language for multimedia and quality of service. Furthermore, some practical issues at application of E2ENP in a protocol framework are presented.

4.1.1 Language Definitions and Optimisation Considerations

There are multiple standardization efforts concerning developments of description models for multimedia. E2ENP reuses concepts belonging to the following specifications:

- Moving Picture Experts Group (MPEG) provides descriptions of user environments and of terminal characteristics within the scope of MPEG-21 Digital Item Adaptation (DIA) specification [72]. E2ENP applies as a motivation for its language optimisation the concept of MPEG-21 DIA, considering that not only the media but also its associated metadata should undergo adaptation in order to satisfy the requirements of different services and users (see also the discussion below). Furthermore, E2ENP reuses concepts of MPEG-7 (see Section 4.1.3.3) to specify univocal naming of the applied multimedia codecs and to provide vector data-types (see MPEG-7 Multimedia Description Schemes (MDS) [109]). Additionally, MPEG-7 has the ability to compress any XML; hence, it can be applied for compression of E2ENP descriptions (see MPEG-7 Binary Format for Metadata (BiM) [254]).
- Multiparty Multimedia Session Control Working Group (IETF MMUSIC WG) [179] develops session description protocols for multimedia communication (i.e. SDP [86] and SDPng [81] with its enhancements for transport QoS [87]). These protocols define description schemes for specifying terminal characteristics in form of codecs. In addition, these protocols apply definitions for audio and video parameterisations for streaming transportation [2], developed within the Audio/Video Transport Working Group (IETF AVT WG) [179]. Initially, E2ENP aimed to be an extension of SDPng [14][20][132][170][173], however, the current development of E2ENP is rather a concurrent specification then an extension of SDPng. Presently, the SDPng progress in MMUSIC experiences temporal cessation as the original SDPng specification [81] exposed multiple shortcomings (see the discussion in Section 4.2) and MMUSIC concentrated its efforts on finalizing SDPng's predecessor SDP. However, the transition to a more flexible format is desired to introduce possibilities for negotiations, QoS, security and other novel system features. An informational draft on this issue is accepted for RFC approval [69].
- International Telecommunication Union (ITU) provides in its media-control specification H.245 [63] descriptions of codecs for defining terminal characteristics. A network-parameter specification for network-performance classes is also available from ITU [255][256]. Additionally, ITU specifies a namespace for defining picture sizes for video information, named Common Intermediate Format (CIF) [159]. E2ENP reuses some of these specifications within its own definition language (see Section 4.1.3.2).
- Video Electronics Association (VESA) [257] and Advanced Television Standards Committee (ATSC) [258] develop also standards for video resolution, i.e. Video Graphics Array (VGA) [259] and High Definition Television (HDTV) [260]. These standards are used together with ITU CIF to specify the E2ENP namespace for video sizes (Section 4.1.3.2).

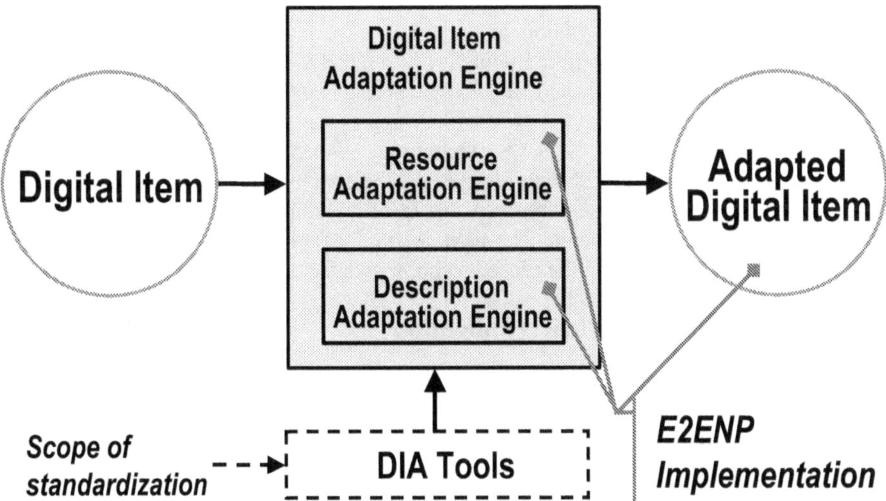

Figure 48: E2ENP and MPEG-21 DIA

- In order to provide security and authentication features within XML the XML-Signature Syntax was introduced [229]. E2ENP reuses concepts and namespaces of this syntax in order to enable the authentication of mentoring entities within the E2ENP PDUs. Furthermore, E2ENP reuses the specification of the XPath [99] technology to point at information added by the mentors to the E2ENP PDUs. Transformation of XML is also considered as a relevant to E2ENP issue. The application of XML standards in E2ENP is shown in Section 4.1.3.4.

The standards presented so far influence the E2ENP descriptions of media capabilities, QoS contracts and streaming configurations. However, the goal of E2ENP is not only to provide a comprehensive description model but also to offer possibilities to negotiate these descriptions in an optimised manner (see also [14][132][173]). The necessity for such optimisation is a requirement of the carrier protocols for E2ENP, especially in cases where unreliable transport and mobile services have to be supported and the respective signalling and negotiations have to be kept as concise as possible (see SIP [62] and Congestion Control Principles [214]). Current implementations of multimedia communication and management systems (e.g. QoS Brokers [17][19], adaptive play-out and congestion management [8] – [10] or content adaptation within the network [90][91][202][221]) consider for adaptation only the perceivable multimedia, as the streaming of the media represents the primary part of the communication. Yet, the communication cannot perform properly in distributed and mobile environments without having accurate controls (e.g. some appropriate management-metadata associated with the perceivable multimedia). The importance of regarding the multimedia as a combination of perceivable information and associated metadata is defined within the MPEG-21 DIA [72], where the adaptation of the data is performed upon Digital Items and each Digital Item (DI) is defined exactly as a combination of media resources and corresponding descriptions that are both adaptable information types (Figure 48). Respecting this MPEG definition, the E2ENP language described in this chapter and its processing tools (see Section 5.2) represent a description-adaptation engine for metadata for optimising it for network transportation at configuration negotiations between the distributed parts of a multimedia service. Furthermore, the language itself (i.e. the generated E2ENP PDUs) corresponds to the result of the metadata adaptation (i.e. Adapted Digital Item in Figure 48) and the whole E2ENP mechanism is the means for controlling the adaptation of the media (i.e. the configurations with E2ENP influence the Resource Adaptation Engine). Consequently, the relevance of E2ENP for MPEG-21 DIA application is that E2ENP can be seen as a metadata-adaptation engine that combines different technologies to provide appropriate controls for Digital Item Adaptation. Note that MPEG-21 is technology independent, hence, it specifies predominantly abstract rules and requirements for the adaptation (i.e. DIA Tools in Figure 48). With this respect, E2ENP can be seen as a technical implementation of the MPEG-21 DIA metadata-adaptation concepts. The next sections detail the practical implementation of E2ENP in consideration with the reused technologies for the E2ENP definition.

4.1.2 E2ENP Practical Application

The application of E2ENP as a protocol requires the registration of its namespaces with a registration authority like the Internet Assigned Numbers Authority (IANA) [261] that is responsible for allocation of globally unique names and numbers (e.g. protocol ports). Furthermore, the name of E2ENP should be registered as a MIME-type in order to apply it as a message body within SIP and other IETF carrier protocols (see Section 2.3.1.1).

Such registrations have not taken place currently, as the contribution of E2ENP [14][132][173] and of similar proposals for SDPng dialects based on MPEG-21 DIA [95] – [97] (i.e. SDPng++) are in very early version of their standardisation requests within IETF MMUSIC WG. Even the major development of SDPng [81] is still not registered at IANA. The exact structure of the SDPng namespace and its correspondent registration depend on the fact how the SDPng work will progress. In this contribution, the shown examples are developed having the assumption that the SDPng and the E2ENP namespaces are registered at IANA. This work sticks to the initial proposal of E2ENP [20][132][170][173], where E2ENP shall become a sub-namespace of SDPng as a dialect of SDPng. Respecting this, the IANA registry for the E2ENP application will occur under the namespace *http://www.iana.org/assignments/sdpng/e2enp* to register the E2ENP XML-namespaces and description values. Additional registry of E2ENP as a MIME type shall be provided under *http://www.iana.org/assignments/media-types/application/e2enp*. Furthermore, it should be considered that E2ENP reuses the IANA namespace *http://www.iana.org/assignments/media-types/multipart/related* [230] [231] in order to provide possibility for transferring multiple E2ENP PDUs simultaneously (see below and Sections 3.2.2 and 3.2.3). The namespaces of MPEG-7 MDS *urn:mpeg:mpeg7:cs:AudioCodingFormatCS:2001* and *urn:mpeg:mpeg7:cs:VisualCodingFormatCS:2001* are also reused to provide codecs identification (see Sections 4.1.3.3 and 4.3.3).

```
<e2enpDesc xmlns="http://www.iana.org/sdpng-e2enp"
xmlns:e2enp="http://www.iana.org/sdpng/e2enp"
xmlns:sdpng="http://www.iana.org/sdpng"
xmlns:rtp="http://www.iana.org/sdpng/rtp"
xmlns:xsi="http://www.w3.org/2001/XMLSchema-instance"
xsi:schemaLocation="http://www.iana.org/sdpng/e2enp e2enp-02.xsd
http://www.iana.org/sdpng/rtp rtp-pkg-sdpng-04.xsd
http://www.iana.org/sdpng-e2enp sdpng-e2enp-01.xsd">
```

Figure 49: Example of XML header definition of E2ENP PDU instances

Taking into account the IANA considerations presented above, the XML header of the E2ENP instance documents is developed as shown with an example in Figure 49[23] (see also Section 4.3.1). The namespace of E2ENP is indicated with *http://www.iana.org/sdpng/e2enp*[24]. Additionally, the namespaces *http://www.iana.org/sdpng* and *http://www.iana.org/sdpng/rtp* indicate the reusing of SDPng and RTP packages of SDPng within the E2ENP specification (see Sections 4.1.3.1, 4.2 and 4.3.3). The namespace *http://www.iana.org/sdpng-e2enp* indicates the integration namespace of E2ENP and SDPng that is used to generate valid XML documents as the basic E2ENP namespace *http://www.iana.org/sdpng/e2enp* provides only the major data types and components of the E2ENP language without fixing them to a specific top-level structure. This approach allows further flexible extensibility and application of the E2ENP namespace for scenarios that go beyond the negotiation purpose (e.g. for service discovery). Further possible namespace that can be reused within the E2ENP XML-header is the definition of the XML-Signature schema (Sections 4.1.3.4, 4.2.3 and 4.3.6) to describe encryption techniques and signing for the E2ENP PDUs. The entry *http://www.w3.org/2001/XMLSchema-instance* specifies that E2ENP is an application of XML schema, i.e. it reuses the basic data types defined through the XML schema (e.g. such types are the descriptions of numbers like integer, double, etc. or the definition of text-based data-types as string or URI) [93][94]. Furthermore, the *XMLSchema-instance* namespace contains the definition of the XPath technology [94]. The indication of the location of the schema (i.e. *xsi:schemaLocation="http://www.iana.org/sdpng/e2enp e2enp-02.xsd http://www.iana.org/sdpng/rtp rtp-pkg-sdpng-04.xsd http://www.iana.org/sdpng-e2enp sdpng-e2enp-01.xsd"*) identifies that currently the E2ENP schemas shall be loaded for validation purposes from the local device and the directory location where the instance XML document is situated (i.e. the schemas and the instance documents are physically co-located). The names *e2enp-02.xsd*, *rtp-pkg-sdpng-04.xsd* and *sdpng-e2enp-01.xsd* are the respective document names of the applied XML schemas and the numbers *02*, *04* and *01* indicate respectively the versions of these documents. Note that the SDPng schema document does not need to be listed in the XML instance document, as it is included in the RTP schema per SDPng definition, consequently the RTP schema inherits the SDPng schema (see Sections 4.2.2 and 4.2.4). Further details on the XML schemas of E2ENP are discussed in Section 4.2.

The application of E2ENP as a MIME type leads to the following definitions within the text-based carrier protocols (like SIP) that transport the E2ENP PDUs:

---

[23]Due to the line wrapping in this document the original indentation as usual for XML layout does not appear correctly in the examples shown in this work, i.e. the bracket pairs "<" and "/>" or "<" and ">", or "</" and ">" always indicate single unbreakable lines. The lines delimited with "<!--" and "-->" brackets are XML comments that are used where appropriate to improve the readability of the XML code. The information between "<!--" and "-->" indicates also a single line.

[24]Note that the XML namespace indication (i.e. the place where the schema is physically located on the Internet) can be different compared to the registration entries of the protocols and the applications within the respective Internet registration authorities. The example shown here represents the typical way according to which XML namespaces belonging to IANA and W3C are indicated in XML-instance documents. Furthermore, different root-domain names can also appear in the instance documents in cases when backup or mirror registries of the major registries are in use, e.g. the initial proposal for E2ENP was to physically locate its schemas under the project domain-name of MIND (http://www.ist-mind.org); however, this domain is not supported any longer.

- In case that a single E2ENP PDU is transported, the *Content-Type* field of the carrier is:

Content-Type: application/e2enp

- In case that multiple PDUs are carried simultaneously, the definition is (see also [231]):

Content-Type: Multipart/Related; boundary=e2enp-pdu

--e2enp-pdu

Content-Type: application/e2enp

<!--THE FIRST PDU XML DATA-->

--e2enp-pdu

Content-Type: application/e2enp

<!--THE SECOND PDU XML DATA-->

...

--e2enp-pdu

Content-Type: application/e2enp

<!--THE N-th PDU XML DATA-->

--e2enp-pdu--

Note that the name of the delimiter line (i.e. *e2enp-pdu*) between the single PDUs can be freely chosen, however, the chosen name for this separator together with the delimiter dashes must be univocal and shall never appear in any of the XML documents representing the E2ENP PDUs (see also the delimiter definition in [217]).

The considerations about the practical application of E2ENP presented here are used for implementing the E2ENP XML schemas and instance documents (Sections 4.2 and 4.3 and Annex C) and for the realisation of the E2ENP XML transportation and management within the E2ENP UA architecture (Section 5.2).

4.1.3 Major External Profiles and Mechanisms in E2ENP

The reusing of standard external profiles and mechanisms is profitable for the application of E2ENP. In order to provide short definitions for media parameterisations, E2ENP applies standardized definitions for audio and video streaming, for video sizes and for codecs naming. Furthermore, E2ENP can directly benefit from reusing the standard for XML binary-representation and compression as defined in MPEG-7. The application of some XML-native technologies to provide linking and digital signing for XML documents is also utilized. All these technologies and the way they are reused in E2ENP are discussed in the next sub-sections.

4.1.3.1 Audio and Video Profiles for RTP

The Real-Time Streaming Protocol (RTP) [1] is the most wide-spread protocol used for developing and deploying real-time multimedia applications like Video on Demand (VoD) or Voice over IP (VoIP). For controlling conferences with minimum efforts, a profile for indicating multimedia codecs and their streaming capabilities (e.g. transmission frequency) using numbers was specified under the name RTP/AVP and a single codec number within this profile is designated as a *payload type* (PT) [2]. The complete set of numbers used for PTs is between 0 and 127, where the numbers from 0 to 23 correspond to static payload types for audio, from 24 to 95 – to static payload types for video or for combined audio/video codecs and from 96 to 127 – to dynamic payloads (see [2] for details on the respective codec assignments). The static payloads are either already assigned to existing codecs or are reserved for special purposes. The dynamic payloads can be freely chosen by the applications when using novel codecs or when defining non-standard streaming capabilities.

Session Description Protocol [86] and Session Description Protocol next generation [81] apply RTP/AVP for describing and negotiating the streaming capabilities of the codecs and for configuring the respective multimedia service. E2ENP directly adopts the RTP/AVP profile definition of SDPng and enhances it with additional data to introduce, for instance, proper handling of the PT numbers [2] and of IPv4 [119] and IPv6 [120] address conventions (see Sections 4.2.2 and 4.3 for examples of the RTP definitions and their application in E2ENP).

4.1.3.2 Video Resolution Standards

There are three major video-resolution standards considered in this work, i.e. Common Intermediate Format (CIF) [159], Video Graphics Array (VGA) [257][259] and High Definition Television (HDTV) [258][260]. All these standards use names to define specific height and width of a video frame in pixel. The most commonly used names (e.g. for displays of mobile telephones, PCs and TVs or for beamer) associated with these standards and applied in later E2ENP examples are shown in Table 1 through Table 3[25]. The complete definition of the CIF, VGA and HDTV video-resolution namespaces can be found in the corresponding standardisations.

Table 1: ITU CIF video resolution names

Format	NTSC-based resolution (width x height)			PAL-based resolution (width x height)		
SQCIF				128	×	96
QCIF	176	×	120	176	×	144
QCIF+	176	×	220	176	×	220
CIF	352	×	240	352	×	288
4CIF	704	×	480	704	×	576
9CIF	1056	×	720	1056	×	864
16CIF	1408	×	960	1408	×	1152

[25]NTSC is the American television standard. PAL is the European television standard.

Table 2: VESA VGA video resolution names

Format	Resolution (width x height)		
CGA	320	×	200
QVGA	320	×	240
EGA	640	×	350
VGA and MCGA	640	×	480
HGC	720	×	348
MDA	720	×	350
SVGA	800	×	600
XGA	1024	×	768
XGA+	1152	×	864
WXGA	1280	×	768
SXGA	1280	×	1024
WSXGA or WXGA+	1440	×	900
SXGA+	1400	×	1050
WSXGA	1600	×	1024
WSXGA+	1680	×	1050
UXGA	1600	×	1200

Table 3: ATSC HDTV video resolution names

Format	Resolution (width x height)		
HDTV 1080i	1920	×	1080
HDTV 720p	1280	×	720
DVD NTSC	720	×	480
DVD PAL	720	×	576
VCD NTSC	352	×	240
VCD PAL	352	×	288

Table 4: MPEG-7 MDS for MPEG-1 Audio

MDS Name for Audio CS: urn:mpeg:mpeg7:cs:AudioCodingFormatCS:2001	
MDS Number	**Codec Name**
3	MPEG-1 Audio
3.1	MPEG-1 Audio Layer I
3.2	MPEG-1 Audio Layer II
3.3	MPEG-1 Audio Layer III

4.1.3.3 MPEG-7 Mechanisms

Three major concepts of MPEG-7 are considered within the E2ENP development, i.e. the possibility to define vectors as data structures [109], the specification of Classification Schemes for codecs within the scope of the Multimedia Description Schemes (MDS) for uniquely identifying media-coding algorithms [109] and the possibility to compress XML using the MPEG-7 Binary Format for Metadata (BiM) [254].

E2ENP uses sets and ranges of data in form of vectors for the specification of its QoS contracts, e.g. to express parameters like potential audio-sampling rates or video-colour qualities. MPEG-7 introduced the processing of complex data structures like vectors and matrices as these definitions were missing in the original XML specifications [109]. However and unlike the MPEG-7 vectors, E2ENP needs simpler and smaller structures for its sets and ranges definitions. Furthermore, E2ENP uses predominantly XML attributes and not elements in order to reduce the sizes of the PDUs that are sent during a negotiation (see also the respective discussion at the beginning of Section 4.3 about the application of attributes and elements when describing capabilities and parameters). Hence, for E2ENP it is inconvenient to directly reuse the element-based MPEG-7-vectors specification. With this respect, E2ENP reuses only the idea of MPEG-7 and it specifies its own convention for vectors. The E2ENP vector-structures are applied in attributes that have names ending with either "*-set*", or "*-range*", e.g.

frame-rate-set="20,25" frame-size-set="QCIF,CIF" color-quality-set="8" overall-quality-range="9900,9960,0.1"

The values in the vectors are comma-separated. The sets can contain just one element (e.g. *color-quality-set="8"*). The ranges contain two values (e.g. *quality-range="9900,9960"*), indicating that the range can be used with any tracing step and is actually discontinuous, or it contains three values (e.g. *overall-quality-range="9900,9960,0.1"*), indicating that the range should be traversed with a tracing step indicated in the last place within the vector. Since this E2ENP convention is non-typical XML definition, the applications of E2ENP should use additional level of parsing to separate the elements of the vectors upon their comma-delimiters. Further examples of E2ENP vectors are also shown in Section 4.3.

MPEG-7 MDS specifies numbered values for uniquely identifying the coding algorithms of different media and it provides names for the respective classification schemes (CS) for the codecs belonging to the same media type [109]. An example of the MPEG-1 audio definition using the audio CS of MDS is shown in Table 4, where *urn:mpeg:mpeg7:cs:AudioCodingFormatCS:2001* is the name of the respective identification scheme for audio. The name of the video scheme is similar: *urn:mpeg:mpeg7:cs:VisualCodingFormatCS:2001*. Since not all available codecs are represented by the MPEG-7 classification schemes (e.g. WAVI [17][162] is not in the video scheme), E2ENP defines also its own classification scheme for audio and video codecs under the names *urn:mpeg:e2enp:cs:AudioCodingFormatCS:2001* and *urn:mpeg:e2enp:cs:VisualCodingFormatCS:2001*, following the same principle as MPEG-7 (examples thereof are given in Section 4.3, see also Annex D). The leading numbers in a classification scheme correspond to the basic algorithms for coding the media (e.g. for *MPEG-1 Audio* this number is 3, see Table 4). Multiple coding algorithms produce different qualities and expose different processing capabilities using the so-called coding layers. Hence, such layers and if available sub-layering of layers is identified with a number preceded by a dot and each level of layers has its own numbering (e.g. the number *4.2.2* points at *MPEG-2 Backward Compatible Multi-Channel Layer II* that is a sub-layer of *MPEG-2 Backward Compatible Multi-Channel* profile, identified with *4.2*. Correspondingly, *MPEG-2 Backward Compatible Multi-Channel* is a sub-scheme of

MPEG-2 Audio that is number *4* in the audio scheme). This rule for layer numbering applies also for different versions of an algorithm (see [109] and Annex D). E2ENP re-uses the audio CS and video CS of MPEG-7 in order to provide univocal identification for audio and video coding capabilities of a peer at negotiations. This consideration is made because some industrial implementations of codecs bear different names then the respective standard algorithms and may incorporate multiple algorithms together with versions and layers thereof. Two typical examples of this fact are the audio coding with MP3 [262] and the video coding with DivX [263]. The MP3 implementation corresponds to *MPEG-1 Audio Layer III* and DivX applies *MPEG-4 Visual Simple Profile* and *MPEG-4 Visual Advanced Simple Profile*. Furthermore, industrial implementations belonging to different manufactures may display incompatibilities due to incomplete or ambiguous definitions in the used standards or due to the application of additional non-standard features in the implementations. With this respect, E2ENP gives the possibility to indicate both the industrial implementation of a codec and the respective coding algorithm to achieve unique identification of the negotiated capabilities, e.g. such identification for MP3 in E2ENP is:

**<e2enp:audio-codec name="MP3"
href="urn:mpeg:mpeg7:cs:AudioCodingFormatCS:2001:3.3"/>**[26]

If the name or the used algorithm for the media coding is univocal, the negotiating party may use only the name or only the CS specification of the codec to negotiate capabilities (see Section 4.3 for further examples).

The third MPEG-7 mechanism considered for utilization with E2ENP is the MPEG-7 Binary Format for Metadata (BiM) that can be used to achieve efficient compression of any XML file [254][264]. The application of BiM can further optimise the sizes of the E2ENP PDUs in addition to the descriptions fragmentation provided through E2ENP (see also Sections 3.1.2, 3.1.4, 4.1.3.4 and 4.3.2). Although not implemented in the current E2ENP UA, the realisation of BiM compression for XML within the UA can be easily achieved due to the modular and configurable structure the UA that can be simply rebuilt and reconfigured to enhance it with additional compression features. The compression modules can be installed on the processing paths for XML in a similar way to the application of the measurement modules within the current UA (see Section 5.2.1.2).

4.1.3.4 XML Linking, Signing and Transformation

The generation and the management of hierarchical QoS specifications with E2ENP require linking between multiple PDUs (see also Section 3.1.4). The typical technologies in XML used for linking are XLink [98], XPath [99] and XPointer [100]. XLink is the technology providing cross-references between different resources (i.e. documents); XPath defines the language for accessing elements and attributes within a single XML document and XPointer is the combination of XLink and XPath that allows to access specific contents enclosed in different independent resources. The direct application of these technologies with E2ENP, however, is problematic as XML was initially designed for document management [251] but not for communication protocols, where different requirements may hold true (see below the discussion about the physical and logical

[26] *href* indicates a reference to an external MDS source, where the complete codec specification is given. If the application has the respective codec specification locally, this parameter is interpreted simply as a name.

object-structures in E2ENP). The possible solutions for the E2ENP linking are detailed in the following:

- The E2ENP documents (the PDUs) are not named XML-documents and it is up to the application if they shall be stored in their original form or shall be transformed and stored in a different form (see also Section 3.1.4). Since E2ENP shall operate in heterogeneous environments, E2ENP does not prescribe how and where the applications store their PDUs. Consequently, this may lead to ambiguous naming and storage specifications of the E2ENP applications, which fact appears counterproductive at utilization of XLink [98] and XPointer [100] for linking between E2ENP PDUs stemming from different negotiation partners. Additionally, the revealing of the exact storage place of the PDUs at negotiation can make the peers vulnerable to attacks of malicious third parties that may try to access the PDUs in different ways then the negotiation procedure and thus manipulate already negotiated PDUs (i.e. manipulate the application's PDU-memory). Respecting the above considerations, the XLink and XPointer technologies are not applied for linking of E2ENP PDUs. E2ENP introduces its own linking mechanism that is applied together with the E2ENP fragmentation mechanism for hierarchical QoS specifications (see beginning of Section 4.3 and Sections 4.3.2, 4.3.4 and 4.3.5). The linking and pointing mechanism between E2ENP PDUs is the E2ENP PDU header (Section 4.3.2).
- The specific negotiation procedures of E2ENP (see Section 3.1) are able to separate hierarchical QoS specifications into fragments and to build these from the bottom of the hierarchy (i.e. starting with the QoS contracts) to its top (i.e. QoS contexts). Furthermore, the QoS specification can be split between different documents. Because of this fragmentation feature, the XML documents of E2ENP are structured differently compared to the QoS specification that may be contained completely or partially in such documents. This means that the physical XML tree of a PDU document can contain the complete logical tree of a hierarchical QoS specification or logical sub-trees of such specification. Hence, these two structures (the physical and the logical) require different addressing when cross-references shall be used, in order to achieve optimisation at definitions of links, e.g. for the short Re-negotiations (see below and Sections 3.1.2 and 4.3.5). Correspondingly, E2ENP supports two kinds of links:

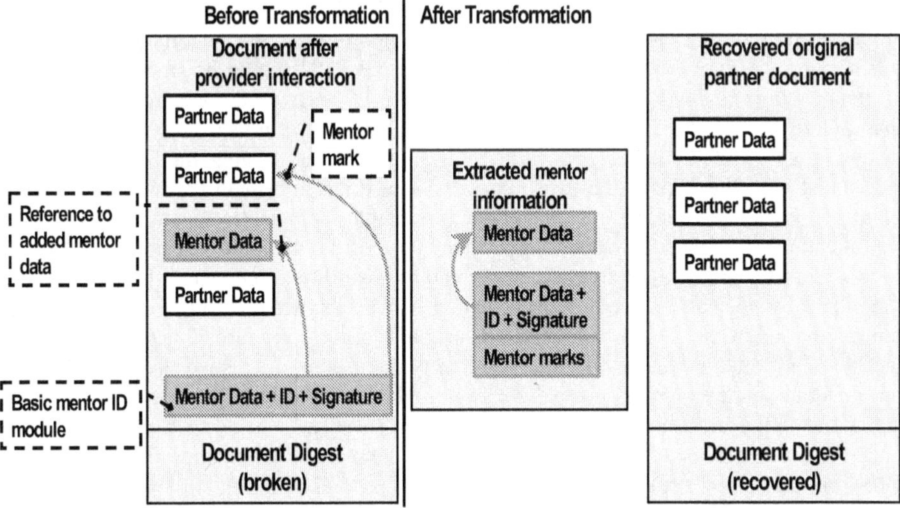

Figure 50: Addition and extraction of mentor information in the PDUs

- o Indicators to the physical XML structure of the documents where the XPath technology [99] is reused to refer to information locally within the same document where the XPath-link is applied. For instance, XPath is used in cases where mentors explicitly add QoS-relevant information to the PDUs to enhance the technical specification of the peers with additional QoS definitions in case media adaptation services shall be applied (see below the considerations on signing and the discussion and examples in Section 4.3.6).
- o Links to the logical QoS specification of the application where the levels of the specification are addressed by their names (see Sections 4.3.4 and 4.3.5), i.e. E2ENP is used to define the names of all the relevant capabilities and QoS levels. E2ENP provides a mechanism for the application to build control contexts at different abstraction levels and to address these contexts at QoS adaptations. The building of the logical links with the names of the QoS-specification levels is an efficient way to refer to the QoS specification. The addressing of a logical structure with its physical representation (e.g. in the way XLink and XPointer do) is non-optimal as on one side the XML definitions contain multiple wrapper elements that are meaningless for the QoS specification, but are necessary for the proper XML-document definition; on the other side the specific fragmentation mechanism would require the composition of several physical links that will result in a too complex link-definitions. The logical links are used in cases of short Negotiations (see Section 4.3.4) and Re-negotiations (see Section 4.3.5) and in cases where the mentors mark the hierarchical QoS specification without introducing changes within the negotiated technical specifications of the peers (see Section 4.3.6).

Consequently, E2ENP defines two novel linking mechanisms. The first one is the linking between PDUs to build and manage hierarchical QoS specifications (see Sections 4.3.2 and 4.3.7). The second one is the logical linking between components of the QoS specification to associate capabilities and to build QoS contexts at different abstraction levels (see Section 4.3.4). Furthermore, the second mechanism is used to refer to contracts/contexts at short Re-negotiations and to enable provider marking of QoS specifications with policy-specific rules (Sections 4.3.5 and 4.3.6).

The mentors' interactions with the E2ENP PDUs require identification and authentication of the mentors as trustworthy supporters of the negotiation procedure. Furthermore, it should be considered that the E2ENP PDUs could be signed by the end-systems using S/MIME digests (see Section 3.2.1). Hence, the mentors should not manipulate the original technical specifications of the terminals (see the respective discussions in Sections 2.3.2.1.2 and 2.3.2.2.2), nor shall they change the S/MIME digests associated with the PDUs in an irreversible way in order to remain trustworthy from the point of view of the negotiating partners and to enable the peers to separate the technical specifications of their communication partners from of the mentors' information. Because of this requirement, E2ENP gives the mentors the possibility to interact with the PDUs in a modular way so that the mentors information can be extracted through recognition and transformation of the XML document to its original form (see Figure 50 and the discussion on the transformation below). In order to identify themselves as trustworthy participants in the negotiations the mentors use a *basic mentor ID module* (Figure 50) that includes the signature of the mentor and information which E2ENP data is being marked within or added to the XML document (see Section 4.3.6). The marked information (i.e. capabilities, QoS contracts or contexts) is indicated with an E2ENP link pointing at the respective data. The link can be physical or logical depending on the fact if the mentor adds technical data (e.g. capabilities) to the technical specification of the peers or it adds only restrictions to the specification (see the above discussion on physical and logical links and the examples in Section 4.3.6). For restricting of control information, the mentors use E2ENP logical links to point at the respective part of the QoS specification. When adding new capabilities, the mentors use XPath to exactly point which information is being added to the document so that it can be extracted from the document to achieve the original QoS specification of the terminal sending the respective PDU. The signature within the "basic mentor ID module" can be an E2ENP-own signature that provides information on what is being marked or added, which algorithm is used for signing of this mentor information and the signature itself. Note that the E2ENP assumes that information on keys and further relevant data for the signing of the PDUs is being exchanged ahead of the negotiation between the mentors and the terminals (see Section 3.2). Furthermore, E2ENP integrates the XML-Signature Syntax [229] that can be used to provide more comprehensive signatures compared to the E2ENP-own optimised signature (see Sections 4.2.3 and 4.3.6 for further details and examples on providing mentor signatures within the E2ENP PDU). The XML-Signature Syntax allows exchanging of complete sets of signature and signature management information compared to the E2ENP-own short signing mechanism. Thus, the XML-Signature Syntax enables the exchanging of keys and key-management information within the negotiation procedure. However, the sizes of the PDUs increase when applying this technique.

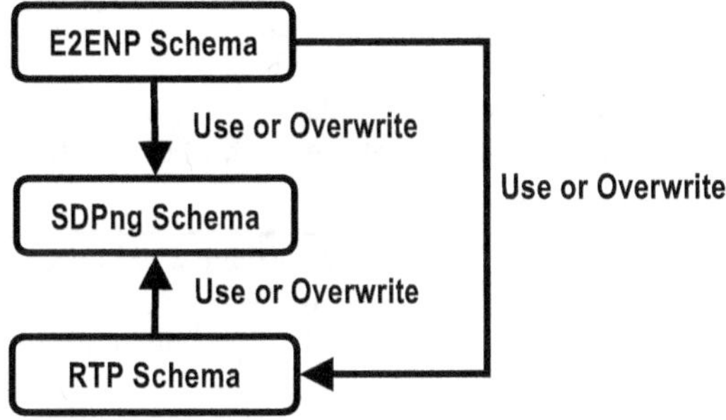

Figure 51: XML schema inheritance problem in initial E2ENP version

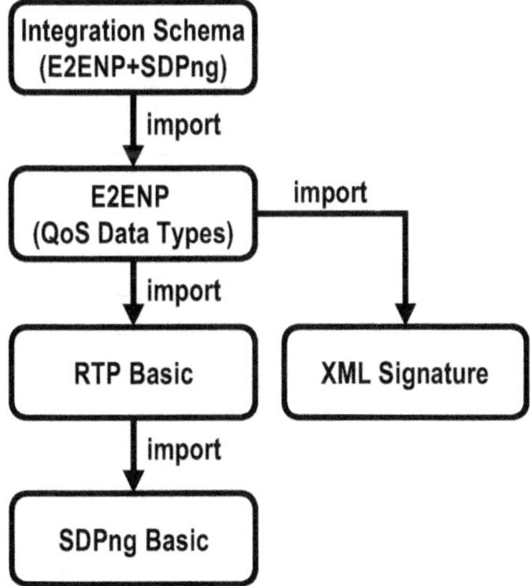

Figure 52: Integration of the namespaces used in E2ENP

Table 5: XML Spy components visualisation convention

 Complex Type	The complex type corresponds to the definition of an object interface. It is used as a definition basis for XML elements with similar characteristics and as a basis for extension or restriction of the content model, i.e. for specification of derived interfaces and objects.
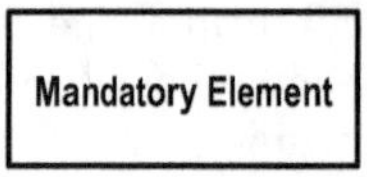 Mandatory Element	The mandatory elements in the content model are those that shall always appear in the XML instances. The mandatory elements and links to such elements (i.e. the integration of the element into another element) are shown in the XML Spy visual model with a uniform line.
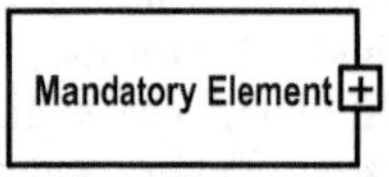 Mandatory Element ⊞	The plus in a square associated with an element indicates that the element contains more elements that are not shown in the visual model, i.e. the element is collapsed.
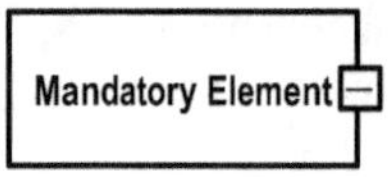 Mandatory Element ⊟	The minus in a square associated with an element indicates that the element contains more elements that are all shown in the visual model, i.e. the element is expanded.
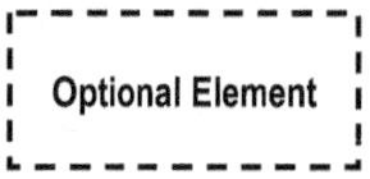 Optional Element	The application of optional elements is non-obligatory and they can be used according to the needs of the generated XML instance. The optional elements and links to such elements (i.e. the integration of the element into another element) are shown in the XML Spy visual model with a dashed line.
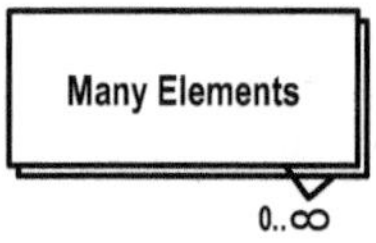 Many Elements 0..∞	The elements indicated with double rectangles and an angle with the symbols "0..∝" or "1..∝" underneath are elements that can be repeated in the content model. The symbols under the angle indicate the cardinality of the element, where "0..∝" is optional or many-times application and "1..∝" is one or multiple.
 Element with Content	The three dashes on the upper-left corner of an element indicate that it contains data of some simple type in the body of the element, e.g. string.
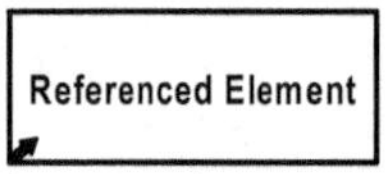 Referenced Element	The arrow at the lower-left corner of an element indicates that it is a reference to a globally defined element.
	The sequential content model is used to define a sequence of elements. The elements grouped with the sequence model should appear in the order of their definition.
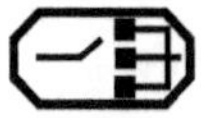	The choice content model is used when one element of the model shall be chosen.
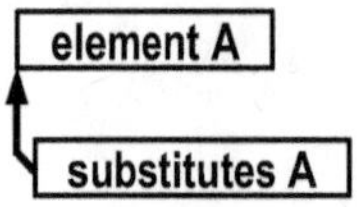 element A substitutes A	The substitution of elements in the content model is indicated with an arrow pointing from the substitute towards the element to be replaced.

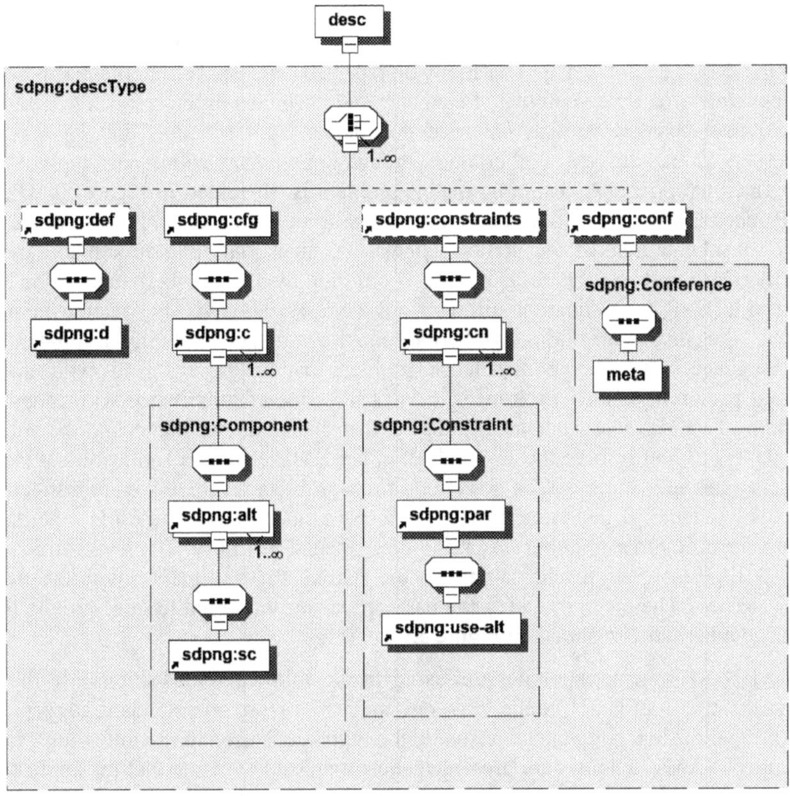

Figure 53: SDPng version 4 [81]

The inclusion of mentor data into the PDUs and the restoration of the original PDUs, respectively the extraction of the mentor data, can be achieved using XML transformation techniques like XSLT [265]. Currently, E2ENP provides only the specification on how the mentors add information into the PDUs and what fields shall be recognised as mentor data and respectively extracted to recover the original peer specifications and if available the corresponding signatures (see Figure 50). The inclusion of PDU transformations is not considered in the E2ENP UA implementation presented in this work. However, since the transformation technology is mature enough, the existing transformation engines can be directly integrated into the E2ENP UA in a similar way and in combination with BiM-Transformers to achieve further processing optimisation for the E2ENP PDUs (see Sections 4.1.3.3 and 5.2.1.2 and [264]).

4.2 XML Schemas

This section discusses the XML schemas applied in the E2ENP language. Compared with the older version of E2ENP [14][20][170][173], the version presented here does not redefine the SDPng namespace in order to avoid the schema-inheritance problem as visualised in Figure 51, where the *Use* relationship indicates the application of the XML schema *import* element to combine description modules belonging to different schemas and the *Overwrite* designates the usage of *redefine* or *include* elements for overwriting or combining certain schema modules. Complications in the instance-document validation of documents stemming from schemas with namespace inheritance where loops or multiple inheritance occur were detected using the validating parsers of XML Spy [252] and of XMLmind XML Editor [253]. The loops or repeated namespace inheritance in schemas and documents affect the visibility of elements and attributes in the respective documents, i.e. the parsers cannot validate instance documents from schemas that have repeated predecessors in the namespace inheritance, though the schemas with repeated predecessors are in principle correct. An example of this problem is illustrated in Figure 51 where the E2ENP schema instances cannot be validated due to detection of multiple inheritance, as the E2ENP schema inherits the SDPng schema two times, one time directly and one time indirectly over the RTP schema. However, if omitting the direct inheritance of the SDPng schema in E2ENP, the validation still does not work, as the visibility of some elements is obscured when using the indirect inheritance. Although not explicitly forbidden in the XML language, the schema-inheritance problem cannot be resolved with the used validating parsers (see also Section 5.2.2.1). The result of inheritance with loops is the generation of valid XML schemas, but the XML-instance documents cannot be validated, as the parsers cannot resolve the ambiguities in the XML-elements application at elements referencing and substitutions. The detection of this problem resulted in some corrections of the E2ENP schema to guarantee proper namespace inheritance where the inheritance is done in a specific order to avoid obscuring of elements' visibility. The re-definition of the SDPng-E2ENP integration schema enables the E2ENP PDU validation with the currently existing parsers. The order in which the namespaces are imported to generate the E2ENP namespace inheritance is shown in Figure 52.

The E2ENP schema examples discussed in the following subsections use the XML Spy visualisation tool [252]. Since, this tool applies its own, non-standard convention for representing the XML components visually, the indications used in the following examples and figures of XML schemas are presented shortly in Table 5. Note that the application of an XML complex type for the definition of elements in the schema tree is indicated with a grey-shaded rectangle bordered with large dashes (larger then the dashes of the optional elements – see Table 5 and compare the optional elements and the grey surrounding boxes of element groups indicating complex-type application in Figure 53). The name of the applied complex type is indicated in the upper-left corner of the shaded rectangle and the element that uses the respective type is one level higher in the visual hierarchy. If the visualisation of a component in XML Spy is done from within its own schema namespace no namespace header is shown in the respective figures, e.g. <signatureType> is shown in the E2ENP schema, where the component is defined, but <e2enp:signatureType> is shown in the integration schema of E2ENP and SDPng. Furthermore, the XML components are displayed with a header on places where globally-defined components are referenced.

The citing of XML elements (respectively types), of attributes and of their values in the later parts of this work (i.e. Sections 4.2 and 4.3) is done using XML-element brackets for the elements and their types (e.g. <purpose>), the attributes are cited with quotations (e.g. "frame-rate-set") and the values are shown in italics (e.g. *audio*).

The next subsections present the foreign XML schemas used in E2ENP and the integration of these schemas in the E2ENP language. The discussed integration deals with the used elements only. The specification of the respective attributes of the elements is detailed in the next Section 4.3 together with the corresponding examples of the usage of the elements and their attributes.

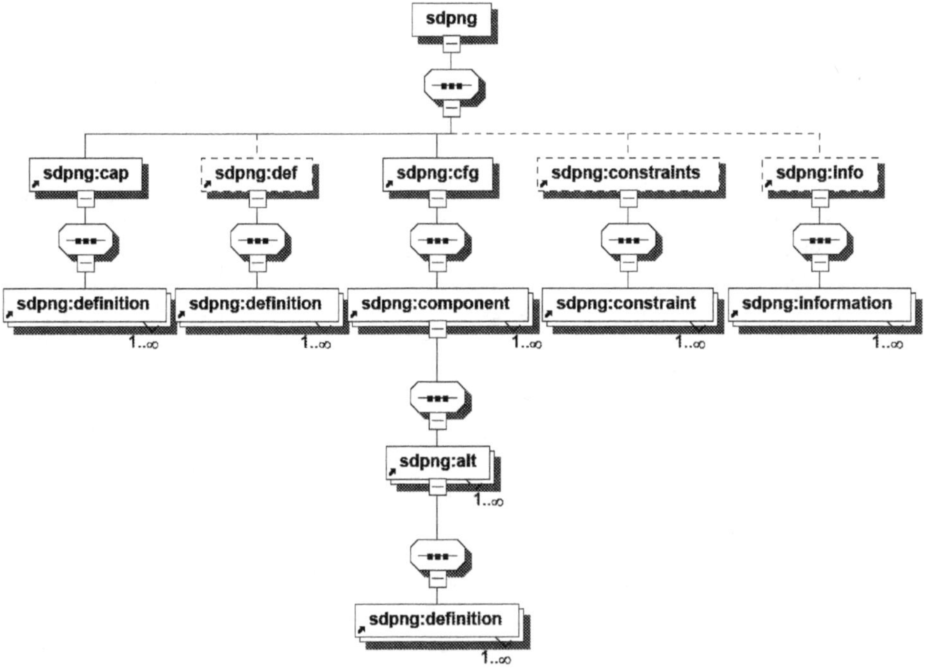

Figure 54: SDPng version 8 [81]

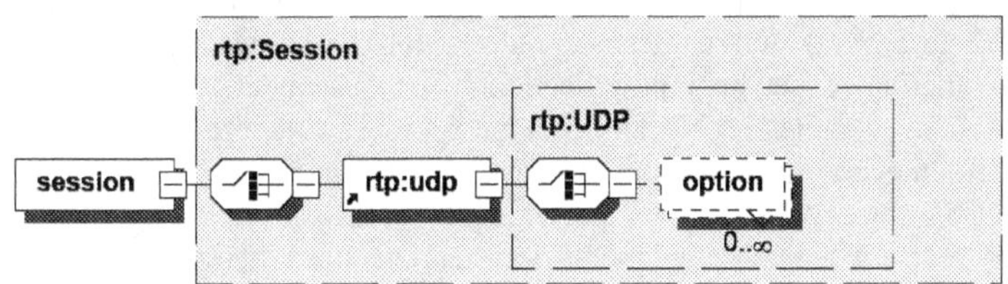

Figure 55: The RTP package of SDPng version 4 [81]

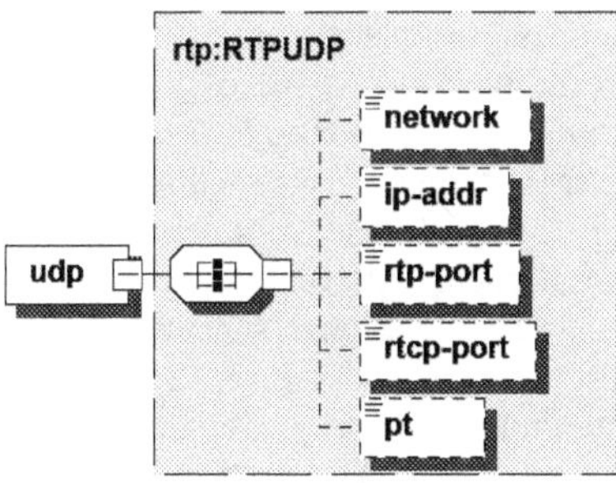

Figure 56: The RTP package of SDPng version 8 [81]

4.2.1 Session Description Protocol Next Generation

E2ENP was initially considered an enhancement of SDPng, but in the meanwhile, it developed as an alternative approach to SDPng that adopts only minimally elements of the original SDPng XML schema. In order to explain the application of SDPng concepts in E2ENP, first the SDPng schemas are introduced (see Figure 53 and Figure 54). SDPng provides in its basic schema [81] the wrapper elements for specifying the top-level structure of the session description documents. The version used within E2ENP is version four of SDPng (Figure 53), where the top-level elements specify: basic media definitions (<sdpng:def>), alternative configurations for the media (<sdpng:cfg>), definition of constraints for the applied configurations (<sdpng:constraints>) and conference configuration (<sdpng:conf>). In version eight of SDPng, the elements <sdpng:def>, <sdpng:cfg> and <sdpng:constraints> have the same meaning as in version four. In addition, the element <sdpng:conf> is removed as SDPng per definition should be applicable for different services, hence, the specific conference definition is replaced by the more general <sdpng:info> wrapper to indicate any additional metadata associated with the session description. Furthermore, the wrapper <sdpng:cap> was introduced to indicate device capabilities that are negotiable, whereas <sdpng:def> in version eight is for non-negotiable parameters like the RTP ports. The versions four and eight of the SDPng schema presented here display two alternative completed versions of the SDPng XML schema corresponding to the numbers of the respective IETF MMUSIC drafts [81].

One major problem with extending the SDPng XML schema is that it prescribes the top-level structure of the documents. Respectively, an extension for adding further top-level elements (like the E2ENP PDU header element – Section 4.3.2) or for removing elements can be achieved applying schema redefinition [170]. However, the SDPng specification contains also other adjoined namespaces and schemas (e.g. for RTP – see Section 4.2.2 and [81]) and namespace-redefinition might lead to problems with the namespaces inheritance (see beginning Section 4.2). Consequently, E2ENP defines its own integration schema for the E2ENP and SDPng top-level structure (see Section 4.2.4) that shall be rewritten every time new top-level elements are introduced, thus, leading to future different E2ENP top-level structures and versions (see also Section 4.3.1 for an example how E2ENP treats different applied namespaces and versions thereof).

The typical practice in XML for defining new elements and for extending elements is through using predefined element types. These types are specified separately to the elements and are used within the schema definition of elements as their extension base. Elements that do not reuse types are of anonymous type and cannot be extended via the type, but for the purposes of schema extension, they can be substituted by other elements. For the application of types belonging to different then the current namespace, it is enough to include the respective namespace in the list of the applied namespaces in the XML header (see Figure 49 the "xmlns" definitions). In order to apply elements of one schema into another schema the XML "import" instruction is used. However, the order of "imports" should not cause the inheritance problem as discussed at the beginning of Section 4.2 to enable XML-document validation. SDPng [81] prescribes that the enhancements of SDPng shall be made through substituting the abstract elements under the top-level elements, i.e. substitution of <sdpng:d>, <sdpng:c> and <sdpng:cn> in version four[27] and of <sdpng:definition>, <sdpng:component>, <sdpng:constraint> and

[27]Note that the SDPng model in version four is not consistent respecting the definition of the abstract elements, as an abstract globally-defined element under <sdpng:conf> is missing.

<sdpng:information> in version eight applies. Consequently, enhancement schemas of SDPng should import the SDPng schema to accomplish the required substitutions. However, SDPng prescribes some additional profiles (e.g. RTP schema – see Section 4.2.2) that also import the SDPng basic schema and apply substitutions of elements. Hence, if enhancements to the SDPng schema wish to reuse all SDPng schemas, they run into a schema-inheritance problem. Thus, E2ENP reuses only the concept of the <sdpng:def> and <sdpng:cfg> elements as there are no type definitions at this level in SDPng (see Section 4.2.4). Furthermore and in order to reuse the RTP profile, E2ENP applies the <sdpng:Component> type (see Figure 54) to specify alternatives for the RTP stream configurations (see Sections 4.2.2 and 4.2.4).

4.2.2 SDPng Profiles

The SDPng specification [81] prescribes three additional XML schemas named profiles, i.e. the RTP profile and the profiles for audio and for video capabilities. However, the audio and the video profiles of SDPng are in very rudimentary development stage. Hence, E2ENP specifies its own XML definitions for audio and for video re-applying where appropriate only the concepts of the SDPng profiles. Furthermore, examples for a profile for transport definitions were presented in [87], but this draft does not provide XML schema definitions. Such definition for transport QoS (based on [87]) was developed for E2ENP. The only profile schema definition inherited in E2ENP is the RTP profile stemming from version four of SDPng. The development of E2ENP introduced some corrections to this profile as some inconsistent or incomplete definitions were detected. The corrected RTP schema adopted for E2ENP matches the XML examples presented in the SDPng specification at version four [81], but it does not correspond to the XML schema of the respective RTP profile to avoid the original ambiguous definitions for RTP transport. The RTP schema after the respective corrections introduced through E2ENP is shown Figure 55 below and examples of corresponding XML code are shown in Section 4.3.4. Furthermore, E2ENP adopts the RTP <PayloadType> complex type and enhances it for its own purposes of RTP streaming (see Sections 4.2.4 and 4.3.3).

For reasons of completeness, Figure 56 shows the RTP profile at version eight of SDPng. The major distinction between the profile definitions in the two different versions of SDPng is that in the later version of SDPng XML elements instead of XML attributes are used. However, as the major task of E2ENP is to provide possibly minimal size of the XML instance documents, the application of attributes in E2ENP is preferable (see the respective discussion and example at the beginning of Section 4.3).

4.2.3 Integration of XML Signature Syntax

E2ENP defines a structure for provider and mentor marking of peers' technical specifications of QoS (Figure 57 and Sections 4.2.4 and 4.3.6). This structure is specified through the complex type <signatureType> in the E2ENP schema. A part of this structure is the E2ENP signature (<e2enp:signature>) that provides a compact possibility for mentor signatures (see Sections 4.1.3.4, 4.2.4 and 4.3.6). In addition to the E2ENP-own signature, the substitution element <fullsignature> is defined that reuses the <SignatureType> as provided through the XML signature definition in the *ds* namespace [229]. E2ENP adopts the XML signature syntax, as it is, without introducing changes to it, as this syntax is already a stable and completed standard.

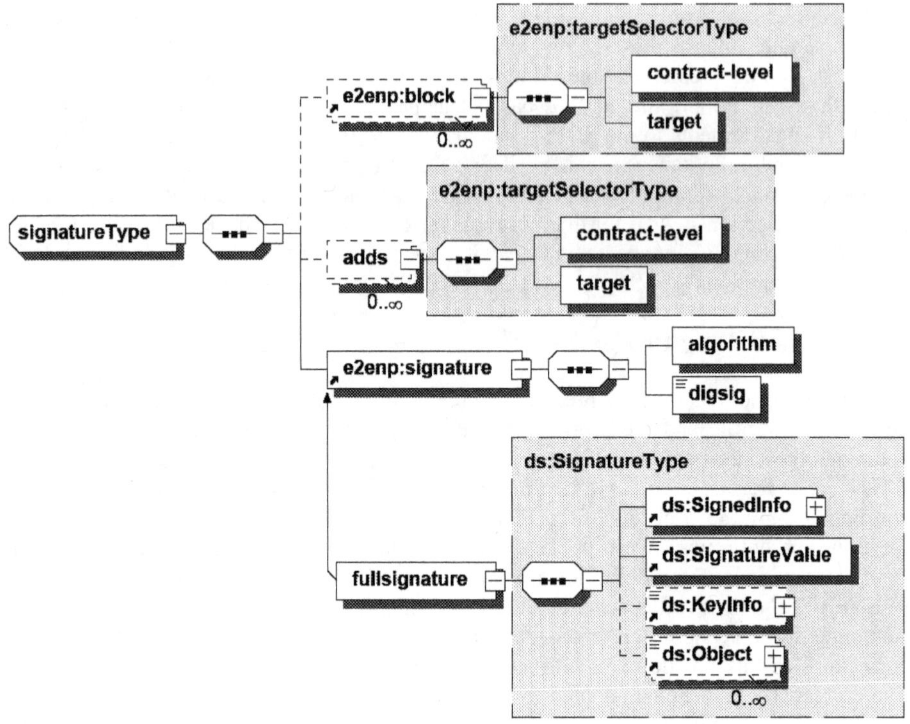

Figure 57: Integration of XML Signature Syntax [229]

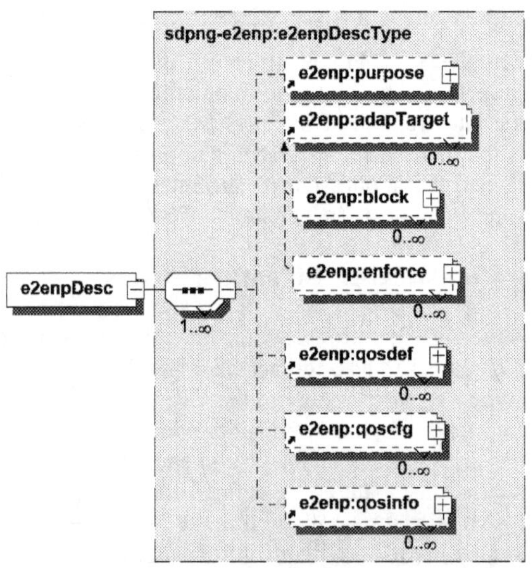

Figure 58: E2ENP and SDPng integration schema

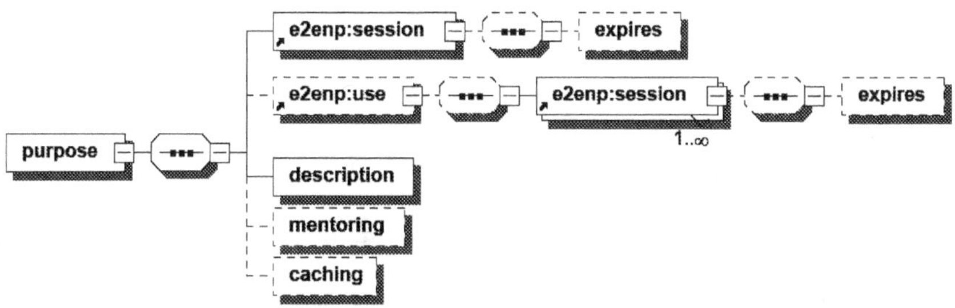

Figure 59: E2ENP <purpose> element

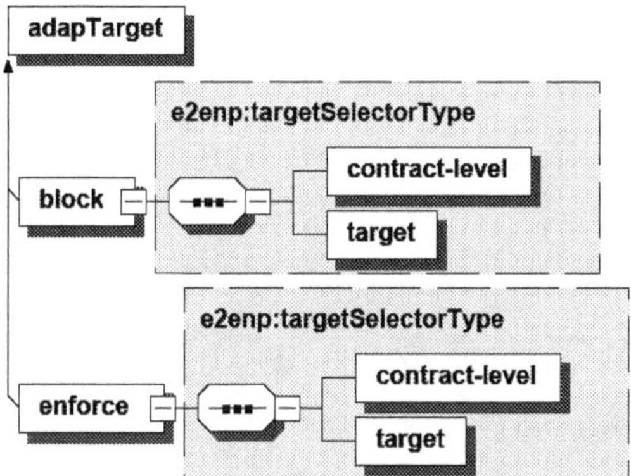

Figure 60: E2ENP <adapTarget> element and its substitutes <block>/<enforce>

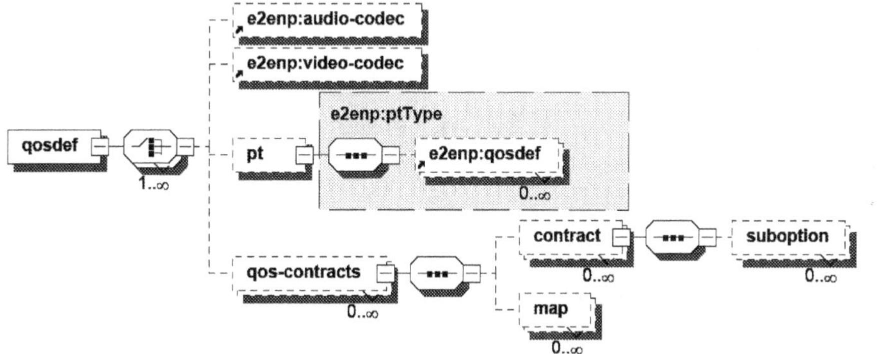

Figure 61: E2ENP <qosdef> element

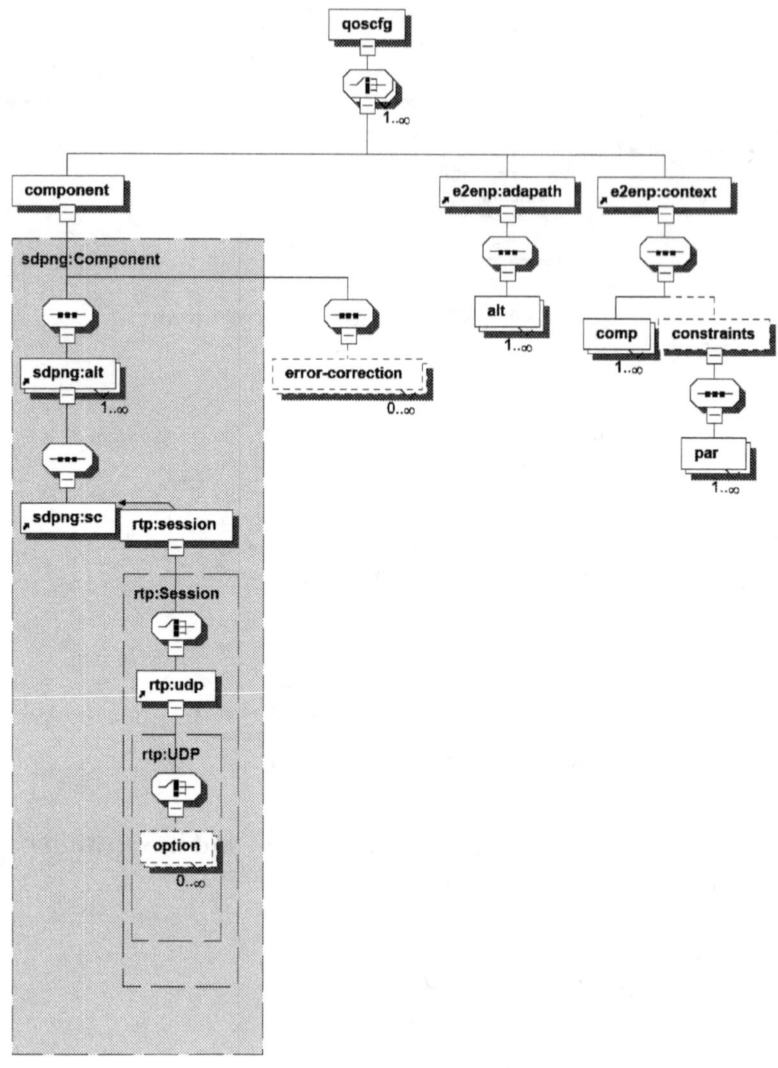

Figure 62: E2ENP <qoscfg> element

4.2.4 E2ENP Schema Definitions

The integration schema of E2ENP and SDPng is shown in Figure 58, in order to visualise the structure applicable for the E2ENP documents (i.e. the PDUs). The top-level elements of the E2ENP schema are described in the basic E2ENP schema and they are associated with each other in the integration schema of E2ENP and SDPng. This feature allows that the top-level elements of E2ENP can easily be referenced in other schemas, different from the E2ENP and SDPng integration. All the top-level elements in the E2ENP schema are defined as optional, in order to cover some specific E2ENP concepts like the negotiation of applicable XML-schema specifications (see Section 4.3.1) or the application of E2ENP without its fragmentation features, i.e. similar to the application of the current state of the art description protocols (see below and Sections 4.3.2 and 6.5 and Annex C). All but the

<purpose> element of the schema can be used multiple times to express negotiation and/or configuration alternatives.

The E2ENP <purpose> element represents the header of an E2ENP PDU (Figure 59). It is used only once per instance document and it carries the unique identification of a PDU. The <session> sub-element of <purpose> represents the *identifier of the PDU* and it carries information about the party that generated the PDU and about the E2ENP session to which the PDU belongs to. The <expires> sub-element of <session> specifies the permanence of the PDU for leasing purposes. If <expires> is missing the leasing of the PDU is permanent, unless some explicit caching directive is available (see below). The <use> sub-element of <purpose> integrates one or multiple <session> elements and applies in cases of associations between PDUs to enhance QoS specifications or to reference certain definitions. The <description> element provides information on the function of the PDU in order to instruct the application to perform specific PDU-processing or negotiation-specific actions, e.g. <description> carries information about the negotiation phase or mode of E2ENP to which the PDU belongs (see Section 3.1). The element <mentoring> indicates mentoring modes associated with the PDU (see Sections 2.4.1, 3.2 and 3.3) and the element <caching> is an explicit storing directive for the PDUs and it indicates if the repeating of the PDU leasing is possible or if the leasing is done just one time. All elements of the PDU header are explained in detail and with extensive examples in Section 4.3 and Annex C.

The <adapTarget> element (Figure 60) is an abstract element for realisation of the E2ENP referencing mechanism at short Re-negotiations (see Section 4.3.5). The substitution elements <block> and <enforce> are applied to define which QoS contracts/contexts shall be blocked through the short Re-negotiation and which other contracts/contexts shall be enforced. The complex type <targetSelectorType> is used for specifying logical links to the hierarchical QoS specification (see also Section 4.1.3.4). It points at the abstraction level of the contexts using the <target> element where the type of the context level is described (e.g. contract, context, session, etc., see Section 4.3.5) and it specifies the path to the respective contents of the hierarchical QoS specification with <contract-level> element. The <targetSelectorType> is used in several E2ENP elements, i.e. <block> and <enforce> for short Re-negotiations, but also for elements for provider and mentor marking of QoS information and for enhancing of hierarchical specifications (see Sections 4.2.3 and 4.3.6).

Figure 61 shows the <qosdef> element that is used for providing definitions of the peer capabilities and for specifying the QoS contracts. The respective definitions are distinguished upon the attribute “name” of the <qosdef> element (see Section 4.3.3). The elements <audio-codec> and <video-codec> are applied for the specification of the audio and video capabilities of the negotiating device. The element <pt> identifies only the RTP payload type numbers used as capability specification (see Section 4.1.3.1). E2ENP gives the peers the possibility to negotiate the configuration of dynamic payload types, thus, the <pt> element integrates the <qosdef> element and the application can specify dynamic configurations of payload types using the E2ENP <audio-codec> and <video-codec> definitions. The <qos-contracts> element specifies the applied QoS contracts at application and at transport level. Its sub-element <contract> contains the definition of the contract and <map> is used to associate the application-level contracts with their respective carrier definitions for network transportation (e.g. RTP). The sub-element <suboption> is used for defining sub-options for the network transport performance and applies only under the definition of transport QoS contracts. The initial E2ENP definition of resource-optimisation policies that belonged to <qosdef> [20] was removed, since this information is redundant to the current specification of QoS contracts and contexts and since this data is technology specific to each of the negotiating peers. Hence, the resource-optimisation policies are not relevant for configuration of heterogeneous environments (see Section 2.3.2.1.2). The exhaustive definition of <qosdef> is provided in Section 4.3.3.

Figure 62 presents the <qoscfg> element that is used for the actual configuration of the multimedia session in terms of QoS contexts. The element <component> utilizes the SDPng <Component> complex-type definition to provide alternative definitions for the media components used in a session (i.e. <sdpng:alt> elements specifies the corresponding alternatives). The RTP structure is applied to specify the actual hosts and ports for the RTP streaming (<rtp:udp> element) and to provide specific configurations for the streaming (<option> element). The <error-correction> element applies in cases that the peer wishes to negotiate specific error-correction mechanisms without going into technical details about the implementation of the error-correction. If available, this element is also associated with the <option> element that is used for the definition of the technical details for the error-correction. Note that the <option> element may freely specify any additional configurations needed for the RTP streaming and it is the means for extension of the RTP-definitions in terms of specific implementation techniques. The element <adapath> specifies possibilities for adaptations involving QoS contexts. It provides alternative context configurations (i.e. <alt>) in a similar manner as the definition of alternatives for the media components configuration (see above <sdpng:alt>). The <context> element enables the associations of contracts and contexts in still higher-level of contexts using the <comp> element (<comp> stands for components of a context). The <constraints> element defines constraints at association of contracts and contacts and at their application, for this purpose the element <par> is applied to identify the exact limitation parameters. Note that E2ENP does not define a top-level element for constraints (compared to SDPng – see Section 4.2.1) as the specific limitations and service rules for the contracts and contexts are applied through the hierarchical QoS specification. Details of the E2ENP-specific configurations and contexts are presented together with examples in Section 4.3.4.

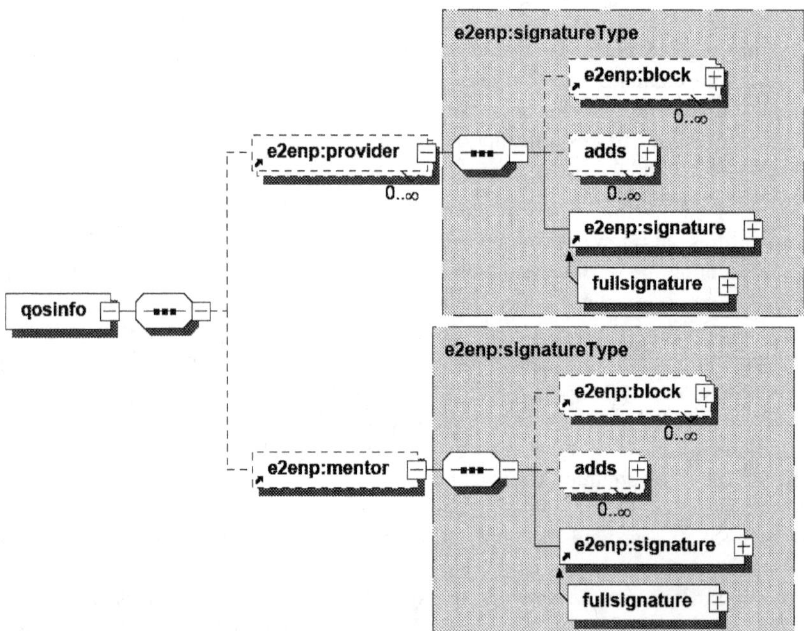

Figure 63: E2ENP <qosinfo> element

```
<xsd:simpleType name="nameContract">
    <xsd:restriction base="xsd:string">
        <xsd:pattern value="(audio|video|data|control|transport)-contract-[0-9]{1,5}"/>
    </xsd:restriction>
</xsd:simpleType>
```

Figure 64: Example of simple type for pattern definition

```
<xsd:simpleType name="contract-type">
    <xsd:restriction base="xsd:NMTOKEN">
        <xsd:enumeration value="audio"/>
        <xsd:enumeration value="video"/>
        <xsd:enumeration value="data"/>
        <xsd:enumeration value="control"/>
        <xsd:enumeration value="transport"/>
    </xsd:restriction>
</xsd:simpleType>
```

Figure 65: Example of simple type for enumeration

```
<key type="type" value="content"/>
```

```
<key type="type">
        <value>content</value>
</key>
```

Figure 66: XML attributes vs. XML elements

```
<e2enp:video-codec frame-rate-set="30" frame-resolution-type="ITU_CIF" frame-size-set="QCIF,CIF" color-type="color" color-quality-set="8"/>
```

```
<m21-dia:Terminal>
    <m21-dia:TerminalCapability xsi:type="m21-dia:DisplaysType">
        <m21-dia:Display>
            <m21-dia:DisplayCapability xsi:type="m21-dia:DisplayCapabilityType" colorCapable="true"
            refreshRate="30">
                <m21-dia:Mode>
                  <m21-dia:Resolution horizontal="176" vertical="120"/>
                </m21-dia:Mode>
                <m21-dia:ColorBitDepth blue="8" green="8" red="8"/>
            </m21-dia:DisplayCapability>
        </m21-dia:Display>
    </m21-dia:TerminalCapability>
    <m21-dia:TerminalCapability xsi:type="m21-dia:DisplaysType">
        <m21-dia:Display>
            <m21-dia:DisplayCapability xsi:type="m21-dia:DisplayCapabilityType" colorCapable="true"
            refreshRate="30">
                <m21-dia:Mode>
                  <m21-dia:Resolution horizontal="352" vertical="240"/>
                </m21-dia:Mode>
                <m21-dia:ColorBitDepth blue="8" green="8" red="8"/>
            </m21-dia:DisplayCapability>
        </m21-dia:Display>
    </m21-dia:TerminalCapability>
</m21-dia:Terminal>
```

Figure 67: E2ENP vs. MPEG-21 DIA description

The <qosinfo> element (Figure 63) can be applied for specifying of any additional meta-information associated with the QoS contracts and contexts. Currently, it is used only for defining of provider and mentor information associated with the QoS information. The providers and the mentors can block peers configurations (element <block>) or add QoS configurations (element <adds>) in accordance with the specific mentoring scenario (see Sections 3.2 and 3.3). Furthermore, mentors provide a signature that they are authorized to perform the respective mentoring actions (see Sections 4.1.3.4 and 4.2.3). The element <qosinfo> can be further enhanced for additional meta-information, e.g. for specific adaptation controls like the MPEG-21 DIA bitstream adaptations as presented in [95] – [97] within the scope of SDPng++. The exact details of the E2ENP mentoring when using the <qosinfo> element are presented in Section 4.3.6.

In addition to all of the elements and complex types, E2ENP defines several simple types for specifying restrictions to the applied parameters and definitions through patterns (using regular expressions, also named *regex* patterns – see [266]) and enumerations. Examples of these components of the E2ENP schema are shown in Figure 64 (pattern definition) and Figure 65 (enumeration). The exact application of the simple types in E2ENP is discussed together with the examples of their utilization in Section 4.3.

4.3 The E2ENP XML Structures and Examples

The development of the E2ENP description language follows two major tasks, i.e. to enable fragmentations of hierarchical QoS specifications of the QoS-aware applications and to apply possibly minimised negotiation-definitions with XML. The accomplishment of these two tasks results in an optimisation for the application signalling as the data units to be transported are reduced in size (see also Chapter 6).

In order to provide fragmentation for any hierarchical QoS specification, E2ENP follows the Application Level Framing (ALF) principle [24] that states:

> *The application should break the data into suitable aggregates and the lower levels should preserve these frame boundaries as they process the data,*

(See also [128] and [129], where similar recommendations about proper application-data preparation and presentation in distributed services are discussed). Additionally, the data carriers may apply restrictions to the sizes of the aggregates [62][214]. Hence, the E2ENP definition language together with the E2ENP negotiation mechanism (see Section 3.1) supplies the application on one side with appropriate means for generating suitable fragments of the hierarchical QoS specification, on the other side the transport requirements for restricting the sizes of the fragments are also satisfied. Consequently, the E2ENP fragmentation can be applied alternatively and in combination with compression techniques within the signalling for multimedia management.

The E2ENP definition language allows the generation of fragments as well-formed and valid XML documents that can be directly handled by the XML processing units in the E2ENP UAs (see Section 5.2.2.1) without the necessity that lower levels of the service are responsible for putting together the application aggregates. For instance, if the transport layer shall fragmentize and put together XML document this would cause extra burden for the transport layer and possible performance faults at the application, e.g. at malformations of the XML structure and/or validity. Consequently, when applying the ALF principle [24] for implementing the E2ENP PDUs such problems can be avoided, as the PDUs are stand-alone and physically independent documents directly manageable in the application. E2ENP gives the possibility to the application to use E2ENP PDUs for different purposes, like fragmentation and construction of QoS specifications, leasing management of such specifications and addressing of known, leased specifications for adaptation purposes. These E2ENP functionalities represent the innovations of E2ENP compared to the state-of-the-art protocols (see Section 1.2.2) that are not able to process and reference fragments of negotiable data-structures. The most important part of the PDUs that serves the realisation of these novel E2ENP features is the E2ENP PDU header presented in Section 4.3.2.

```
<e2enpDesc xmlns="http://www.iana.org/sdpng-e2enp" xmlns:e2enp="http://www.iana.org/sdpng/e2enp"
xmlns:sdpng="http://www.iana.org/sdpng" xmlns:rtp="http://www.iana.org/sdpng/rtp"
xmlns:xsi="http://www.w3.org/2001/XMLSchema-instance"
xsi:schemaLocation="http://www.iana.org/sdpng/e2enp e2enp-02.xsd
http://www.iana.org/sdpng/rtp rtp-pkg-sdpng-04.xsd
http://www.iana.org/sdpng-e2enp sdpng-e2enp-01.xsd">
    <e2enp:purpose>
        <e2enp:session user="Mary" session-id="2890843112" version="2890841393" nettype="IN"
        addrtype="IP4" addr="43.196.180.1"/>
        <e2enp:description type="request" name="pre-negotiation" style="full" mode="push"/>
        <e2enp:caching cache-results="true"/>
    </e2enp:purpose>
</e2enpDesc>
```

Figure 68: Negotiation of the applicable XML namespaces

```
<e2enp:purpose>
    <e2enp:session user="Mary" session-id="2890844526" version="2890842807" nettype="IN"
    addrtype="IP4" addr="43.196.180.1">
        <e2enp:expires time="3600"/>
    </e2enp:session>
    <e2enp:use>
        <e2enp:session user="Mary" session-id="2890843112" version="2890841393" nettype="IN"
        addrtype="IP4" addr="43.196.180.1"/>
    </e2enp:use>
    <e2enp:description type="request" name="negotiation" style="short" mode="push"/>
</e2enp:purpose>
```

Figure 69: Example 1 – Purpose header

```
<e2enp:purpose>
    <e2enp:session user="Mary" session-id="2890843112" version="2890841393" nettype="IN"
    addrtype="IP4" addr="43.196.180.1">
        <e2enp:expires time="36000"/>
    </e2enp:session>
    <e2enp:use>
        <e2enp:session user="Billy" session-id="3452134576" version="23417645908" nettype="IN"
        addrtype="IP6" addr="243f:6a88:85a3:08d3:1319:8a2e:0370:7344"/>
        <e2enp:session user="Billy" session-id="12345678543" version="9875456789" nettype="IN"
        addrtype="IP6" addr="243f:6a88:85a3:08d3:1319:8a2e:0370:7344"/>
    </e2enp:use>
    <e2enp:description type="request" name="taken-over" style="full" mode="push" migration="true"/>
    <e2enp:mentoring mode="external-negotiation" migration="session-split"/>
</e2enp:purpose>
```

Figure 70: Example 2 – Purpose header

Since E2ENP's task is to provide minimisation of the definition sizes for QoS-relevant contents, E2ENP applies XML-attributes for the necessary key-value-pair descriptions to reduce the sizes of the instance documents. An example illustrating the fact that the application of XML-attributes shrinks the parameter definitions compared to the utilization of XML-elements is shown in Figure 66. Furthermore, the reduction of the sizes of the E2ENP documents is associated with E2ENP's negotiation capabilities to provide optimised definitions of spaces for the expected performance of the service through specification of sets and ranges values. In order to demonstrate this feature, a comparison example of E2ENP and MPEG-21 DIA definitions is presented in Figure 67, where the equivalent definitions are shown in italics (i.e. E2ENP "frame-rate-set" and MPEG "refreshRate" are analogous, E2ENP "frame-size-set" and MPEG <Resolution> also and the definition of the colour capabilities in both of the description formats expresses similar configuration). Although the definitions of MPEG-21 DIA [72] are a very powerful tool for specifying of user environments and their performance upper bounds, the DIA descriptions are not optimised for protocol and negotiation purposes as they use multiple wrapper elements and need to repeat definitions for specifying spaces of the service performance in form of contracts/contexts (i.e. MPEG needs two <TerminalCapability> blocks, where E2ENP applies just the attribute "frame-size-set" for the same purpose – Figure 67). The E2ENP language applies strongly optimised format for the purpose of protocol communication, as evident from Figure 67, where the E2ENP description comprises just one XML line compared with the multiple elements of the MPEG-21 DIA description.

The following sub-sections present examples of the E2ENP description language and discuss the meaning and the application of the specified parameters and XML definitions. Examples of complete E2ENP PDUs are also shown in Annex C.

Note that the E2ENP XML implementation presented in this work concentrates on the special innovative features of E2ENP as a protocol for providing fragmentations of hierarchical QoS specification and for optimising the signalling and the negotiation of multimedia sessions (see also Chapters 5 and 6). The language specification of E2ENP is open for future developments and re-definitions. At its current development stage, E2ENP does not claim to achieve the completeness of industrial standards (e.g. MPEG-21 DIA), which have taken decades of specification and which concentrate on the specification of media in general, compared to E2ENP that treats only multimedia streaming. The streaming of media is considered to be most appropriate for the demonstration of the innovative E2ENP features at management of hierarchical QoS specifications in mobile and dynamic environments (see also Sections 1.2.2.2 and Chapter 2 for the respective discussions on the shortcomings in the state-of-the-art mechanisms).

4.3.1 The XML Header Negotiation

The peers implementing E2ENP may negotiate the applicability of XML namespaces utilizing similar PDUs to the one shown in Figure 68. The negotiation of the used XML namespaces is done with the XML header definition (i.e. <e2enpDesc> and the definitions of "xmlns" attributes, and the names of the schema documents where appropriate – Figure 68) before an actual negotiation of capabilities and QoS takes place. Such XML header negotiation has the advantage that the peers may prepare themselves what XML descriptions should be expected during a capabilities and QoS negotiation so that the peers can provide customised validation of the PDU documents.

The customised validation is a possibility to optimise the XML processing. The argumentation for this E2ENP negotiation is that not all the applicable XML namespaces are necessary for all the application scenarios (e.g. in ad-hoc scenarios there are no providers, hence, the provider signatures with XML-Signature Syntax are not applicable in such scenario). Furthermore, only a part of the namespaces is obligatory to provide functioning technical specifications. Considering the possibility for customised validation, E2ENP implements SAX parsing [267] that traces sequentially the XML document and uses a customised application-internal model to validate and to store the content model of the PDUs. E2ENP applies for the purpose of customised validation configurable XML processing architecture, omitting the complete document validations as typical for XML and skipping for further processing those XML components that are not understood by the applications in consideration with corresponding configuration presetting (see Section 5.2.2.1).

4.3.2 The PDU Header

The E2ENP PDU header is the <e2enp:purpose> section and it has several negotiation and PDU-management functions (see Figure 68, Figure 69 and Figure 70, and the summary at the end of this section). It is used as a unique identifier of one fragment of the hierarchical QoS specification and/or as an identifier of a specific management action (e.g. leasing of PDUs – see Sections 4.3.5 and 4.3.7).

The <e2enp:session> element applies for the purposes of *PDU identification*, where the information about the user ("*user*") and the device that issued the PDU ("*nettype*", "*addrtype*" and "*addr*") are given as attributes, together with some additional univocal identifiers (attribute "*session-id*" as a *unique name of the PDU*, associating the PDU with the E2ENP session to which the PDU belongs to and attribute "*version*", identifying different versions of the same PDU). The names of the attributes of <e2enp:session> are based on the similar names for identifying a negotiation bid in the SDP [86], where the correspondent to <e2enp:session> description is the origin line ("o=").

The <e2enp:session> element can have a child element <e2enp:expires>, where the attribute "time" indicates the expiry time of the PDU in seconds. The absolute time that indicates the starting time for counting of the expiry time is provided through the carrier protocol (e.g. SIP Date header [62]). This association of the E2ENP relative time and the absolute time of the carrier protocol is possible, as the carrier protocols apply usually time information to provide ordering of the messages for transport purposes (see also [50][52]). Furthermore, exact application of time for E2ENP is not that crucial, due to the possibility to hold back expired data in a Zombie-state in case this data is actively used by some application (see Section 3.1.4).

For purpose of *association of PDUs*, one or multiple <e2enp:session> elements are applied as children of the <e2enp:use> element. In cases when the fragments of the hierarchical QoS specification are exchanged sequentially, the application of E2ENP may point only at the last known fragment when referencing the QoS specification with the <e2enp:use> element, as the sequential ordering of the PDUs' cross-references enables the applications to resolve the chain of associated PDUs stating with the last pointer to the QoS specification (see Figure 69). In case that multiple QoS specifications are referenced and managed simultaneously with a single PDU (e.g. at joining of QoS information belonging to different specifications or for invalidation purposes – see Figure 70 and Section 4.3.7) those PDUs are listed within the <e2enp:use> construct. The application understands how the PDUs shall be managed upon the respective contents of the <e2enp:session> elements and upon the <e2enp:description> and <e2enp:mentoring> elements.

The <e2enp:description> element provides information about the actions associated with the respective PDU. It has the following attributes:

- "type" indicates the type of the message , i.e. *request* or *response* (These values are defined using XML enumeration).
- "name" indicates the negotiation action associated with the PDU. Possible values defined via enumeration are:
  - *standard* – for SDP/SDPng-like negotiations without fragmentations of the technical specifications of the applications, i.e. the capabilities and the QoS information is send as a single piece of information.
  - *pre-negotiation*, *negotiation* and *re-negotiation* – for the E2ENP-specific negotiation phases that apply associations of and references to QoS information (see Section 3.1.2)
  - *start-reservation*, *ready-reservation*, *cancel-reservation* and *cancelled-reservation* – to manage the network reservation procedures (see Section 3.1.3)
  - *expire* – invalidation PDU (see Sections 3.1.4 and 4.3.7)
  - *taken-over* – a PDU or a group of PDUs is taken over by a mentoring party (Figure 70 – the taken-over PDUs are listed in the <e2enp:use> section, see also Section 3.3)
- "style" indicates if the peers perform a *full* or a *short* Negotiation/Re-negotiation (see Section 3.1.2). Upon this field, the application knows how to treat the referenced PDUs and how to resolve the references to QoS information contained in the PDUs (see also Sections 4.3.3, 4.3.4 and 4.3.5). The children of <e2enp:use> element indicate the referenced PDUs at short E2ENP phases
- "mode" defines the applied negotiation modes (see Section 3.1.1). Its possible values are specified through XML enumeration as *push*, *pull* and *push-pull*. The application uses the value of "mode" to launch different sequences of SIP transactions to execute the required E2ENP mode (see Section 3.1.3)
- "coordination" indicates if network-reservation coordination shall be applied. The "coordination" attribute is a Boolean value
- "migration" is a Boolean value that specifies if the PDU is used for session migration purposes (see Section 3.3.1.2). The application understands how to proceed upon the referenced PDUs (i.e. the children of <e2enp:use> element indicate the references) and the rest of the <e2enp:description> and/or <e2enp:mentoring>attributes
- "coupling" indicates that a coupling scenario for independent multimedia sessions shall be performed (see Section 3.3.1.2) and is defined as a Boolean value. The application understands how to proceed upon the referenced PDUs (i.e. the children of <e2enp:use> element indicate the references) and the rest of the <e2enp:description> and/or <e2enp:mentoring> attributes

The <e2enp:mentoring> element indicates the communication with mentoring entities and has two attributes "mode" and "migration". The attribute "mode" is applied to designate how the mentor is involved in the communication process of the peers. If the mentor only helps for the negotiation without being involved in the media session of the peers, the pre-defined enumeration value *third-party-assisted* applies, e.g. negotiations for alternative services (see Section 3.3.1.1). If the mentor is involved in both the negotiation and the media session, the enumeration value *external-negotiation* applies, e.g. in session migration and transcoder scenarios (see Sections 3.3.1.2 and 3.3.2). The attribute "migration" of <e2enp:mentoring> indicates the mode for performing of a session migration, i.e. *session-split* or *session-migration* are possible values defined per XML enumeration (see Section 3.3.1.2). Note that the mentoring performed through provider proxies for managing the network resources and reservations is not indicated explicitly in the E2ENP header, as this information is a part of known and pre-agreed profiles exchanged between the provider and the terminals at registration (see Section 3.2).

The <e2enp:caching> with its attribute "cache-results" (defined as a Boolean) provide information on the leasing modes for the PDU that carries the caching directive, e.g. if no <e2enp:expires> element is available in association with the PDU and <e2enp:caching cache-results="true"/>, this is an indication of permanent leasing of the PDU or when a PDU as associated <e2enp:expires time="36000"/> element and <e2enp:caching cache-results="false"/> element, this indicates that no time reset for the PDU is allowed.

Considering the definitions of the elements and their attributes within the <e2enp:purpose> section, the functionality of this element in E2ENP is to provide PDU associations for fragmentising and for building of hierarchical QoS specifications and for referencing of elements and structures within such specifications. Furthermore, <e2enp:purpose> indicates multimedia-session and PDU-management functions and communicates information about the leasing of a PDU.

```
<e2enp:qosdef name="capabilities">
   <e2enp:audio-codec name="PCMU" scope="applicable"/>
   <e2enp:audio-codec name="G729" scope="applicable"/>
   <e2enp:audio-codec name="MP3"
   href="urn:mpeg:mpeg7:cs:AudioCodingFormatCS:2001:3.3"
   scope="possible"/>
   <e2enp:pt name="rtp-avp-0" pt="0" format="PCMU"/>
   <e2enp:pt name="rtp-avp-18" pt="18" format="G729"/>
   <e2enp:pt name="rtp-avp-97" pt="97" format="MP3">
      <e2enp:qosdef name="capabilities">
         <e2enp:audio-codec sampling-rate-set="8000" channel-set="2"/>
      </e2enp:qosdef>
   </e2enp:pt>
   <e2enp:video-codec name="H263" scope="applicable"/>
   <e2enp:video-codec name="H261" scope="applicable"/>
   <e2enp:video-codec name="WAVI"
   href="urn:mpeg:e2enp:cs:VisualCodingFormatCS:2001:11"
   scope="possible"/>
   <e2enp:pt name="rtp-avp-34" pt="34" format="H263"/>
   <e2enp:pt name="rtp-avp-31" pt="31" format="H261"/>
   <e2enp:pt name="rtp-avp-100" pt="100" format="WAVI">
      <e2enp:qosdef name="capabilities">
         <e2enp:video-codec frame-rate-range="5,30"
         frame-size-set="QCIF,CIF" color-quality-set="8,16"
         overall-quality-range="9100,10000"/>
      </e2enp:qosdef>
   </e2enp:pt>
</e2enp:qosdef>
```

Figure 71: Capabilities definition

```
<e2enp:qosdef name="contracts">
    <e2enp:qos-contracts for="audio">
        <e2enp:contract type="audio" name="audio-contract-1" sampling-rate-set="4000,8000"
        channel-set="1"/>
        <e2enp:contract type="audio" name="audio-contract-2" sampling-rate-set="8000,12000"
        channel-set="1,2"/>
        <e2enp:contract type="audio" name="audio-contract-3" sampling-rate-set="12000,16000"
        channel-set="1"/>
        <e2enp:map contract="audio-contract-1" format="rtp-avp-0" peer-role="receiver"/>
        <e2enp:map contract="audio-contract-2" format="rtp-avp-18" peer-role="receiver"/>
        <e2enp:map contract="audio-contract-3" format="rtp-avp-97" peer-role="receiver"/>
    </e2enp:qos-contracts>
    <e2enp:qos-contracts for="video">
        <e2enp:contract type="video" name="video-contract-1" frame-rate-set="10,15"
        frame-resolution-type="ITU_CIF" frame-size-set="CIF" color-quality-set="8,16,32"
        overall-quality-range="9500,9800"/>
        <e2enp:contract type="video" name="video-contract-2" frame-rate-set="15,20"
        frame-resolution-type="ITU_CIF" frame-size-set="QCIF" color-quality-set="8,16"
        overall-quality-range="9800,9900"/>
        <e2enp:contract type="video" name="video-contract-3" frame-rate-set="20,25"
        frame-resolution-type="ITU_CIF" frame-size-set="QCIF,CIF" color-quality-set="8"
        overall-quality-range="9900,9960"/>
        <e2enp:map contract="video-contract-1" format="rtp-avp-34" peer-role="receiver"/>
        <e2enp:map contract="video-contract-2" format="rtp-avp-31" peer-role="receiver"/>
        <e2enp:map contract="video-contract-3" format="rtp-avp-100" peer-role="receiver"/>
    </e2enp:qos-contracts>
    <e2enp:qos-contracts for="transport">
        <e2enp:contract type="transport" name="transport-contract-1" pbr="320" sbr="320"
        mps="72" mbs="72" qtp="guaranteed-service">
            <e2enp:suboption name="sensitivity-1" type="SI-parm" me2ed="80" me2edv="10"
            mplr="2E-2"/>
            <e2enp:suboption name="sensitivity-2" type="SI-parm" me2ed="120" me2edv="20"
            mplr="2E-2"/>
        </e2enp:contract>
        <e2enp:contract type="transport" name="transport-contract-2" pbr="96" sbr="96"
        mps="72" mbs="72" qtp="guaranteed-service">
            <e2enp:suboption name="sensitivity-1" type="SI-parm" me2ed="80" me2edv="10"
            mplr="2E-2"/>
            <e2enp:suboption name="sensitivity-2" type="SI-parm" me2ed="120" me2edv="20"
            mplr="2E-2"/>
        </e2enp:contract>
        <e2enp:contract type="transport" name="transport-contract-3" pbr="29" sbr="29"
        mps="72" mbs="72" qtp="guaranteed-service">
            <e2enp:suboption name="sensitivity-1" type="SI-parm" me2ed="80" me2edv="10"
            mplr="2E-2"/>
            <e2enp:suboption name="sensitivity-2" type="SI-parm" me2ed="120" me2edv="20"
            mplr="2E-2"/>
        </e2enp:contract>
    </e2enp:qos-contracts>
</e2enp:qosdef>
```

Figure 72: QoS contracts definition with E2ENP

4.3.3 Capabilities and QoS Contracts

The application of the <e2enp:qosdef> element is dedicated for specifying of terminal capabilities and of QoS contracts (see Figure 71 and Figure 72, and Sections 2.2.1 and 2.2.2). The "name" attribute of <e2enp:qosdef> indicates if its child elements are the capabilities or the QoS contracts (the respective values are fixed with XML enumeration).

Figure 71 shows a definition of system capabilities in form of codecs according to the RTP/AVP definition [2]. The two major sections <e2enp:audio-codec> and <e2enp:video-codec> specify the supportable codecs with their names and applicability scope. The "href" attribute is an indication of a univocal specification of the codec (see in Section 4.1.3.3 MPEG-7 MDS and Annex D). The "applicability" attribute denotes if the negotiated capabilities are immediately applicable after the negotiation (value *applicable*), are assumed to be applicable (value *possible*), but currently the application does not wish to use them (e.g. due to resources shortage) or the application does not have such capabilities (value *not-applicable*). The partner application can remember the capabilities indicated as *possible* and use them in future negotiation rounds. All of the "applicability" values are specified with XML enumeration.

The <e2enp:pt> (Figure 71) specifies RTP-transport for the codecs, referencing the already defined with <e2enp:audio-codec> and <e2enp:video-codec> descriptions. For this purpose, the name of the codec is used (i.e. "name" attribute in the audio and video codec definitions) and respectively referenced, using the "format" attribute in the <e2enp:pt> element. The aim of the separate mapping of the codec definitions to the payload-types is to optimise the PDU processing. The application can first prove if a given codec is supported and if the codec is not supported the respective transportation format (in this case RTP payload types) can simply be ignored. Additionally, for some codecs it is possible to associate dynamic payload types, which in fact is a complementary information to the codec, therefore it is presented separately from the codec itself.

Dynamic RTP payload types are characterised through a definition of parameterisation sub-element of <e2enp:pt>. The sub-element is an additional <e2enp:qosdef> element containing correspondingly an <e2enp:audio-codec> or an <e2enp:video-codec> elements (Figure 71). For the purpose of audio and video parameterisation, the respective codec definitions contain several attributes that are the same as the definitions for audio and video QoS contracts (see below and Figure 72). However, the parameterisations of the dynamic RTP payloads represent actual technical specification of a codec, while the definitions of E2ENP contracts represent possible configurations required from the users of the application.

The numbers of the applicable payloads are fixed for the "pt" attribute of <e2enp:pt> using the following regex pattern *([0-9]{1,2})|[1][0-1][0-9]|([1][2][0-7])* that corresponds to the RTP/AVP definition (see also Section 4.1.3.1). Furthermore, the "name" attribute is fixed with a similar pattern (i.e. *rtp-avp-(([0-9]{1,2})|[1][0-1][0-9]|([1][2][0-7]))*) to uniquely identify the specified capabilities for future referencing at complete specification of the RTP-stream configurations, i.e. with hosts and port numbers (see Section 4.3.4). The two regex specifications are parts of the RTP <PayloadType> definition (see Section 4.2.2) that is inherited from and enhanced within the <e2enp:ptType> (Section 4.2.4).

Figure 72 shows the structure of the QoS contracts at transport level and at application level (i.e. audio and video QoS contracts). The values of the "for" attribute of <e2enp:qos-contracts> are fixed with an enumeration to *audio*, *video*, *data*, *control* and *transport* and indicate the wrapper for the respective type of contracts. The same enumeration is also used for the "type" attribute of the underlying <e2enp:contract> element to associate the respective contracts with their types. Currently, E2ENP specifies definitions for only audio, video and transport contracts. The identifiers *data* and *control* are considered for possible extensions for non-streaming and for interactive multimedia applications (see also [20][132][173]).

Within the audio and the video QoS contracts, the <e2enp:contract> describes the user's QoS preferences as application QoS parameters. However, as the user's preferences and the actual codecs parameterisation (see above) are expressible with similar definitions, E2ENP defines reusable groups of attributes for the audio and the video definitions. Inside the audio contracts the attributes are (see also [2]):

- "sampling-rate-set" and "sampling-rate-range" (see also the definition of E2ENP vectors in Section 4.1.3.3) with sampling-rates specified in Hz and
- "channel-set" that identifies the number of channels (e.g. Mono=1, Stereo=2, Dolby=5.1, etc.)[28].

Within the video contracts the meaning of the parameters is (see also [2][63][72][159]):

- "sampling-rate-set" and "sampling-rate-range" with sampling-rates specified in Hz
- "frame-rate-set" and "frame-rate-range" define number of video frames in Fr/s
- "frame-resolution-type" identifies the standard types for frame sizes as *ITU_CIF*, *VESA_VGA*, *ATSC_HDVT* or *custom* (there values are fixed with XML enumeration, see also Section 4.1.3.2). The "frame-resolution-type" attribute is associated with the application of the "frame-size-set" attribute (see below)
- "frame-size-set" are the displayable frame sizes, defined in Height[Pixel]xWidth[Pixel] (e.g. *176x120*) or as standardised video sizes (see Section 4.1.3.2 and above)
- "color-type" is used for the specification of the colour scheme as *greyscale* or *color* (the values are fixed with XML enumeration) and it is associated with the usage of the "color-quality-set" attribute (below)
- "color-quality-set" identifies the quality of the colour in bits, e.g. 8bit colour or 16bit colour. If it is applied with the *greyscale* "color-type" the bit value represents the possible variations of grey, otherwise the value applies for the three colours of the RGB (red-green-blue) scheme
- "overall-quality-set" and "overall-quality-range" are percentage values, where 1% is equal to 100 units of the corresponding descriptions. This definition is used to express the overall visual perception of the user, e.g. in terms of punctual defects or other malformations that represent specific combinations and relations of the above-mentioned technical parameters for video (see also [135])

The mapping between the QoS contracts and the technical specifications of codecs to form a technically-accomplishable QoS contract is done using <rtp:map>. The QoS contract is identified through the attribute "name" within the <e2enp:contract> element and these "name"-s are fixed correspondingly with a pattern to identify that the element is a contract and to associate the element with its media type (e.g. audio, video or transport).

---

[28]Note that the definition of audio channels is only as a set as the channels represent discrete values with irregular intervals between the values. This consideration holds true also for other discontinuous parameter definitions.

The pattern applied for the definition of "name" in <e2enp:contract> is the regex *(audio|video|data|control|transport)-contract-[0-9]{1,5}*. The association of the <e2enp:contract> with <e2enp:map> is done using the "contract" attribute of <rtp:map> to point at the "name" of the respective <e2enp:contract>.

The definition of the transport QoS is adopted as a concept from [87]. The technical specification in [87] is derived from ITU standards like [255][256], where the transport performance parameters are initially defined (e.g. bandwidth, delay, jitter, etc. and the transport-service classes). The most important parameters for understanding of the association of the contracts at application and at transport level done in E2ENP are the "name" attributes of the <e2enp:contract> (see the regex definition in the previous paragraph) and <e2enp:suboption> (regex pattern: *sensitivity-[0-9]{1,5}*) used in the transport definitions. These names are used later for referencing the respective elements by their "name" attribute. The meaning of the single attributes used in the transport-contract and traffic-sensitivity definition is:

- "pbr" – peak bit rate expressed in kilobit per second
- "sbr" – sustainable bit rate also in kilobit per second
- "mps" – maximum packet size in byte
- "mbs" – maximum burst size in kilobit
- "qtp" – network quality type with values *guaranteed-service*, *controlled-load* and *best-effort* that are fixed via enumeration

```
<e2enp:qoscfg>
    <e2enp:component name="audio-stream-1" media="audio">
        <sdpng:alt name="AVP-audio-0">
            <rtp:session format="rtp-avp-0">
                <rtp:udp role="receive" addr="43.196.180.1" rtp-port="7800"
                rtcp-port="7801"/>
            </rtp:session>
        </sdpng:alt>
        <sdpng:alt name="AVP-audio-9">
            <rtp:session format="rtp-avp-9">
                <rtp:udp role="receive" addr="43.196.180.1" rtp-port="7840"
                rtcp-port="7851"/>
            </rtp:session>
        </sdpng:alt>
        <e2enp:error-correction name="FEC" method="fec_redundancy"/>
        <e2enp:error-correction method="arq"/>
    </e2enp:component>
    <e2enp:component name="video-stream-1" media="video" reservation="receiver">
        <sdpng:alt name="AVP-video-34">
            <rtp:session format="rtp-avp-34">
                <rtp:udp role="receive" addr="43.196.180.1" rtp-port="7900"
                rtcp-port="7901" rtp-port-offset="2"/>
            </rtp:session>
        </sdpng:alt>
        <sdpng:alt name="AVP-video-31">
            <rtp:session format="rtp-avp-31">
                <rtp:udp role="receive" addr="43.196.180.1" rtp-port="7920"
                rtcp-port="7921"/>
            </rtp:session>
        </sdpng:alt>
    </e2enp:component>
</e2enp:qoscfg>
```

Figure 73: RTP streaming configuration

- "me2ed" – maximum end-to-end delay expressed in milliseconds, i.e. jitter
- "me2edv" – maximum end-to-end delay variation in milliseconds
- "mplr" – maximum packet-loss ratio in percent, e.g. 2E-2 = 0.02 % see Figure 72

Furthermore, the parameter "type" of <e2enp:suboption> expresses the kind of the applied sensitivity parameter. The sensitivity is freely definable using the value *SI-parm*, or is bound to the application of transport classes, values *SI-class* or *SI-flav* (the three values are fixed per enumeration, see also [87]).

The mapping between the application and the transport contracts and definitions is achieved at session establishment time, since only then it is known how many streams would be opened for the respective media session and which network resources are necessary for these streams (see Section 4.3.4). The provider restrictions on the QoS contracts are expressed with additional elements and are discussed in Section 4.3.6. The capabilities (Figure 71) and the QoS contracts (Figure 72) are usually exchanged between end-systems during a Pre-negotiation (see Section 3.1.2).

```
<e2enp:qoscfg level="stream">
   <!-- adaptation path for single streams -->
   <e2enp:adapath name="audio1" ref_component="audio-stream-1">
      <e2enp:alt default="true" name="nominal" ref_contract="audio-contract-1"
      ref_transport-contract="transport-contract-3" ref_suboption="sensitivity-1"
      scope="applicable"/>
      <e2enp:alt name="choice1" ref_contract="audio-contract-3"
      ref_transport-contract="transport-contract-3" ref_suboption="sensitivity-2"
      scope="applicable"/>
   </e2enp:adapath>
   <e2enp:adapath name="video1" ref_component="video-stream-1">
      <e2enp:alt default="true" name="nominal" ref_contract="video-contract-1"
      scope="applicable"/>
      <e2enp:alt name="choice1" ref_contract="video-contract-2" scope="possible"/>
   </e2enp:adapath>
   <!-- Possible associations of streams between user A and B -->
   <e2enp:context name="association1-1" scope="applicable">
      <e2enp:comp name="element1" ref_adapath="audio1"/>
         <e2enp:comp name="element2" ref_adapath="video1"/>
      <e2enp:constraints>
         <e2enp:par name="lipsync-delay" ref_adapath="audio1" max="2"/>
         <e2enp:par name="aggregated-bw" max="64000"/>
      </e2enp:constraints>
   </e2enp:context>
   <e2enp:context name="association1-2" scope="possible">
      <e2enp:comp name="element1" ref_adapath="audio1"/>
   </e2enp:context>
   <e2enp:adapath name="associations-A-B">
      <e2enp:alt default="true" name="nominal" ref_context="association1-1"/>
      <e2enp:alt name="choice1" ref_context="association1-2"/>
   </e2enp:adapath>
</e2enp:qoscfg>
```

Figure 74: QoS contexts definition with E2ENP

4.3.4 Media Session Configuration – QoS Contexts

The <e2enp:qoscfg> element is used to specify stream configurations and QoS contexts. The application of <e2enp:qoscfg> without attributes corresponds to the stream configurations (Figure 73). The <e2enp:qoscfg> with "level" attribute identifies QoS contexts and the level of the hierarchy at which this specification is applied (Figure 74).

Figure 73 shows the RTP-specific stream configuration (see also Sections 2.2.1). This description is partially adopted from SDPng (see Section 4.2.4) and is extended to provide some additional definitions, like

- The reservation mode – attribute "reservation" of <e2enp:component> is adopted as a concept from [36] and its values are defined using the regex pattern *receiver|sender|both* to express which party reserves the network resources
- The fixing of the IP addresses with a pattern, which pattern is also used in the address definition of the <e2enp:purpose> section, i.e. *((([:]{0,2})([0-9a-fA-F]{0,4})){0,8})(([1]?[0-9]{1,2}|[2][0-4][0-9]|[2][5][0-5])([.]([1]?[0-9]{1,2}|[2][0-4][0-9]|[2][5][0-5])){3})?* covers both the specification of IPv4 and IPv6 addresses
- The introduction of RTP and RTCP port-offsets with the attribute "rtp-port-offset" in <rtp:udp> according to the specification of SDP [86] that corresponds to the definition of stream flows in case such flows are applicable for the medium (see Section 2.2.1).
- The negotiation of the error-correction techniques associated with the RTP components is defined with the <e2enp:error-correction> element.

Each <e2enp:component> element is used to specify a single media stream, e.g. Figure 73 shows the definition of one audio and one video stream. The definitions important for the specification of QoS contexts (see below) are the "name" attributes of the <e2enp:component>. These name definitions are referenced within the QoS contexts later. The <sdpng:alt> element specifies an alternative configuration of the respective media stream that is applicable within the application's resource-management as a transparent to E2ENP adaptation mechanism (i.e. fine-grained re-negotiation, see Section 3.1.2).

Figure 74 shows the logical configuration of the streams in association with QoS contexts (see also Sections 2.2.2.2). First adaptation paths for every single medium are described and the adaptation displays an association between application and transport contracts. A QoS context is defined as the association of adaptation paths that express the nominal and alternative QoS specifications to be enforced during adaptation and the applicable time-synchronization and/or QoS-correlation constraints.

The <e2enp:adapath> (Figure 74) defines QoS adaptation rules for single streams, stream groups and multimedia sessions as a whole. The <e2enp:alt> element specifies possible alternative QoS contracts used within the adaptation process. The first <e2enp:alt> element in a group identifies a nominal contract the end-systems should first apply to reserve resources at multimedia-session establishment. For example, Figure 74 shows two alternative QoS contexts, which can be applied for a multimedia session, i.e. *association1-1* with audio and video streams and *association1-2* with only audio stream. The application of these two contexts from the perspective of the whole session is specified with <e2enp:adapath name="associations-A-B">. Streams grouping and synchronization rules for an association of media streams are defined over the <e2enp:context> element where the <e2enp:constraints> element is dedicated for defining of restrictions associated with the adaptation paths or with the whole session. The attributes of <e2enp:constraints> are: "name" to define the name of the constraint (the application can freely define this value); "ref_adapath" to express a restriction upon an adaptation path and the attributes "min", "max" and "avg" to associate minimum, maximum and average performance values to the constraint type defined under the "name" attribute of <e2enp:constraints>. The importance and the applicability of adaptation paths and contexts is defined through the attributes "default", "name" and "scope" of <e2enp:alt> and the attribute "scope" of <e2enp:context>, where the attributes "scope" have the same meaning as in the definition of capabilities (see Section 4.3.3 and Figure 71).

The audio, the video and the transport configurations and the QoS contracts and contexts are identified and referenced in E2ENP by their names (consider Figure 72 through Figure 74). For example, the audio information described in the RTP specific stream configuration (Figure 73) is referenced via its name *audio-stream-1*, a video QoS contract information defined in <e2enp:qos-contracts for="video"> (Figure 72) is identified by its name *video-contract-1*. The referencing of already-defined components is done using the attributes named "ref_" (e.g. "ref_component", "ref_contract", "ref_adapath", etc.). For instance, *ref_component="video-stream-1"* (Figure 74) points at the element *<e2enp:component name="video-stream-1" media="video" reservation="receiver">* (Figure 73) or *ref_contract="video-contract-1"* (Figure 74) points at *<e2enp:contract type="video" name="video-contract-1" frame-rate-set="10,15" frame-resolution-type="ITU_CIF" frame-size-set="CIF" color-quality-set="8,16,32" overall-quality-range="9500,9800"/>* (Figure 72). Thus, giving names to the defined components and referencing later these components by their names enables the creation of higher-level definitions like the QoS contexts. The application of attributes starting with "ref_" is one part of the E2ENP referencing mechanism that serves the building of the hierarchical QoS specification. The mechanism for addressing of branches of the QoS specification is explained in Section 4.3.5 (see also Section 4.1.3.4).

The examples in Figure 73 and Figure 74 correspond to the information exchanged between the end-systems during a short Negotiation. If a full Negotiation is applied, the contents of Figure 71 through Figure 74 are put in a single PDU (see also Section 3.1.2).

```
<e2enp:purpose>
    <e2enp:session user="Mary" session-id="2890844526" version="2890842807"
    nettype="IN" addrtype="IP4" addr="43.196.180.1">
        <e2enp:expires time="3600"/>
    </e2enp:session>
    <e2enp:use>
        <e2enp:session user="Mary" session-id="2890843112"
    version="2890841393" nettype="IN" addrtype="IP4"
    addr="43.196.180.1"/>
    </e2enp:use>
    <e2enp:description type="request" name="re-negotiation" style="short"
    coordination="true"/>
</e2enp:purpose>
<e2enp:enforce>
    <e2enp:contract-level path="associations-A-B.association1-1.video1.video-contract-2"/>
    <e2enp:target name="$contract"/>
</e2enp:enforce>
<e2enp:block>
    <e2enp:contract-level path="associations-A-B.association1-1.video1.video-contract-1"/>
    <e2enp:target name="$contract"/>
</e2enp:block>
```

Figure 75: Short Re-negotiation PDU, Re-negotiation of contracts

```
<e2enp:enforce>
    <e2enp:contract-level path="associations-A-B.association1-1"/>
    <e2enp:target name="$stream-context"/>
</e2enp:enforce>
<e2enp:block>
    <e2enp:contract-level path="associations-A-B.association1-2"/>
    <e2enp:target name="$stream-context"/>
</e2enp:block>
```

Figure 76: Short Re-negotiation of contexts

4.3.5 QoS Adaptation Based on Memory

The applications of E2ENP can use a memory for hierarchical QoS specifications. Therefore, the applications refer to already negotiated data in order to perform QoS adaptation at Re-negotiation time. Figure 75 and Figure 76 show the PDU contents of a short Re-negotiation (Section 3.1.2). The <e2enp:enforce> element is used to point at the new QoS contract/context to be applied and the <e2enp:block> defines, which currently used QoS contract/context should be blocked at adaptation. For instance, Figure 75 shows the enforcing/blocking of contracts, where the *video-contract-2* is enforced and the *video-contract-1* is blocked. The links to these contracts are specified with their complete paths in the hierarchical QoS specification (i.e. *associations-A-B.association1-1.video1.video-contract-2* and *associations-A-B.association1-1.video1.video-contract-1*) in order to avoid ambiguities at naming, e.g. *association1-2* may also contain a *video-contract-1* but the corresponding configuration of this association is different from the association *association1-1*. Figure 76 shows an example of enforcing/blocking of contexts, i.e. *associations-A-B.association1-1* is enforced and *associations-A-B.association1-2* is blocked. The respective full definitions of the components of the hierarchical QoS specification shown as references in this section may appear in different PDUs as described with examples in Sections 4.3.3 and 4.3.4 and in Annex C.

The application of a request with single or multiple <e2enp:enforce> sections only (i.e. without any <e2enp:block>) is used to test at the partner if *possible* definitions have turned *applicable* (see Section 4.3.3). The answer to a request containing only <e2enp:enforce> is with an <e2enp:block> definition with the same contents as an <e2enp:enforce> in the request if the descriptions shall remain *possible* or with <e2enp:enforce> if the contents have turned *applicable*. The purpose of a request with single or multiple <e2enp:block> sections is to indicate that some *applicable* configurations have turned *possible*, e.g. at peer's migration to other provider or technology. The application understands if the Re-negotiation concerns only the applicability of configurations or corresponds to a QoS adaptation request upon the "coordination" attribute of the <e2enp:description>. If "coordination" is not available the peer wishes only to test the applicability of *possible* configurations or it wishes to block configurations permanently and thus to mark the hierarchical QoS specification branches as *possible.*

The referencing mechanism for QoS specifications is applied via the "path" attribute of <e2enp:contract-level> and identifies a complete branch of a hierarchical QoS definition that was produced during previous Pre-negotiation and Negotiation phases (see Sections 2.2.2.3 and 3.1.2 and the respective examples and figures in Sections 4.3.2, 4.3.3 and 4.3.4). The identification of the branch is done using the names of the specified definitions (see below and Section 4.3.4). The <e2enp:target> element defines the abstraction level of the QoS information. Thus, the "path" attribute (of <e2enp:enforce> and of <e2enp:block>) and the <e2enp:target> build the second part of the E2ENP logical-linking mechanism for QoS definitions (see Sections 4.1.3.4 and 4.3.4). The definition of the "path" values is not fixed in the XML specification of E2ENP (it is just a string definition), as it is used for both E2ENP logical paths and XPath definitions (see Sections 4.1.3.4 and 4.3.6). However, the specification of E2ENP [20][132][173] fixes the value of the logical path in a way that the single names of the elements of the hierarchical QoS specification are dot-separated (see Figure 75 and Figure 76).

In Figure 75, the stream representation of *audio1* changes between the currently active *audio-contract-1* and the alternative configuration *audio-contract-2.* The application of <e2enp:enforce> and of <e2enp:block> for enforcing/blocking QoS contexts is shown in Figure 76, e.g. if the association of an audio and video stream is no longer applicable this association can be blocked and the association with a single audio stream can be enforced (see the corresponding definition of associations in Section 4.3.4 and Figure 74).

The XML structure of a short E2ENP Re-negotiation for configuration-adjustment or for media-adaptation purposes (as evident from the examples in Figure 75 and Figure 76) is always fixed irrespective of the complexity of the QoS specification, i.e. irrespective of how many alternative audio/video streams or how may associations of streams and contexts are defined. At short Re-negotiations, the peers use only <e2enp:enforce>/<e2enp:block> elements to point at the respective change. Such feature for compact addressing of configurations is currently not available within the state-of-the-art mechanisms, where full re-negotiation of parameters is always applied in cases of media adaptation. Furthermore, the state-of-the-art mechanisms have no notion of potential and actual configurations, like the E2ENP indications of *possible* and *applicable* configurations. Hence, the current techniques for negotiations are always bound to communicate complete configurations.

A full Re-negotiation PDU is similar to the full Negotiation PDU (see Section 4.3.4) and applies in E2ENP only in cases that the quality of the application's service is violated and the application has to switch to new and still-not-negotiated configurations. Consequently, the short Re-negotiation compared to the full one and to the state-of-the-art mechanisms (see above) is very compact and leads to efficient signalling at adaptation (see the respective measurements in Chapter 6).

```
<e2enp:qosinfo for="mentoring">
    <e2enp:provider id="e2enp://myfunnyprofider:123V#gP@acme.com/">
        <e2enp:block>
            <e2enp:contract-level path="transport-contract-1"/>
            <e2enp:target name="$contract"/>
        </e2enp:block>
        <e2enp:signature>
            <e2enp:algorithm method="http://www.w3.org/2000/09/xmldsig#dsa-sha1"/>
            <e2enp:digsig type="certificate">
                <!-- the certificate contents-->
            </e2enp:digsig>
        </e2enp:signature>
    </e2enp:provider>
</e2enp:qosinfo>
```

Figure 77: <qosinfo> of provider entity at Pre-negotiation

```
<e2enp:qosinfo for="mentoring">
    <e2enp:provider id="e2enp://myfunnyprofider:123V#gP@acme.com/">
        <e2enp:block>
            <e2enp:contract-level path="associations-A-B.association1-1.video1.video-contract-2"/>
            <e2enp:target name="$contract"/>
        </e2enp:block>
        <e2enp:signature>
            <e2enp:algorithm method="http://www.w3.org/2000/09/xmldsig#PGPData"/>
            <e2enp:digsig type="digest">
            <!-- the digest contents-->
            </e2enp:digsig>
        </e2enp:signature>
    </e2enp:provider>
</e2enp:qosinfo>
```

Figure 78: <qosinfo> of provider entity at Negotiation

```
<e2enp:qosinfo for="mentoring">
    <e2enp:mentor id=" e2enp://mentor:vG$I809@acme.com/">
        <e2enp:adds>
            <e2enp:contract-level path="//qos-contracts[@for='audio']/contract[@name='audio-contract-3'"/>
            <e2enp:target name="$contract"/>
        </e2enp:adds>
        <e2enp:signature>
            <e2enp:algorithm method="http://www.w3.org/2000/09/xmldsig#dsa-sha1"/>
            <e2enp:digsig type="certificate">
            <!-- the certificate contents-->
            </e2enp:digsig>
        </e2enp:signature>
    </e2enp:mentor>
</e2enp:qosinfo>
```

Figure 79: <qosinfo> for transcoding and XPath application

4.3.6 External Parties Interactions with the PDU

The external-parties interactions with the E2ENP PDUs are identified within the <e2enp:qosinfo for="mentoring"> (Figure 77 through Figure 79). This element with its attribute “for="mentoring"” shall be recognised by the PDU management mechanism at the receiver of the PDU, so that the receiver distinguishes the mentor’s information and can extract it from the PDU to achieve the initial technical specification of the sender of the PDU (see Section 4.1.3.4 and Figure 50).

The identification of the providers and the mentors (elements <e2enp:provider> and <e2enp:mentor>) is done using the E2ENP URI (Section 2.2.2.1) and respectively inserting this information in the “id” attribute of the corresponding elements.

The element <e2enp:block> (Figure 77 and Figure 78) has similar meaning and application as within the short Re-negotiations (see Section 4.3.6). The mentoring entities, however, utilize this denotation to inform the users that the respective contracts or contexts are not supportable within the domain of influence of the mentor that issued this limitation. Additionally, if the complete branches of the hierarchical QoS specifications are still not communicated between the peers, the mentors that listen in the peers negotiations mark the QoS specification only upon the known information (see Figure 77), e.g. at Pre-negotiation only the contracts are known, hence the provider uses only the names of the contracts to mark QoS specification as blocked.

The element <e2enp:adds> (Figure 79) identifies a branch of the XML-tree of the current PDU, where the interaction with the provider has taken place, to identify an XML-element added by the mentor to the hierarchical QoS specification. The mentor can insert multiple elements and each inclusion of an element is identified with the "path" attribute of <e2enp:contract> using an XPath definition. The XPath definition points at the physical place of the added element (see Section 4.1.3.4) so that the receiver of the PDU can filter the QoS specification to separate the initial-sender technical information from the technical data added by the mentors (see above, Section 4.1.3.4 and Figure 50). The "target" attribute, in this case, helps the receiver to logically distinguish to what part of the hierarchical QoS specification the added information belongs. The mentors may add information to the hierarchical QoS specification at application of alternative management entities like transcoders to enhance the hierarchical QoS specifications of the negotiating partners (Figure 79).

When the mentors interact with the PDUs of the peers, mentors have to provide information about their authenticity as trustworthy entities. For this purpose, they apply the <e2enp:signature> element for signing of the blocked/added QoS information with either the E2ENP's short signature (Figure 77 through Figure 79) or with complete XML signature (see Section 4.2.4 and Figure 57 and the standard examples of XML signatures in [229]). The element <e2enp:algorithm> provides information about the used algorithm for signing the PDU data, i.e. the attribute "method" uses the predefined namespace for signature algorithms as defined in [229]. The element <e2enp:digsig> contains the signature (i.e. the place of the signature in Figure 77 through Figure 79 is shown with an appropriate XML comment) and it has the attribute "type" to identify the type of the signature, e.g. as *digest* or *certificate*. The type of the signature depends on the requirements of the application for security and these requirements are exchanged between the terminals and the provider during the registration of the devices (see Sections 3.2.1 and 4.1.3.4). Correspondingly, *digest* means that the mentor uses parts of the PDU to sign the data and *certificate* is just a cookie-like identifier of the mentor.

```
<e2enp:purpose>
    <e2enp:session user="Mary" session-id="2890844526" version="2890842807" nettype="IN"
    addrtype="IP4" addr="43.196.180.1">
        <e2enp:expires time="0"/>
    </e2enp:session>
    <e2enp:use>
        <e2enp:session user="Mary" session-id="2890843112" version="2890841393" nettype="IN"
        addrtype="IP4" addr="43.196.180.1">
            <e2enp:expires time="18000"/>
        </e2enp:session>
    </e2enp:use>
    <e2enp:description type="request" name="pre-negotiation" style="short"/>
</e2enp:purpose>
```

Figure 80: PDU's time resetting

```
<e2enp:purpose>
    <e2enp:session user="Mary" session-id="2890844526" version="2890842807" nettype="IN"
    addrtype="IP6" addr="::43.196.180.1">
        <e2enp:expires time="0"/>
    </e2enp:session>
    <e2enp:use>
        <e2enp:session user="Mary" session-id="2890843112" version="2890841393" nettype="IN"
        addrtype="IP6" addr="::FFFF:43.196.180.1">
            <e2enp:expires time="0"/>
        </e2enp:session>
        <e2enp:session user="Mary" session-id="2890843112" version="2890841393" nettype="IN"
        addrtype="IP6" addr="1080::8:800:200C:417A">
            <e2enp:expires time="30"/>
        </e2enp:session>
        <e2enp:session user="Mary" session-id="2890843112" version="2890841393" nettype="IN"
        addrtype="IP6" addr="FEDC:BA98:7654:3210:FEDC:BA98:7654:3210">
            <e2enp:expires time="50"/>
        </e2enp:session>
    </e2enp:use>
    <e2enp:description type="request" name="expire" style="short"/>
</e2enp:purpose>
```

Figure 81: PDUs' invalidation

4.3.7 PDU Time Set and Invalidation

The E2ENP leasing management (Section 3.1.4) uses short or full PDUs to signal the repetition of the leasing of PDUs (Figure 80) or the cancellation of the leasing (Figure 81). The short PDUs have similar contents as the examples in Figure 80 and Figure 81 and contain no further elements then the <e2enp:purpose> element (see also Section 4.3.2). The full PDUs can contain combinations of the capabilities- and QoS-definition elements of E2ENP (see Sections 4.3.3 and 4.3.4).

The time-resetting PDU in Figure 80 signals that it immediately expires as it contains no further technical information that would influence the QoS specification of the peers, i.e. <e2enp:expires time="0"/>. The PDU resets an already existing PDU that is referenced in the <e2enp:use> element of the <e2enp:purpose>. The time for the repeated leasing is indicated with the <e2enp:expires time="18000"/>. The <e2enp:description> element identifies the PDU as *pre-negotiation* as it does not influence the technical specification of the peers (see the definition of leasing Pre-negotiation in Section 3.1.3 and Section 3.1.4).

The invalidation PDU in Figure 81 also signals its immediate expiration with <e2enp:expires time="0"/> as it does not carry further technical specifications. The PDUs that shall expire are referenced within the <e2enp:use> element of <e2enp:purpose> (see also Section 4.3.2) and their final-expiration time is indicated within the corresponding <e2enp:expires> element. The expiration function is indicated within the attribute "name" of <e2enp:description>. The receiver of such PDU can act according to the application logic of the service that will result in the following actions:

- Putting the expired PDUs into Zombie state if these PDUs are in use within a multimedia session (see Section 3.1.4) and delete the PDUs as soon as the session ends or delete the PDUs immediately if they are not used in a media session

- Try to continue the leasing of the PDUs also in combination with resetting of Zombie states (see above and Section 3.1.4)
- Re-negotiate the technical specification, e.g. as an E2ENP Pre-negotiation in case that the peers are not in a multimedia session that uses the expired PDUs or as a short/full E2ENP Re-negotiation if the peers are involved a media session (see Sections 3.1.2 and 3.1.3). Note that the expiration of PDUs always starts as a leasing Pre-negotiation (Section 3.1.3), no matter if the PDU contains technical data or not. This is necessary because multiple actions from the side of the receiver may take place as result of the expiration. Hence, these actions have to be planned through an E2ENP Pre-negotiation, in order not to cause violations in the resource-management systems of the affected service (see Section 2.3.2). A Pre-negotiation PDU in such case contains the indication of a corresponding negotiation mode (i.e. *pull* or *push-pull*) in order to signal that the Pre-negotiation continues with exchange of technical data or that a subsequent Re-negotiation is expected (see Sections 3.1.1, 3.1.3 and 4.3.2).

This chapter presented the XML language of E2ENP. Specific features like the PDU referencing and leasing are implemented within the E2ENP User Agent together with E2ENP's negotiation, consistency-management and resource-management concepts (see Chapter 5 where details on these concepts are presented). Furthermore, Chapter 6 presents an evaluation of the E2ENP UA in its Offer/Answer Model implementation to show that the concepts for reducing of the application signalling through PDU recollection and corresponding referencing mechanism hold true for the E2ENP UA. The evaluation contains also comparison with the state-of-the-art protocols and this comparison is provided through emulation of SDPng functionality.

5 Negotiation Engines

This chapter presents the implementation concepts of the negotiation engines for the purpose of multimedia-session management. Furthermore, it discusses the actual realisation of the End-to-End Negotiation Protocol User Agent (E2ENP UA). This chapter summarises also the experience at developing the E2ENP negotiation engine and at integrating foreign software modules into its architecture.

The E2ENP UA is a Java application based on the Java Standard Edition (J2SE), version 1.4.x [222]. The implementation of the E2ENP UA and of further concepts of the negotiation are presented using the Unified Modelling Language (UML) [268][269] and the corresponding figures were developed applying partially the UML design-software *Together*[29] [216].

The general architecture of the E2ENP UA is presented in Section 5.1, where the major design principles at developing and implementing of an E2ENP-based negotiation engine are discussed. Details of the E2ENP UA components and their implementation are given in Section 5.2, highlighting the innovations provided through the E2ENP UA implementation in comparison with SIP and the Offer/Answer model in terms of phases-based negotiation, multi-application management of concurrent services and abstraction of the applications from the actual signalling protocol. The E2ENP UA architecture was already presented by the author of this book (in co-authorship) in [132] and a detailed model and API of the E2ENP UA at its Offerer/Answerer-performance mode were published (also in co-authorship) as European patent (EP1414211) [170] (see also Sections 3.1.3 and 3.3.1.2). In addition to the Offerer/Answerer mode (Section 5.2.2.4.2), this work presents also the application of the E2ENP UA at External-Negotiation mode (see Section 5.2.2.4.3) that is suitable for performing of session migrations and of third-party call-control scenarios (see also Sections 3.3.1.2 and 3.3.2). Finally, Section 5.3 discusses some additional implementation aspects of E2ENP that concern the simultaneous application of the E2ENP UA and the local/network resource-management systems.

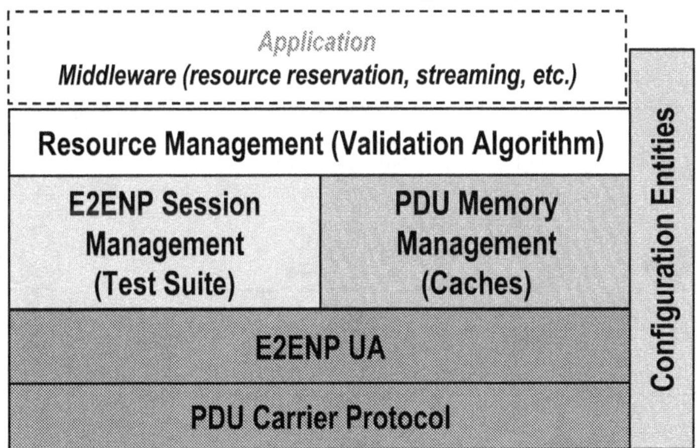

Figure 82: E2ENP general architecture

[29]Note that due to the application of different versions of Together some of the similar UML model-types display slightly different appearance.

5.1 General Architecture and Design

The general architecture of an E2ENP-based engine for management of multimedia applications is shown in Figure 82. According to their level of implementation, the different components of this architecture are presented in different shades. The dark-grey shaded boxes in Figure 82 indicate fully designed and developed components. In light grey are the components that are completely designed but are not fully implemented. Nevertheless, these components display adequate functionality for performing tests with the E2ENP UA (see below and Section 5.2.1). In white are the components that are only conceptually planned for reasons of clarity and completeness, e.g. if the E2ENP UA is applied as a session manager of a QoS Broker, the components shown in white are other then the session manager components of the Broker (see also Section 2.4.2).

Correspondingly, the E2ENP UA is fully developed and is presented in detail in Section 5.2. The PDU-carrier protocols considered in this work are SIP and RMI implementations that are based on existing applications and/or were re-engineered to fit the E2ENP UA plug-in requirements (see below and Section 5.2.2.3). The PDU memory management is implemented in form of PDU caches associated with the E2ENP UA (Sections 5.2.1 and 5.2.2.2). The E2ENP Session Management is implemented as a test suite for the E2ENP UA performance and a modification of this suite is applied within the performance tests in Chapter 6 (see also Section 5.2.1). The tests of the validation algorithm require the additional development of an appropriate multimedia application, but since the goal of this work is to provide concepts for improving the support of multimedia applications at their session management and the work does not aim to develop such applications, the models concerning the validation algorithm and the application-specific management are presented only theoretically (see Sections 2.3.2.2.2 and 5.3, and Annex E).

The E2ENP design envisions that the components belonging to the E2ENP-engine are configurable and pluggable within the architecture, so that the performance of the engine can be adapted to different devices that operate within different network technologies (i.e. mobile or stationary user terminals, automatic services or provider/mentoring devices shall be supportable). With this respect, the E2ENP-engine contains a layer of configuration entities (Figure 82). The implementation of this layer includes the application of the *factory method* design pattern [146] to activate the generation of the single components in the architecture (see also Section 5.2.1) and of the *abstract factory* pattern to generate the specific structure of the E2ENP-engine (see Section 5.2.1.2).

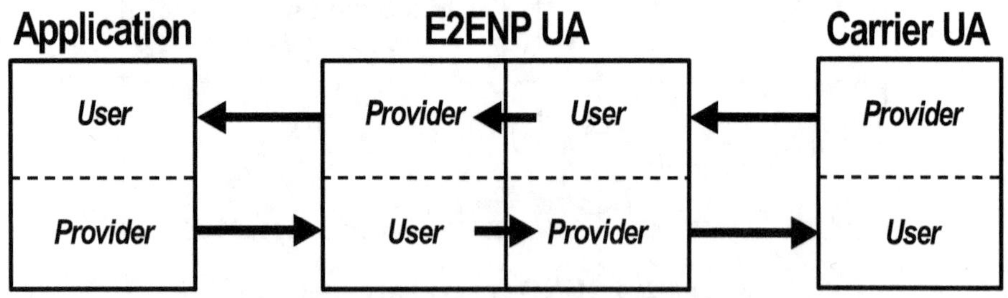

Figure 83: Asynchronous subsystem communication

Since some of the components of the architecture have to be kept unique in order to guarantee consistency for their performance (see the respective requirements in Section 2.3.2.1.1), these components are designed using the *singleton* pattern [146] to allow their instantiation just once (e.g. the PDU caches apply this pattern – Section 5.2.2.2). The uniform access to such objects is then managed over the E2ENP state machines and the MUTEX (see Section 5.2.2.4).

The communication between the components inside the E2ENP UA is asynchronous (see Figure 83) in order to allow the application of versions of these components, exposing different complexity and performance, and to enable the communication with the E2ENP UA within networks with different throughput (see also Sections 5.2.2.3 and 6.4). Consequently, the adjacent communicating components display user-provider behaviour that is characterised through the following activities, i.e. the user requests the service and utilises the results of the service and the provider makes the service available and prepares the expected results. In terms of the implementation, the entity using the services offered by a given API is a *Service User* and the component implementing the services of such API is the *Service Provider*. This user-provider behaviour is facilitated through the application of two software patterns, i.e. the *observer* (also *publish-subscribe*) and the *producer-consumer* [146]. The *observer* pattern allows the separation of the E2ENP-UA-component interfaces into activation and listener interfaces, i.e. as soon as an entity activates an action that is associated with an adjacent component and that shall return results, the initiator (i.e. the observer) starts to listen to incoming events from its respective neighbour-components. For this purpose, the observer registers its listeners at its observed adjacent entity at E2ENP UA starting and configuration procedure (i.e. binding – see also Section 3.1.3). The *producer-consumer* pattern is associated with the synchronisation of the activation and the listening mechanisms implemented through the *observer* pattern, so that the listening entities do not wait actively and may generate failure conditions in cases they get no answer from the observed components after certain periods. The user-provider behaviour is implemented on the boundary between the E2ENP UA and the carrier protocol implementation to separate between the transport functions of the carrier and the PDU management performance of the E2ENP UA. Furthermore, the E2ENP UA also exposes such user-provider behaviour towards the application to separate its PDU and negotiation processing from the application-specific actions. The respective interfaces for implementing the Offerer/Answerer mode and the External-Negotiation mode of the E2ENP UA are discussed in Sections 3.1.3, 3.3.1.2, 5.2.1, 5.2.2.4.2 and 5.2.2.4.3.

The PDU content management within the E2ENP UA applies also several patterns for implementing the Parser, the Generator and the Content Proxy (see also Section 5.2.2.1). At transformation of the XML code into Java objects and vice versa the *visitor* design pattern [146] is applied. The Parser and the Generator apply this pattern to navigate through the XML object-tree for the purposes of the transformation. Additionally, at this low level of simple transformation, the XML elements are defined as objects of the same type to ease the access of the visitors (i.e. the Parser and the Generator) to the contents of the objects (i.e. the visited objects). The Content Proxy provides the E2ENP UA and the applications of E2ENP with an intelligent access to and with a generalised view upon the information contained in the PDUs using the *proxy* and the *façade* patterns [146]. At instantiation, the *proxy* can be loaded with default information in cases some PDUs cannot or shall not provide certain data that is necessary for the performance of the application. Note that the contents of the PDUs are specified according to the requirements of the application for negotiations and in consideration with provider restrictions upon the session management (see the respective discussions in Chapters 3 and 4). Consequently, the Content Proxy is a configurable object. Furthermore, the *façade* simplifies the access to the

PDU information, abstracting the entity using the PDU from the specific XML structure (e.g. the wrapper components in the XML structure are invisible to the application). Further details on the E2ENP UA content management are presented in Section 5.2.2.1.

The E2ENP UA exposes an innovative feature at scheduling the access of multiple applications to the carrier-protocol environment. The current implementations of the SIP specification assume a single application using a single SIP UA (see [62][184][185]). The E2ENP UA makes the uniform access of multiple applications to a single SIP-carrier possible, thus enabling the uniform control of the system resources. For this purpose the E2ENP UA implements a MUTEX as a state machine, applying the design patterns *lock*, *scheduler* and *thread pool* (see [146]), i.e. to lock the access to the carrier and achieve exclusive handling per task, to schedule the access of multiple applications to the carrier environment and to regulate the number of tasks actively waiting for transportation of their negotiation data. Details of the MUTEX implementation are given in Sections 3.1.3 and 5.2.2.4.1.

5.2 E2ENP User Agent

This section presents the implementation of the E2ENP UA and its components. The specifications of the basic structure of the architecture, of the extensibility of the architecture and of the exact realisation of the individual entities within the E2ENP UA architecture are extensively discussed.

5.2.1 User Agent Architectures

The E2ENP UA is presented in this section at its basic and extended versions.

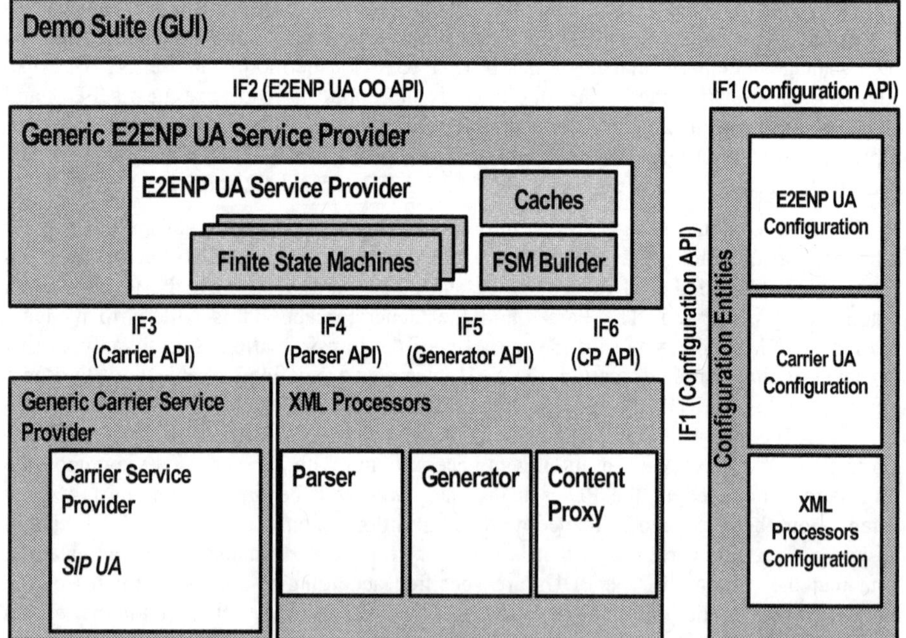

Figure 84: E2ENP UA architecture

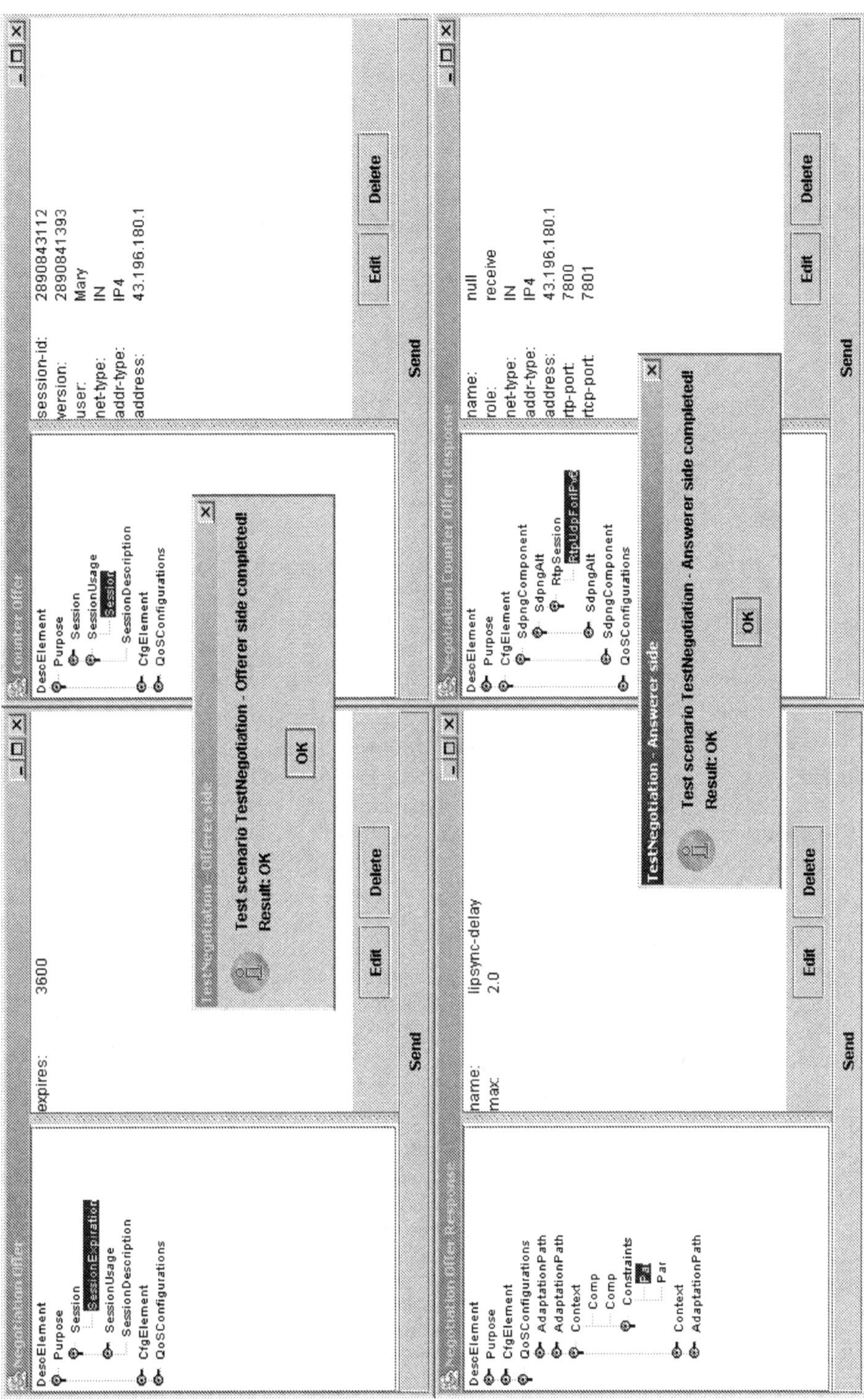

Figure 85: E2ENP UA – Demo screenshot

5.2.1.1 Basic Architecture

Figure 84 shows the basic architecture of the E2ENP User Agent (E2ENP UA) which consist of the following entities:

- The XML Processors are responsible for the E2ENP content management (see also Section 5.2.2.1)
    - The Parser translates E2ENP PDUs into system-internal representation (i.e. Java objects);
    - The Generator translates system-internal objects into E2ENP XML descriptions
    - The Content Proxy wraps up the PDU information into an API that is suitable for the specific needs of the E2ENP application (see Sections 5.1 and 5.2.2.1)
- The PDU caches are parts of the E2ENP *Service Provider* sub-system (see also below and Sections 5.1 and 5.2.2.2) and represent a memory on the available PDUs that is shared between the application and the E2ENP UA (see also Sections 2.3 and 3.1.4). The caches are used to enable consistent exchange and referencing of negotiated information.
- The Carrier Service Provider is the application for transporting of the E2ENP control messages, e.g. a SIP User Agent and the corresponding SIP stack (see Section 5.2.2.3). The specific implementations of the carrier protocol are wrapped up into the Generic Carrier Service Provider that represents an adaptor for associating the specific call control of the carrier application with the predefined call-control between E2ENP UA and the carrier as presented in Sections 3.1.3 and 3.3.1.2.
- The E2ENP Finite State Machines (FSMs) implement the specific E2ENP communication and are parts of the *Service Provider* sub-system (see also Section 5.2.2.4). Each instance of a FSM corresponds to one specific role of the UA (e.g. Offerer or Answerer) and to one E2ENP control session that is created by the application of E2ENP (e.g. the E2ENP UA Demo application is used to demonstrate E2ENP control sessions – see the GUI of this application for an E2ENP Negotiation session in Figure 85). The E2ENP UA uses the FSMs to manage the access of multiple E2ENP control sessions to the carrier application (see Sections 2.3.2.1.1, 3.1.3 and 5.2.2.4.1). The E2ENP UA controls the consistency of the E2ENP information, i.e. the E2ENP UA keeps track of mutually-dependant E2ENP sessions saving into the caches (see above and Section 5.2.2.2) the PDU identification data and verifying this information in association with other E2ENP sessions. Consequently, the E2ENP FSMs are concerned only about the PDU headers as these headers carry the information on mutual dependencies and on specific actions required from the FSMs (see Section 4.3.2). The FSMs together with their specific builder (see also below the discussion on the configuration components) and with the caches are wrapped up into a Generic E2ENP UA Service Provider component that adapts the access to the FSMs and to the caches to the generalised API for application access and negotiations, which API is partially presented in Sections 3.1.3 and 3.3.1.2.
- The Configuration Entities component is used to build the specific structure of the E2ENP UA according to the needs of the application (see also Sections 5.1 and 5.2.1.2 and the example of the API for building the Continent Proxy in Section 5.2.2.1). This component contains three major configuration entities that correspond to the building of one specific E2ENP UA subsystem, i.e. the carrier protocol, the E2ENP UA Service Provider and the language processors (e.g. for XML). At starting of the system, the

application/middleware configures first the E2ENP UA before using the negotiation engine. The Configuration Entities are also applicable to dynamically reconfigure the E2ENP UA during the system execution, e.g. in cases of changing technology support in different networks or provider domains. However, the practicability of the dynamic E2ENP UA reconfiguration to plug in or out sub-components and systems is limited, due to the specific persistent realisation of some of the re-used components (e.g. SIP stacks [184][185] and XML SAX parser [267][270][271]) that do not allow dynamic re-plugging without restarting of the complete system.

The re-usable APIs specified for the E2ENP UA are indicated in Figure 84 with *IF1* through *IF6*. Correspondingly, the functions of these APIs are (see also [170]):

- IF1 is the configuration interface of the E2ENP UA that shall be used by the application/middleware to create a specific instance of the E2ENP UA. The realisation of this interface is asynchronous.
- IF2 is the interface that the E2ENP UA exposes towards the respective application in order to provide the application with generalised negotiation logic. This interface consists of methods corresponding to the E2ENP's negotiation control procedures, e.g. phases, modes, etc. (see Sections 3.1 and 3.3.1.2). The realisation of this interface is asynchronous.

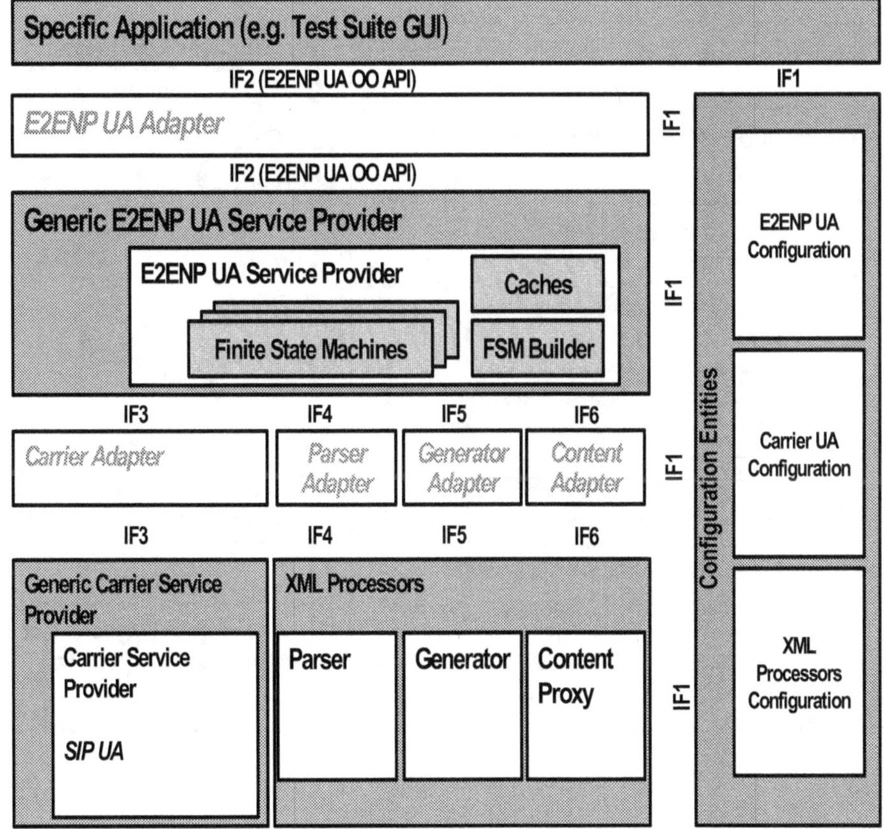

Figure 86: E2ENP UA architecture extensibility

- IF3 is used for specific call management and transportation of the E2ENP data, e.g. it imitates SIP methods (see Sections 3.1.3 and 3.3.1.2). The realisation of this interface is asynchronous.
- IF4, IF5 and IF6 are the access interfaces for the Parser, the Generator and the Content Proxy and their realisation is synchronous (see also Section 5.2.2.1).

These interfaces are also applied for the implementation of the advanced E2ENP UA architecture discussed in the next Section 5.2.1.2.

5.2.1.2 Advanced Architecture – Measurements and Enhancements Modules

The advanced E2ENP UA architecture enables the inclusion of additional processing components on the paths where the predefined E2ENP UA interfaces are situated (see Section 5.2.1.1 and Figure 86). In order to include additional components on the processing paths of the E2ENP UA, the processing components need to implement the interfaces of the E2ENP UA so that they can configure the basic E2ENP UA sub-systems and be able to forward the messages between the basic E2ENP UA sub-components.

At configuration of the advanced E2ENP UA architecture, the Configurations Engines access the Adapter components to configure them and the Adapters configure correspondingly the basic components applying the IF1 interface.

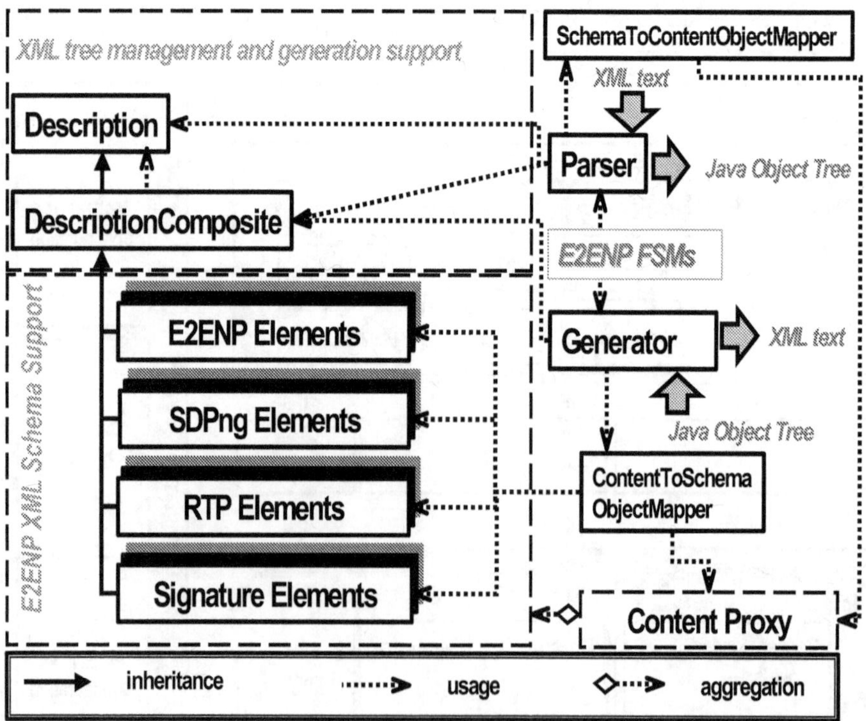

Figure 87: E2ENP processing – association between processed XML code and its processors

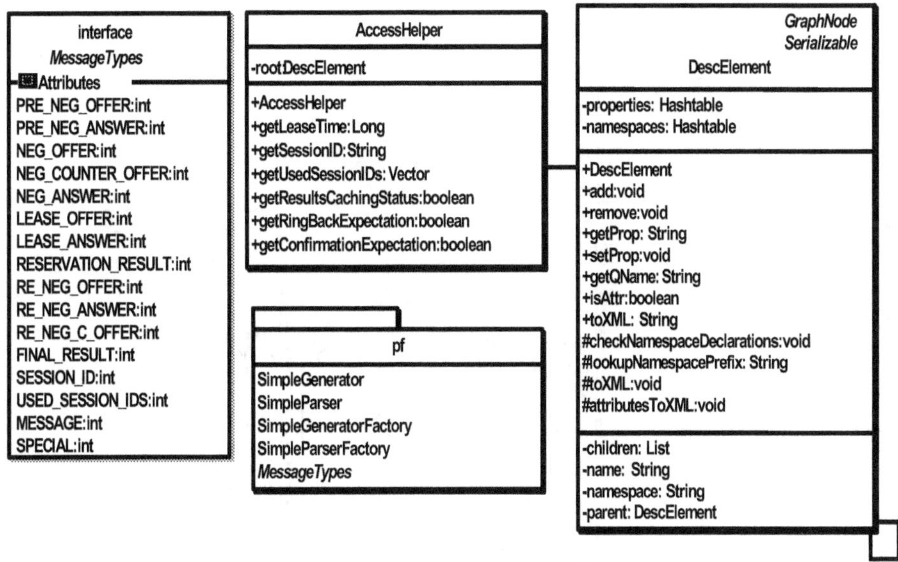

Figure 88: E2ENP content access and the parser/generator package

Figure 89: Relation between the objects representing <e2enp:purpose>

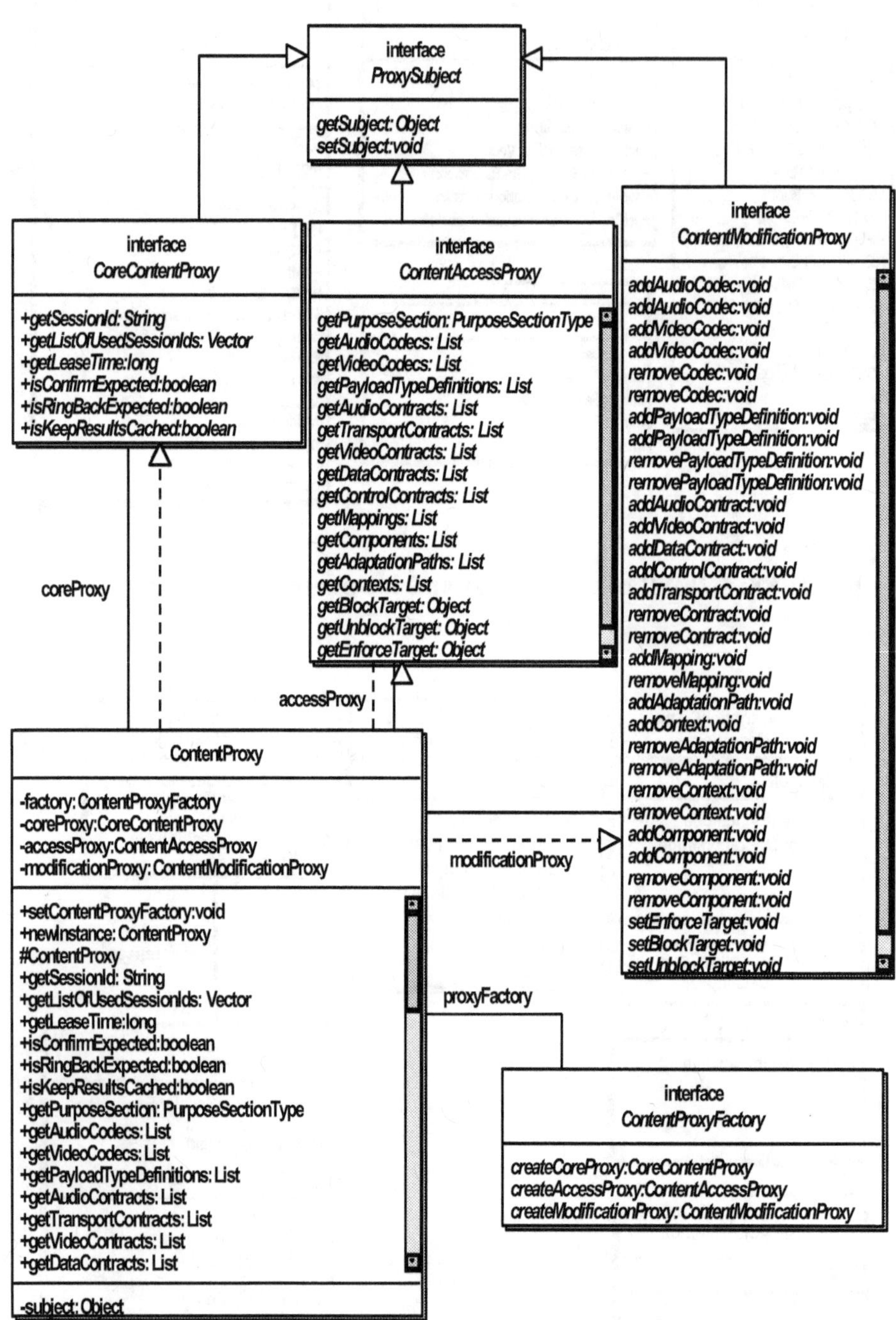

Figure 90: Content-Proxy interface design (classes relation)

The forwarding is enabled because of the fact that some of the potentially variable parameters of the API methods are defined as *java.lang.Object* (e.g. for the PDUs). Hence, the API does not need to be changed in cases that the contents and the structure of such object parameters change due to some additional treatment of the data inside the Adapters. This API feature allows generalised processing for inclusion of components like BiM compression and/or XSLT transformation (see also Sections 4.1.3.3 and 4.1.3.4).

Adapter components are also applied for the purpose of E2ENP UA internal synchronisation and performance stabilisation at communication between the E2ENP FSMs and the SIP implementations used as data carriers (see Section 5.2.2.3). Furthermore, Adapters are used for performing of measurements (see Chapter 6). For the measurement case, the Adapters generate a time event (i.e. a microsecond mark) before forwarding a message. The time events are visible in the console implementation of the E2ENP UA. The evaluation of the time marks enables the estimation of the performance speed of the E2ENP UA and its components in local and in distributed scenarios.

5.2.2 Architectural Components

This section discusses the realisation of the E2ENP UA sub-systems.

5.2.2.1 XML Processing – Parser, Generator and Content Proxy

The architecture of the XML-processing sub-system in E2ENP UA is displayed in Figure 87, where a class-like diagram of the E2ENP XML code is presented together with the corresponding XML-processing components responsible for transforming the textual XML representation into java objects (suitable for the application) and vice versa from objects to XML (for transportation purposes). The current implementation of the XML processing for E2ENP applies SAX parsing with the SAX interface for Java [267]. This interface was implemented for the E2ENP UA using Oracle XML Developers Kit [270] and Apache SAX parser Xerxes-J [271]. The tests of the XML validation with the validation function of these two parsers confirmed the definition problems within the initial version of the E2ENP language that are discussed in Section 4.2. The parsing of E2ENP with SAX that does not apply the classical XML validation can speed up the application processing (see below the explanation of the partial, customised validation used in E2ENP). Although the construction of the E2ENP application objects is similar to the DOM XML representation [251] (see below the definitions of *Description* and *DescriptionComposite*), the difference between the SAX-based XML parsing in E2ENP and the classical DOM parsing is the customised validation (see below) that results in optimised internal representations of objects. The E2ENP processing is configurable to omit pieces of the XML code that are non-understandable for the application, thus producing an optimised representation for the application. Since the E2ENP code is used together with session-control protocols that may imply stringent time limitations, the SAX parsing together with customised validation can improve the performance of the Parser (see also the measurements in Section 6.3).

The Parser generates non-specific data objects named *DescriptionComposite* (see Figure 87). Each instance of a *DescriptionComposite* represents any XML element with its value and attributes. The class *DescriptionComposite* inherits the abstract class *Description* that serves as an accumulator for the information within a single XML element (i.e. element's name and value as well as its attribute names and values). The *Description* provides abstract methods for the management of the information contained in an XML element. These methods are fully implemented within the *DescriptionComposite*. The Parser builds an object tree out of the generated Composite objects. This tree corresponds to the XML tree contained in an E2ENP PDU document. The linking of the XML objects is provided through the *DescriptionComposite* techniques for accessing its direct parent and children objects (i.e. a *DescriptionComposite* object has only one parent that is also a Composite and one or many children Composites). For the linking purposes, a *DescriptionComposite* is able to accumulate one or multiple *Description* objects. Correspondingly, the Parser represents only low-level processing of the XML code that is still not understandable from the E2ENP FSMs and the application.

In order to make the E2ENP code manageable in the application the Parser applies the *SchemaToContentObjectMapper* that consists actually of several access-helper classes (see Figure 88 showing the access-helper used at accessing the information of the PDU header together with the parser/generator package and the interface used for the content-access management). The access-helpers are used within the mapping process between the low-level XML representation objects and the high-level application objects. The relevant *DescriptionComposite* objects are then wrapped up in a specific structure corresponding to the XML schema of E2ENP (see Section 4.2 and Figure 87). An example of such specific objects and their relations is shown in Figure 89 and it corresponds to the contents of the <e2enp:purpose> element and its sub-elements (see Section 4.3.2). Each of the E2ENP specific elements is implemented as a single class that extends the *DescriptionComposite* in order to inherit the object-tree building mechanisms provided by the *DescriptionComposite* and to enable easy element processing at low level of transformation at the Parser and at the Generator. The wrap-up process takes place in parallel to the generation of the *DescriptionComposites* and that is where the *visitor* pattern discussed in Section 5.1 is implemented. The Parser visits each *DescriptionComposite* through its low-level access interface, delegates the element to the *SchemaToContentObjectMapper* and the *SchemaToContentObjectMapper* transforms the *DescriptionComposite* into a specific element.

In case that *customised validation* is done, the *SchemaToContentObjectMapper* simply omits the transformation of those elements that are not expected from the application. The customised validation considers at transformation two important constraints. First, XML wrapper components remain invisible for the application, as they do not contain relevant for the application information, however, this information remains in the object-tree in order to enable easy back-transformation of the objects into XML code (see below the description of the Generator). Secondly, complete blocks that are not understood by the application (e.g. the transport contracts in case the application cannot reserve resources on its own) can be removed from the object tree in case the application has configured the Parser to do so. The *SchemaToContentObjectMapper* is pre-configured when the Parser's configuration takes place to perform customised validation. The respective configuration interfaces of all the XML processing components used in the E2ENP UA are shown in Figure 88 and Figure 90 and are indicated at the end of their names with *Factory*. Note that some of the interface methods visible in Figure 89 and Figure 90 are repeated (e.g. *addComponent* in *ContentModificationProxy* in Figure 90) as they represent the same action with different configuration parameters. However, at UML representation of classes the parameters of the methods are omitted and only the return types of the respective functions are shown.

When the transformation of an element is ready, the *SchemaToContentObjectMapper* links the respective element to the Content Proxy that is the aggregator of the specific XML-element representations (see Figure 87). The Content Proxy corresponds to the interface visible to the application (Figure 90).

The Content Proxy uses the *MessageTypes* interface (Figure 88) to give the application the possibility to recognise the PDU's purpose only upon the short marks that indicate the type of the message (e.g. pre-negotiation, lease-offer, etc.). Furthermore, the Content-

Proxy gives the possibility to the application to manipulate the contents of a PDU (Figure 90), e.g. the application can make a copy of the PDU and manipulate its information to produce an answer to a received PDU. The ready answers of the application are sent to the XML Generator.

The Generator (Figure 87) is a customized generator for XML, designed to produce any XML output document out of the *DescriptionComposites* aggregated into the Content Proxy. The Generator uses the specific Java representations for the XML elements that correspond to the XML schema definitions of E2ENP (Section 4.2) to provide corresponding code validation at transforming the application's representation of the PDUs into XML code. For the purpose of validation, the Generator applies the *ContentToSchemaObjectMapper* that is also configurable like the *SchemaToContentObjectMapper*. The public method *toXML* of the specific-object representations returns a string of the elements' contents and of the underlying XML elements' contents (i.e. the procedure is recursive). If this method is executed by the Generator at the top-level container of the PDU's code (i.e. the *DescElement* – see Figure 88), a complete XML representation of the PDU is generated. The role of the *ContentToSchemaObjectMapper* in the current E2ENP implementation is applied for debugging purposes to prove if the instances of the specific elements generate well-formed XML snippets. The function of the Generator to produce XML out of the object contents through executing the *toXML* method of the objects corresponds to the implementation of the *visitor* pattern discussed in Section 5.1. The Generator visits each *DescriptionComposite* through its *toXML* method for proving and serialising its contents to a string form.

The so presented XML-processing components (Figure 87) enable three different views on the PDU's contents, i.e. low-level for XML-to-object and vice-versa transformations at the Parser and at the Generator, middle-level view for validation and debugging purposes and high-level view for the E2ENP FSMs and the application. The middle-level view enables customised application- and protocol-specific validation of the XML contents that is different from the classical XML validation. The classical validation proves all the XML elements in a document for correctness, while the customised validation proves only those elements that are relevant for the application. Consequently, processing acceleration can be achieved.

Each specific Parser and Generator started by the application is implemented as a singleton object (see Section 5.1) to guarantee the unique applicability of the correspondent components, i.e. one Parser and one Generator dedicated for a specific session-description language are re-usable objects and are needed only once in the system. Processing parallelisation can also be achieved through running of multiple singleton Parsers and Generators of the same type. The number of Parser/Generator pairs of the same type run in the system simultaneously depends on the number of simultaneously runnable Carriers and FSMs. The E2ENP UA system discussed in this work represents only one negotiation role per instance of the system, i.e. one Parser, one Generator, one Carrier and one FSM role are simultaneously runnable. However, the configurability of the system would allow future multiprocessing versions of the E2ENP UA (see also the discussion about the carrier application in Section 5.2.2.3).

Further details of the API implementation of the XML processing components for E2ENP can be found also in the E2ENP UA patent application [170].

The model for processing of XML contents in terms of a protocol like E2ENP is also reused for another dialect of SDPng based on MPEG-21 DIA (i.e. SDPng++). This dialect was published in co-authorship by the author of this book in [96][97][202][221].

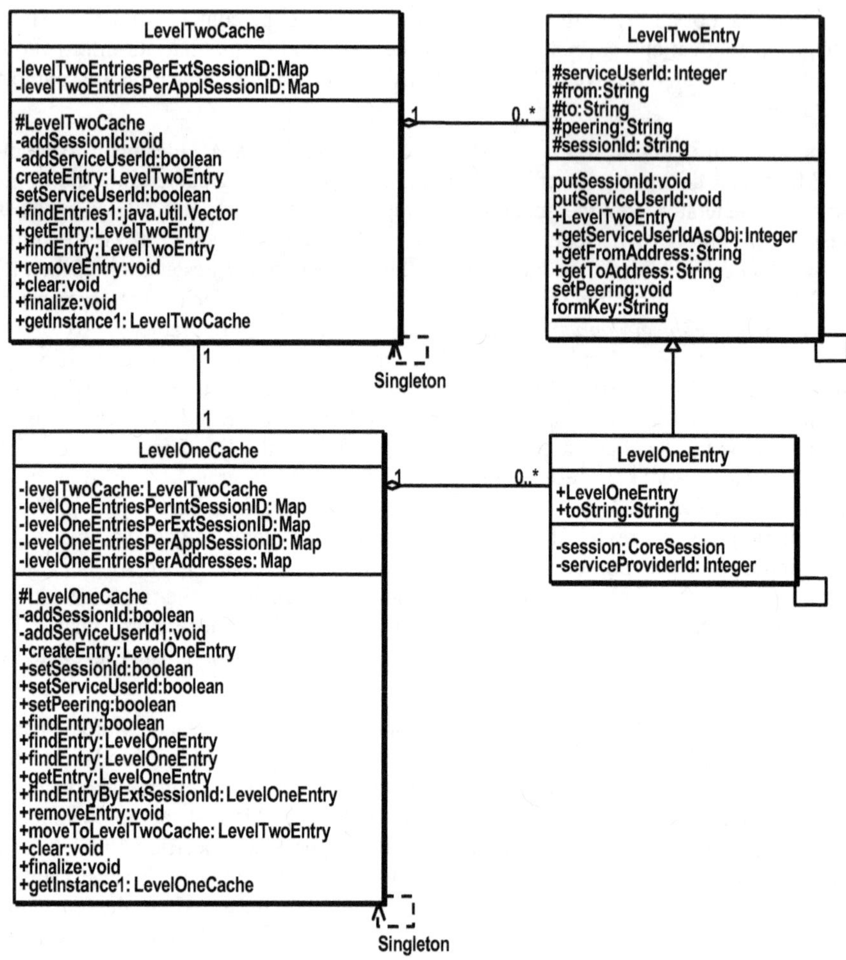

Figure 91: PDU caches

5.2.2.2 PDU Caches

The E2ENP caches (Figure 91) represent a data-structure for saving of the object identifiers associated with the PDUs (see also beginning of Section 2.3 and Sections 2.3.2.1.1, 3.1.4, 5.1 and 5.2.1.1). This data-structure is applied for the negotiation procedures at PDU-management and at multimedia-session-control actions. The part of the PDU memory associated with the PDU contents is a part of the application (see also beginning of Section 2.3), as the E2ENP UA does not understand any other contents of the PDUs but only their headers as identifiers (see Section 4.3.2). The PDU headers are used for generating of the E2ENP-session identifiers (see below) that are *unique* for the complete E2ENP-based system.

The caching concept of E2ENP is similar to the multi-level-caching paradigm [272] used in the modern CPUs for optimising the access to the memory. However, the purpose of the *LevelOneCache* and the *LevelTwoCache* is slightly different. The *LevelOneCache* represents the memory for PDUs used only during the negotiation process to detect inconsistencies at negotiation and to resolve deadlocks at collisions of applications (see Sections 2.3.2.1.1 and 5.2.2.4.1). The *LevelTwoCache* is the long term memory of the application used at referencing of control data (e.g. at short Re-negotiations) and applicable to resolve inconsistencies in cases of lost or malformed PDUs. A PDU can be lost or malformed due to unreliable transport or due to misbehaviour of some of the negotiating peers resulting in failure generation at the *LevelOneCache* and not saving of the PDU in *LevelTwoCache* (see below). A peer missing its negotiation results after a time-out or receiving malformed data may try to repeat the negotiation process to achieve consistency. Alternatively, at subsequent referencing of an inconsistently saved PDU, the requesting peer will notice the absence of a referenced PDU at its partner, as the partner will generate unavailability failure. Consequently, the peers will try to repeat the lost information in an additional negotiation round, thus, assuring the consistency of the long-term saved data.

The application-specific identifier of a PDU is decoupled from the E2ENP UA identification scheme used within the FSMs. The E2ENP engine is the one responsible for the cache management as a provider of the caching service. Correspondingly, the application accesses the caches through a *serviceUserId* (i.e. the application uses the services of the E2ENP UA) and the FSMs access the caches through the *serviceProviderId* (see the respective methods and variables in Figure 91). The *sessionId* is then the full identifier of a PDU that contains the PDU-header data and it is accessible over its short identifiers *serviceUserId* and *serviceProviderId.*

The caches keep two types of data according to the fact if this data is the short-term or the long-term memory of the application (see Section 3.1.4):

- The pre-cached data is marked with *serviceProviderId* as primary key and with *serviceUserId* as secondary key during the execution of an E2ENP negotiation process. The indexing of the pre-cached data is accomplished in the E2ENP UA that takes the *serviceUserId* of the managed application and generates correspondingly a *serviceProviderId* and a *sessionId.* Once a negotiation round succeeds, the pre-cached data moves to the permanent cache and can be used within subsequent E2ENP negotiations. The pre-cached data is modelled as an object named *LevelOneEntry* and this entry is saved into a data-structure named *LevelOneCache* (see Figure 91).
- The cached data is the long-term memory for PDUs shared between the application and the E2ENP UA. This memory is proven by the E2ENP FSMs at any following negotiation round that references already known sessions to achieve consistency of the negotiated and memorised information. The verification of the state of a PDU is done using the *sessionId* of the respective cache entry. Consequently, the *sessionId* is associated with information about PDUs that are in Zombie state but that are still in use in a multimedia session (see Section 3.1.4). Each record in the long-term cache is leased between the negotiation partners and it can be refreshed according to the given rental conditions signalled between the partners (see Sections 3.1.4, 4.3.2 and 4.3.7). The cached data is modelled as an object named *LevelTwoEntry* and this entry is saved into a data-structure named *LevelTwoCache* (see Figure 91).

The PDU-identifier databases *LevelOneCache* and *LevelTwoCache* are implemented using the *java.util.TreeMap* class that is accessed over the *java.util.Map* interface (see also Section 3.1.4). The caches provide methods for accessing, searching, manipulating and removing of the cache records at PDU-management and multimedia-session-control actions. The two levels of the E2ENP memory are implemented as singleton objects (see also Section 5.1) that are maintained by the E2ENP UA. For reasons of management simplification of the caches, their objects run within the instance of the E2ENP UA and display a synchronous interface towards the FSMs.

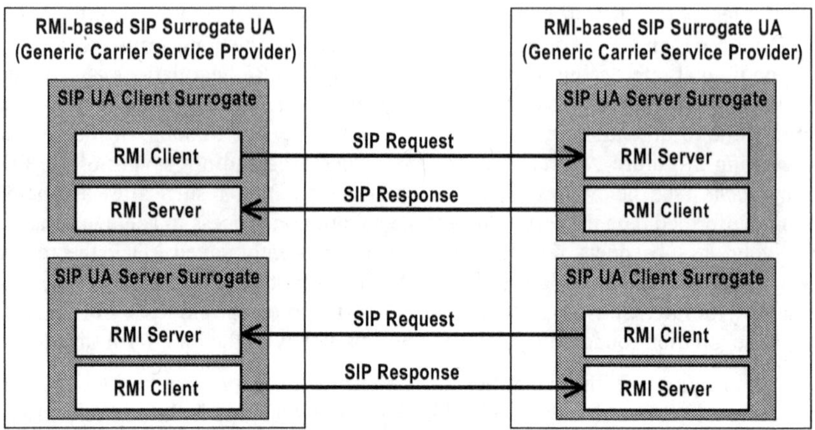

Figure 92: RMI-based SIP surrogate

The caches aggregate correspondingly the *LevelOneEntry* and *LevelTwoEntry* objects. *LevelOneEntry* inherits from *LevelTwoEntry,* since the two kinds of entries contain similar information, but the *LevelOneEntry* includes also temporary information about the running session. The information contained in the both records is the information about the PDU and its associated E2ENP and SIP sessions. The information contained in the *LevelOneEntry* only is temporary information used from the E2ENP UA to identify the PDUs and the sessions univocally until the corresponding application identifiers (e.g. *serviceUserId*, *sessionId*) are generated.

Further details of the cache APIs are also presented in the E2ENP UA patent application [170].

5.2.2.3 Data Carriers

The Session Initiation Protocol [62] is chosen as a representative carrier for the E2ENP PDUs (see Sections 3.1.3 and 3.3.1.2). Multiple open-source and freeware applications of SIP were tested for suitability for integration into the E2ENP UA architecture, e.g. jSIP [184], NIST SIP [185], BT SIP [273] and BonePhone [274]. Furthermore, two additional SIP implementations are also considered for future E2ENP tests but are not further regarded in this work, i.e. oSIP [275] and TTsuite-SIP [276].

The jSIP [184] and the NIST SIP [185] implementations proved to be most suitable for integration into the E2ENP UA. They display the user-provider behavioural schema applied in the E2ENP UA model (this model is discussed in Section 5.1), i.e. jSIP and NIST SIP provide transport services in form of an integrated provider module and the E2ENP UA is the user thereof. However, the cross-test between the jSIP and the NIST SIP User Agents showed that the two implementations are incompatible. The SIP specification [62] prescribes UDP [123] as default carrier for SIP and message repetition (i.e. similar to the *at-least-once* communication semantic [39][50][52]) and/or DiffServ-like QoS assurance for the SIP messages for unreliable transport. However, the implementations of SIP do not stringently follow these prescriptions. Correspondingly, the disregarding or the misinterpretation of the SIP prescriptions for message management in lossy networks leads to incompatibilities of the SIP UAs. jSIP implements only *at-least-once* communication model as prescribed for UDP and NIST SIP applies only *at-most-once* model as prescribed for TCP (see also SIP [62]). This leads to synchronisation problems at communication of the Offerer and the Answerer implemented with two different SIP UAs and the result can be inconsistency in the PDUs exchange (see also Section 5.2.2.2). However, the SIP Stacks (i.e. the language interpreters) of both jSIP and NIST SIP are compatible.

Considering the implementation problems with SIP, the first E2ENP UA implementation was developed using an RMI-based [277] SIP surrogate to guarantee simple and stable implementation of the carrier for E2ENP UA debugging purposes and for initial E2ENP UA tests [65], [131] – [133] (see also Chapter 6). However, RMI (as based on TCP) is generally unsuitable for networks, where transport interruptions may occur, as the interruption and the re-invoking of the TCP threads is a very expensive process [180]. The RMI-based SIP surrogate (Figure 92) imitates the SIP UA behaviour over an accordingly defined asynchronous API that simulates SIP methods and responses, i.e. this API applies partially the JAIN SIP API [178]. The SIP-UA client is implemented as a combination of an RMI-client and an RMI-server application. The RMI client at the SIP-UA surrogate-client invokes the SIP requests. The RMI server at the SIP-UA surrogate-client receives the SIP responses. The SIP-UA surrogate-server is also a combination of an RMI client and an RMI server. Correspondingly, the RMI server at the surrogate-server receives the SIP requests and the RMI client responses to these requests. The two RMI client-server couples enable thus the simulation of asynchronous communication as defined for SIP. The two RMI client-server couples are integrated in an adapter application that imitates the SIP UA behaviour and that represents the Generic Carrier Service Provider defined for E2ENP (Section 5.2.1.1). The adapter exposes towards the E2ENP UA the predefined interface for SIP communication (i.e. IF3 defined in Section 5.2.1.1).

The implementation of the Generic Carrier Service Provider with jSIP and NIST SIP is also done applying an adapter application that converts the jSIP and the NIST SIP communication APIs (i.e. the full or the partial implementations of the JAIN SIP API [178] within jSIP and NIST SIP) to the predefined API (i.e. IF3) for E2ENP (see also Sections 3.1.3 and 3.3.1.2). Furthermore, the Generic Carrier Service Provider is a practical solution that enables a configurable artificial delay for the SIP UAs to adjust them to the performance of different networks. The SIP specification [62] does not consider that SIP may be applied in different kinds of networks. Consequently, the implementations thereof are in their basic performance unsuitable for low-throughput and mobile networks. The debugging tests of the jSIP and NIST SIP showed that the UAs (especially the NIST SIP UA) are designed to operate predominantly under conditions of local area networks (LAN) and that the UAs start to misbehave (to block or to generate time-out failures) as soon as they are launched within slower networks, as the jSIP and NIST SIP User Agents are implemented with fixed time-outs. The jSIP and NIST SIP implementations allow the access to the SIP UA state machines on the level of states (see Section 2.3.1.2) to launch the respective SIP messages over the SIP UA provider API and to send SIP responses towards the application (the respective communication messages are sent asynchronously). The Generic Carrier Service Provider uses this mechanism to introduce artificial delays (configurable thread blocking of the SIP UA state machines) at launching the corresponding methods towards the SIP UA and towards the E2ENP UA thus stabilising the performance of the UAs in low-throughput networks. In the current implementation of the Generic Carrier Service Provider, the artificial delays are configured manually, as data about the necessary time-outs in different networks are still not available (see also Section 6.4). If having such data, future versions of the Generic Carrier Service Provider and of SIP UAs may apply automatic adaptation procedure for different network technologies.

All SIP-adapter applications (i.e. for RMI-based SIP surrogate, for jSIP and for NIST SIP) implement a configuration procedure (see Sections 5.1 and 5.2.1.1) to guarantee that the SIP UA is run as a singleton application to enable the coordination and the synchronisation as defined for E2ENP (see Section 2.3.2.1.1). Currently, the E2ENP UA guarantees the exclusive access of the multimedia applications to just one SIP carrier (see also Sections 3.1.3 and 5.2.2.4.1). Parallel implementation of multiple SIP stacks is also possible due to the configuration mechanisms of the E2ENP UA (see also the respective discussion about the Parser/Generator implementation in Section 5.2.2.1). However, the tested implementations of SIP are not thread-save and contain persistent components which may cause coordination and synchronisation problems at simultaneous running of multiple SIP UAs in one application. This problem with SIP has a historic background, as SIP was initially specified for just VoIP applications, where each application runs independently of the other ones. Correspondingly, current SIP applications that follow the initial SIP concept may not be suitable for multitask middleware as considered for E2ENP and such SIP realisations need to be re-engineered for multi-application purposes. The latest SIP specification [62] does not explicitly exclude multipurpose and parallel implementations. Hence, future development of SIP implementations has to consider also middleware applications.

5.2.2.4 Finite State Machines

This section discusses the state machines implemented within the E2ENP UA. The state machines are designed using UML State Charts and Activity Diagrams [216][269]. The major difference between these charts is that the Activity Diagram provides the possibility to indicate incoming and outgoing signals that steer the state switches and it enables generalisation of states in form of activities. Thus, the Activity Diagrams are more suitable for expressing hierarchical state-machine performance compared to the simple State Charts that express the relation of the states on a single abstraction level.

The E2ENP UA is implemented in consideration with the fact that a more generic then SIP session management protocol might be in use. Respectively, state machines with different complexity and configuration requirements may apply. Consequently, the interface towards the application or the middleware (i.e. IF2 see Section 5.2.1.1) is asynchronous and loosely coupled to the respective *Service User* part of the interface. The state machines communicate with the application/middleware utilizing message queues. These queues guarantee the stability of the E2ENP UA performance at configuration of the UA so that no *Service User* tries to launch the E2ENP UA before the UA has reached a stable configuration state. Furthermore and in cases that a *Service User* blocks infinitely, thus blocking the complete communication for all other registered *Service Users*, the loose coupling guarantees that the E2ENP UA remains available for the rest of its users.

The execution of a communication session with the E2ENP UA is associated with an object representing the session. Such object contains the information related with the given state-machine instance, i.e. what phase is being launched (e.g. Pre-negotiation, Negotiation, Re-negotiation), what role the session has (e.g. Offerer, Answerer), what state the call has reached, etc. The capacity of the E2ENP UA to accept certain amount of sessions simultaneously is managed through a thread pool where each communication session corresponds to a single-thread execution (see also [170]).

The state machines presented in the following sections summarise the E2ENP performance at management of multiple communication sessions (Section 5.2.2.4.1), the application of E2ENP based on SIP for an Offer/Answer Model with resource-reservation coordination (Section 5.2.2.4.2) and the application of E2ENP based on SIP for an external-action model that is suitable for session mobility (Section 5.2.2.4.3).

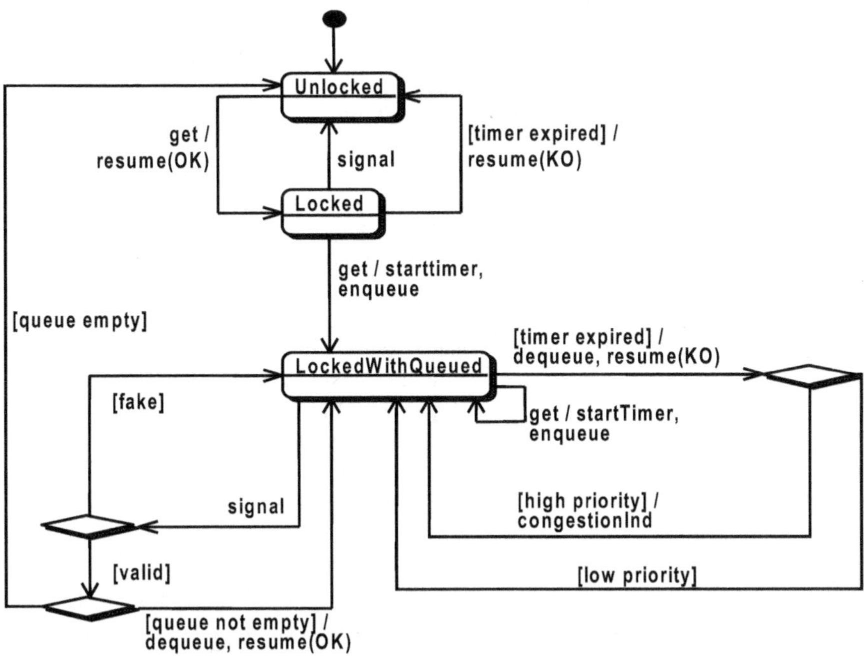

Figure 93: E2ENP UA coordination MUTEX

5.2.2.4.1 Priorities-based MUTEX

The MUTEX (Figure 93) is the mechanism that enhances the basic SIP state-machines (see Section 2.3.1) with multi-application-management capabilities, like the priorities-based resources-reservation coordination. The MUTEX provides transaction-like processing of the resource-reservation coordination performed by the peers, enabling also failure management and deadlock resolution at collisions (see also Sections 2.3.2.1.1 and 3.1.3). Consequently, the MUTEX implements not only mutual exclusion of concurrent services but it is also offers scheduling and garbage collection to its applications (see below).

An interaction with the MUTEX is indicated in the subsequent text with *_ResvMux* and this indication is shown in all of the following E2ENP UA state diagrams that are associated with the specific negotiation scenarios (Sections 5.2.2.4.2 and 5.2.2.4.3). The E2ENP scenario-specific state machines communicate with the MUTEX applying the primitives *get* and *signal*. With *get* the state machine requests to obtain processing time for a period specified within *get*. The *signal* releases the MUTEX lock, respectively it releases a task locked in the queue for starting its negotiation processing over the network (see the paragraph below). The specification of a period for acquiring the MUTEX enables the MUTEX to forcibly stop misbehaving applications, generating a failure indication in case the application does not succeed with its processing within the limits of the specified time. Correspondingly, the tasks in the MUTEX are started/stopped explicitly, using the *resume* primitive, where the parameters *OK* or *KO* indicate the ability of a task to continue processing or the necessity for forcible clear-up (see below). The getting of the MUTEX for Pre-negotiations and other non-critical transactions is done directly. If such transactions cannot succeed to obtain processing time, they may simply be repeated according to the requirements of the respective application. Even the leasing of PDUs is not critical as the application is in position to set expired PDUs in Zombie state and wait until it gets a negotiation slot (see Sections 3.1.4 and 5.2.2.2). The transactions associated with resource-reservation coordination (i.e. Negotiation and Re-negotiation phases) *book* the MUTEX (see also Section 3.1.3). The booking enables such transactions to try repeatedly to *get* the MUTEX, before they end with a failure notification and the respective handling is returned to the application logic. The getting of the MUTEX is associated with the restricted size of the MUTEX queue, respectively with the restricted number of threads within the thread pool of the E2ENP UA and the restricted number of active tasks waiting for processing in the MUTEX queue. This restriction is necessary as the negotiation engine and the respective network communication for negotiations represent also application resources that shall be coordinated together with the rest of the system resources. The limitation of the number of tasks actively waiting for negotiation corresponds to the limitations of the system at acquisition of CPU, memory and network-device capacity resources for processing a negotiation round. Note that at getting of the MUTEX each negotiation task is associated with a thread from the MUTEX pool. Correspondingly, the size of the MUTEX queue has one position less then the size of the thread pool. The one additional thread of the pool is the actively working task in the MUTEX (i.e. the application that makes its negotiation round at that point of time).

If there is no communication session pending on the MUTEX, the MUTEX is in the state *Unlocked* (Figure 93). If an application succeeds to *get* and to lock the MUTEX (state *Locked* – Figure 93), its session is directly processed when there is no other equal or higher priority task actively working, otherwise the session is blocked and its task is queued (i.e. *LockedWithQueued* state in Figure 93). The MUTEX artificially increases the priority of the currently actively working session (compared with the queued sessions), in order to avoid the resetting of the state of the session each time a higher priority task tries to acquire the MUTEX. The actively working tasks are not stopped to avoid resource re-setting (except when they are affected by timer expiry), since the application of SIP and the associated *economy principle* are too expensive to allow intermediate interruptions (see also Sections 2.3.2.1.1 and 3.1.3). Correspondingly, the priorities of the sessions are valid only for the MUTEX queue to achieve the priority-based handling. If a session cannot finish within the specified period declared at MUTEX getting, the MUTEX indicates the application about the occurred failure. The MUTEX throws *congestionInd* (Figure 93) that is mapped by the E2ENP UA to a *failureInd* dedicated to the corresponding *Service User* (a negotiating application). However, if the expired task is still in position to continue, it is not pre-empted by another task automatically. The MUTEX starts an additional timer within the period of which the affected application (affected from the initial timer-expiry) can start a clear-up process (i.e. send *_ResvMux.signal*) and return to a previous stable state or it can decide to continue. In case that the MUTEX receives no response from the application after the second timer is expired, the MUTEX clears the corresponding tasks forcibly. Due to the timer-expiration problem associated with the MUTEX the Negotiation and the Re-negotiation phases are designed as history states (see Section 5.2.2.4.2). This gives the application the possibility to roll back to a previous stable state associated with a moment before the application's timer expired in case that the Negotiation/Re-negotiation is cancelled and the media session cannot start or adapt. The decision which task continues after a time-expiration event is not made by the MUTEX, as the MUTEX has no notion about the applications' task prioritisations and the respective system conditions. Hence, it is supposed that the application takes the resolution decision applying an application-internal arbitration unit [170]. The MUTEX provides such arbitration unit only with information about the state of an occurred congestion and with a mechanism for resumption of an expired task. A simplified version of the arbitration unit is implemented for test purposes within the E2ENP test suite (see Section 5.1), which version works with statically predefined priorities for a fixed test scenario (see Section 6.2).

The tasks are ordered in the MUTEX queue according to their priorities. If the sessions have equal priorities, they are scheduled in a first-in-first-served order (corresponding to sequence of their arrival at the MUTEX). When an active task ends, the MUTEX re-arranges the queue and ejects those tasks which timer is expired, indicating the respective *Service Users* about the occurred problem (see also above). When the MUTEX retrieves a new task from the queue for active processing, it proves its validity (i.e. the specified processing/waiting time for the task is proven), if this time is still valid the task is processed (guard condition *valid* – Figure 93), otherwise a new queue re-arrangement takes place (guard condition *fake*). The efficiency of the queue re-arrangement vs. the active processing time of the MUTEX is proportional to the number of tasks specified for the MUTEX queue, i.e. if the MUTEX allows a larger queue the re-arrangement time may take precedence. This problem is avoided through the configurability of the MUTEX queue, respectively through the configurability of the thread pool. For the measurements in Chapter 6, a configuration with a dozen of threads in the pool is selected. The initial debugging and performance tests of the E2ENP UA provided the information about the suitability of this selection. Furthermore, the initial tests delivered the times necessary for the applications' timer-configuration at negotiations and for the MUTEX operations.

The MUTEX is also used in combination with the PDU caches (Section 5.2.2.2) for resolving of double seizures of E2ENP calls, i.e. the distributed management of the applications' access to the negotiation environment. As the caches store information on the *From*/*To* SIP-header couples and additional information from the respective PDU headers, this information is used to detect the collision (see Section 2.3.2.1.1). The E2ENP UA can allow processing of the concurrent calls in a push-pull mode, where the winner of the bully process for obtaining the signalling slot makes its call first and the looser follows with its call after the winner has finished (see also Section 2.3.2.1.1). The E2ENP UA decides who is the winner through evaluating random fields from the PDU header (e.g. *"session-id"* attribute, see Section 4.3.2) at *booking* the MUTEX for lock granting. However, the E2ENP resolution mechanisms should not interfere with the SIP-own resolution process, where at similar collisions *500 Server Internal Error* [62] can be sent. The solution of this problem is provided through the E2ENP configuration mechanism, where the application is allowed to configure the E2ENP engine for managing conflict resolutions. Consequently, the application may also apply its own congestion-resolution process (as discussed above) or in combination with the SIP-native mechanism.

The current E2ENP MUTEX is designed to administrate the access of multiple applications to a single SIP UA implementation. Such solution enables uniform service coordination using SIP. The current SIP specification and implementations of the SIP state machines allow only a single application per SIP stack [62]. Consequently, in a native SIP environment alone, the global coordination of applications that share system resources is not possible. The E2ENP MUTEX mechanism can be extended further to enable regulated access to a shared pool of SIP UAs, applying simultaneous access to multiple locks which number corresponds to the number of the available SIP UAs.

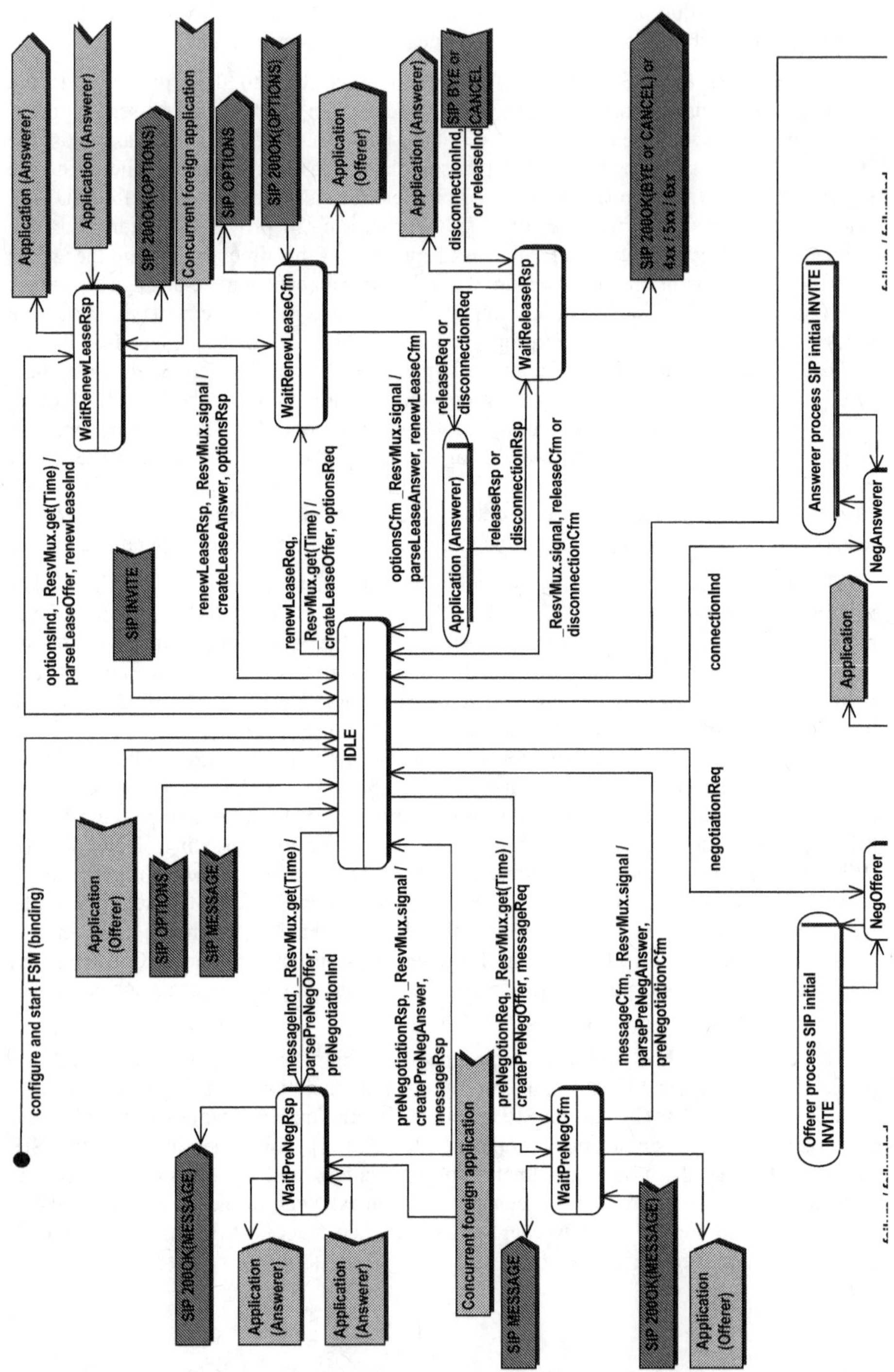

Figure 94: E2ENP UA Offerer/Answerer root state (A)

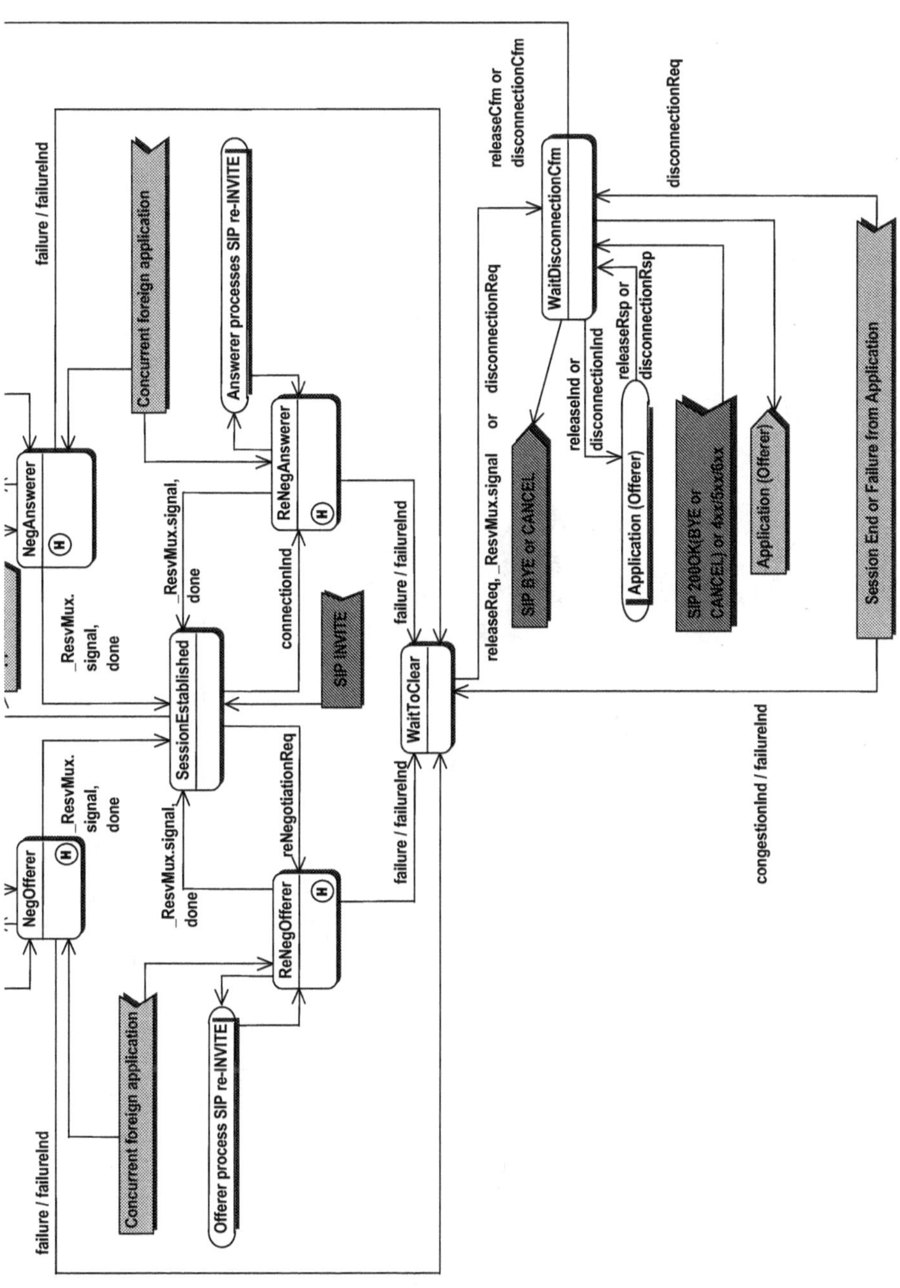

Figure 94: E2ENP UA Offerer/Answerer root state (B)

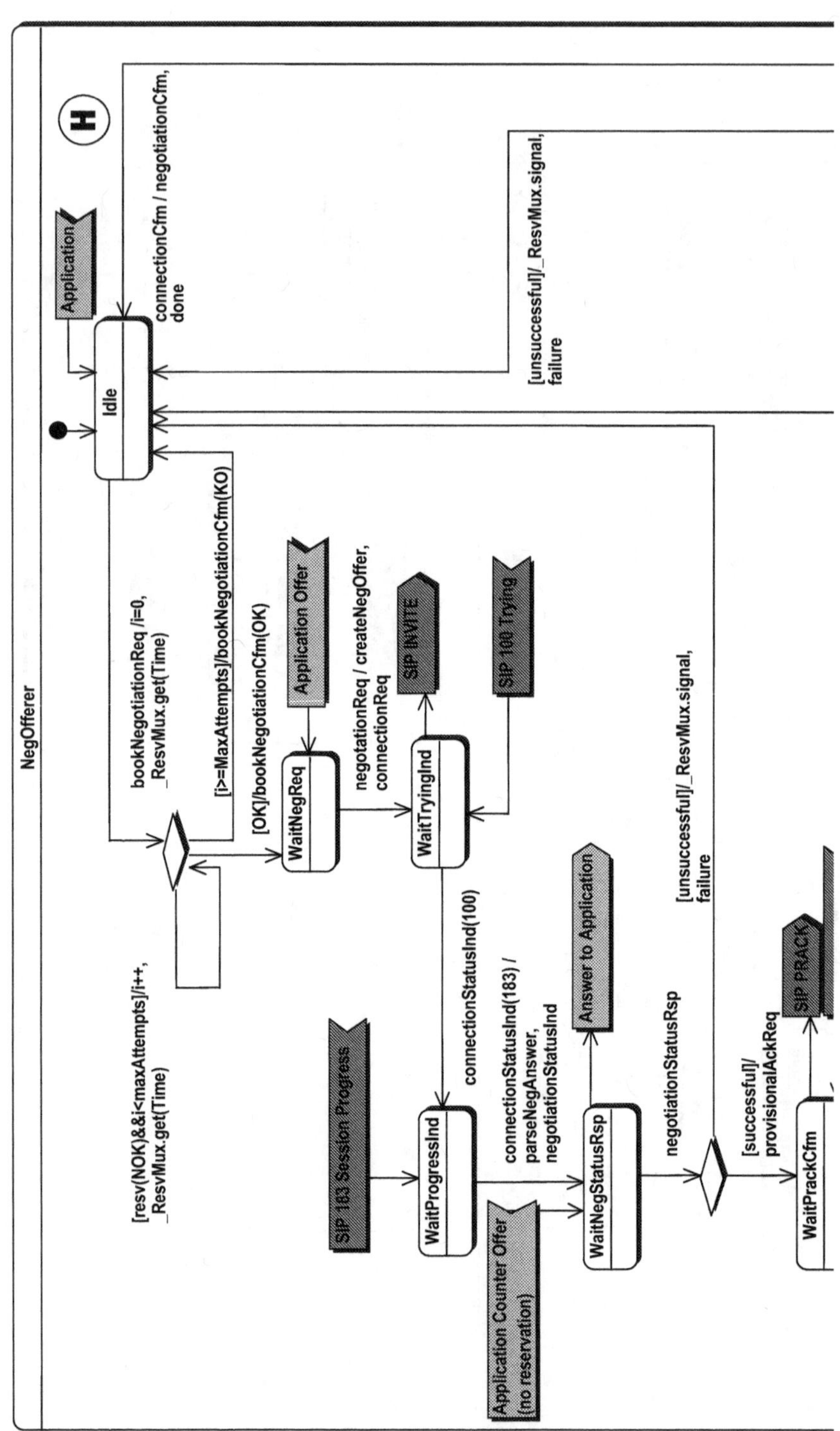

Figure 95: E2ENP UA Offerer-Negotiation history state (A)

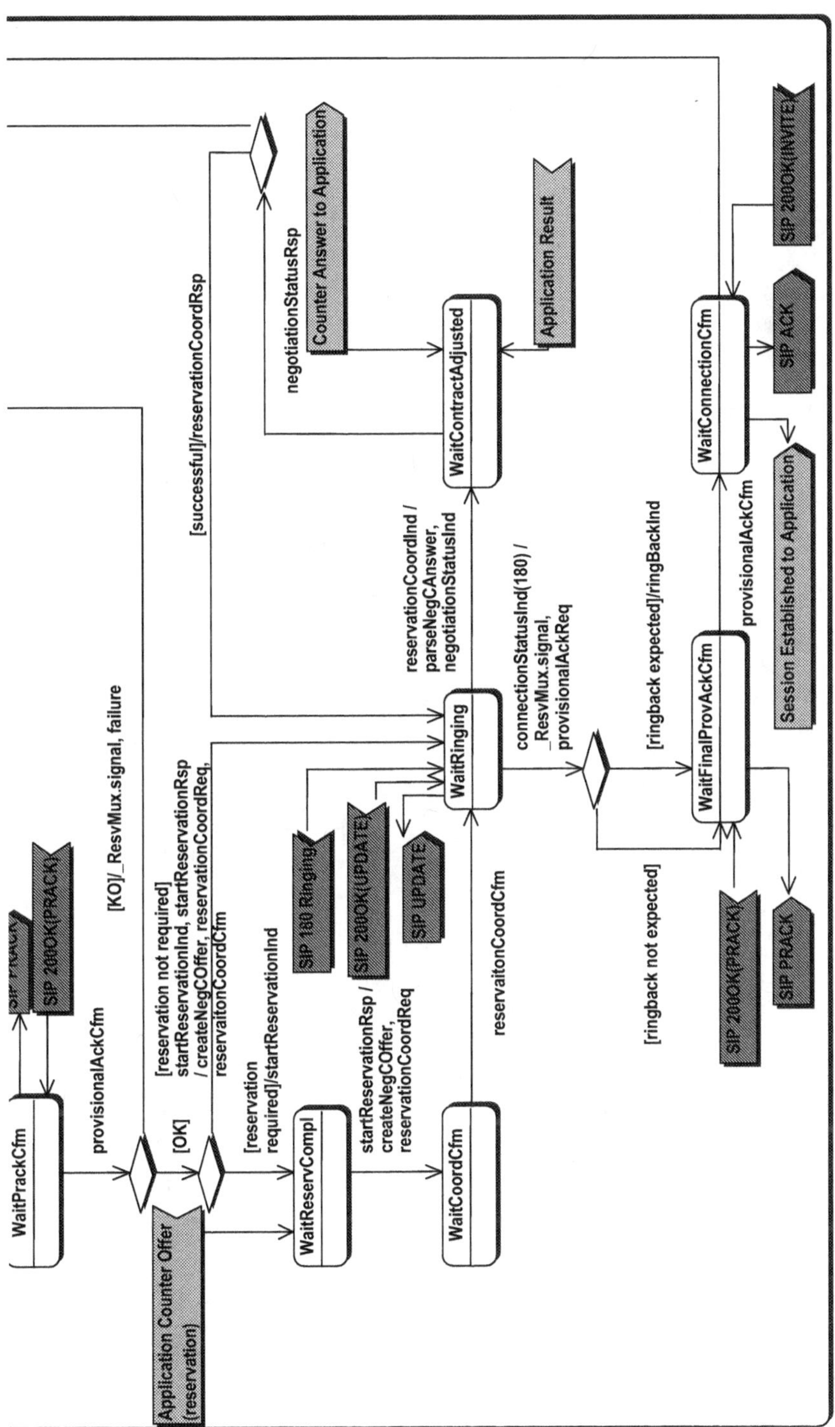

Figure 95: E2ENP UA Offerer-Negotiation history state (B)

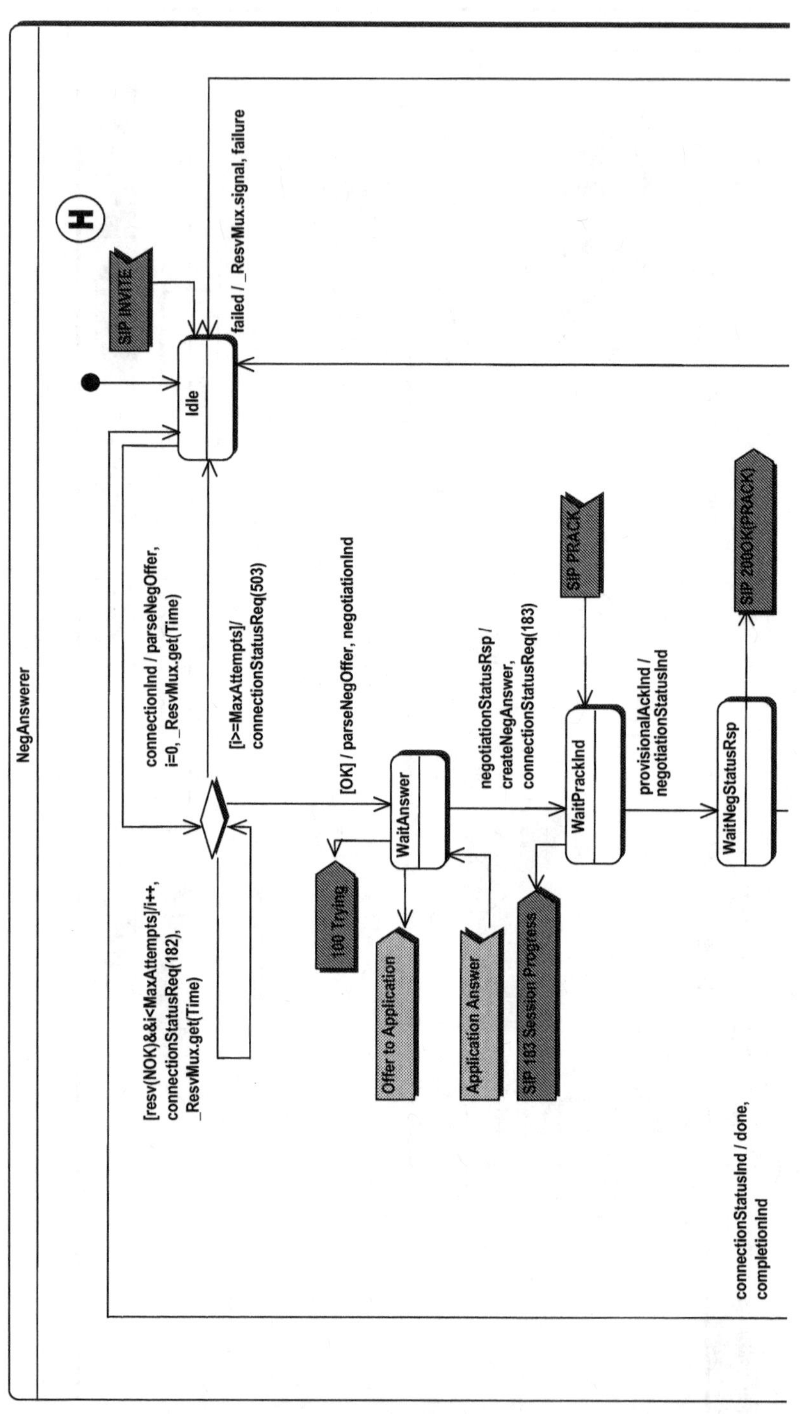

Figure 96: E2ENP UA Answerer-Negotiation history state (A)

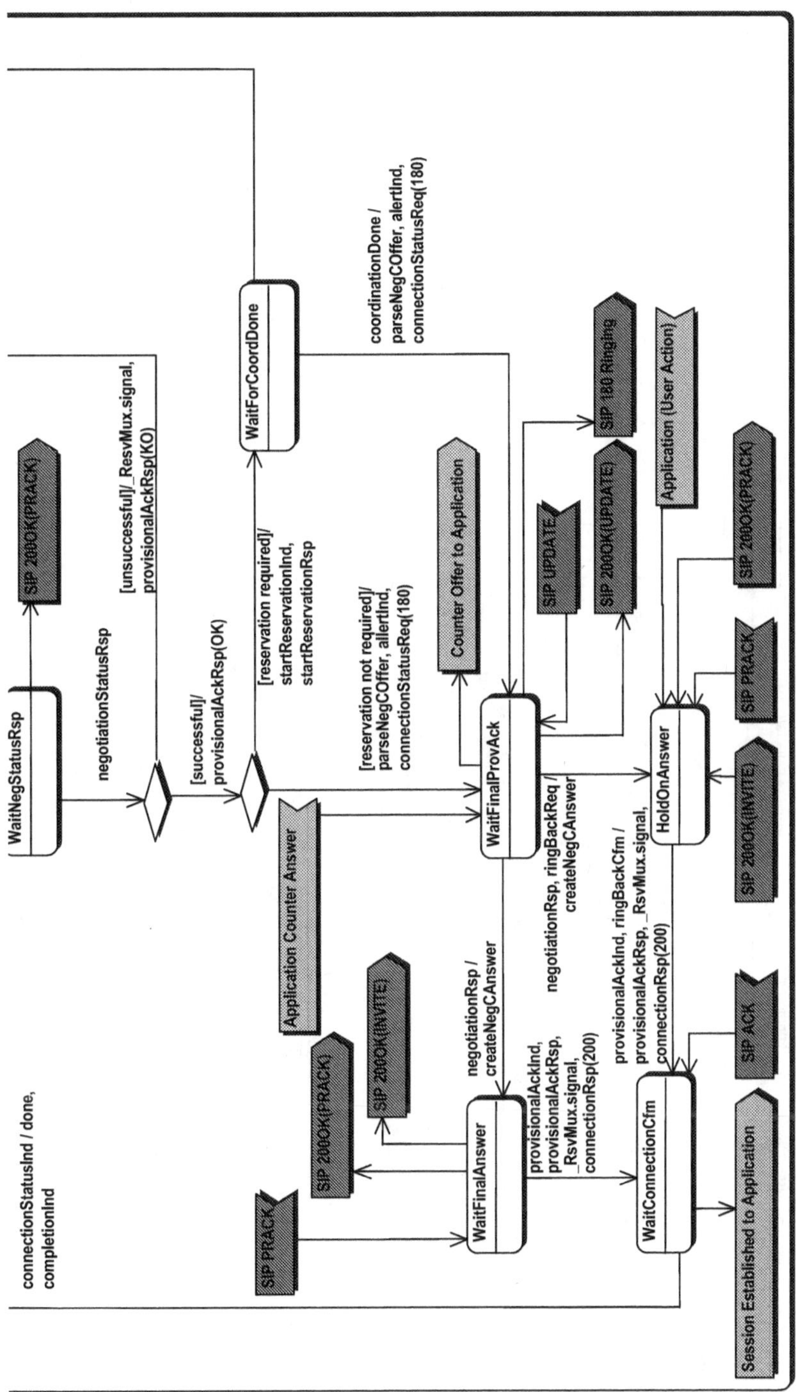

Figure 96: E2ENP UA Answerer-Negotiation history state (B)

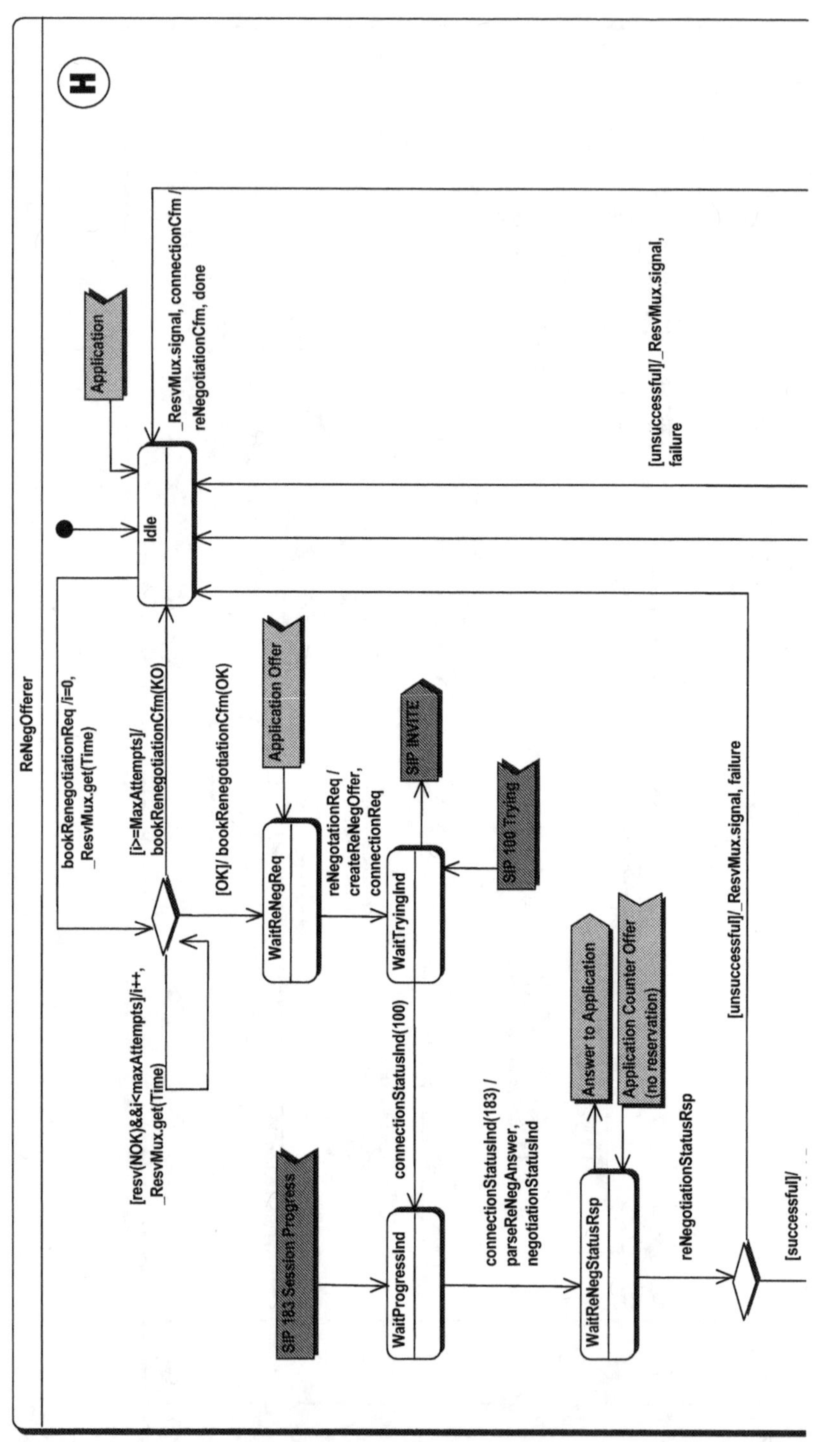

Figure 97: E2ENP UA Offerer-Re-negotiation history state (A)

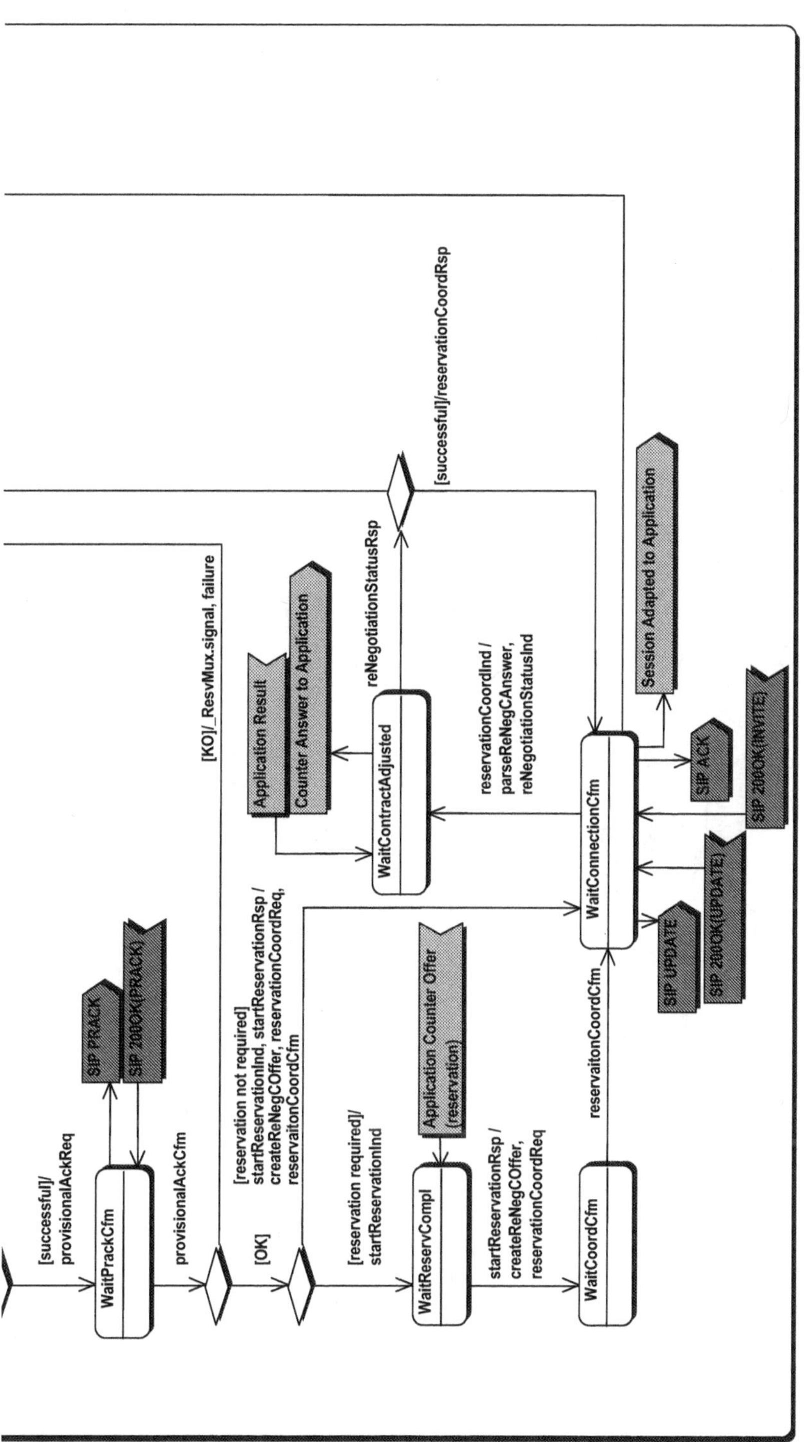

Figure 97: E2ENP UA Offerer-Re-negotiation history state (B)

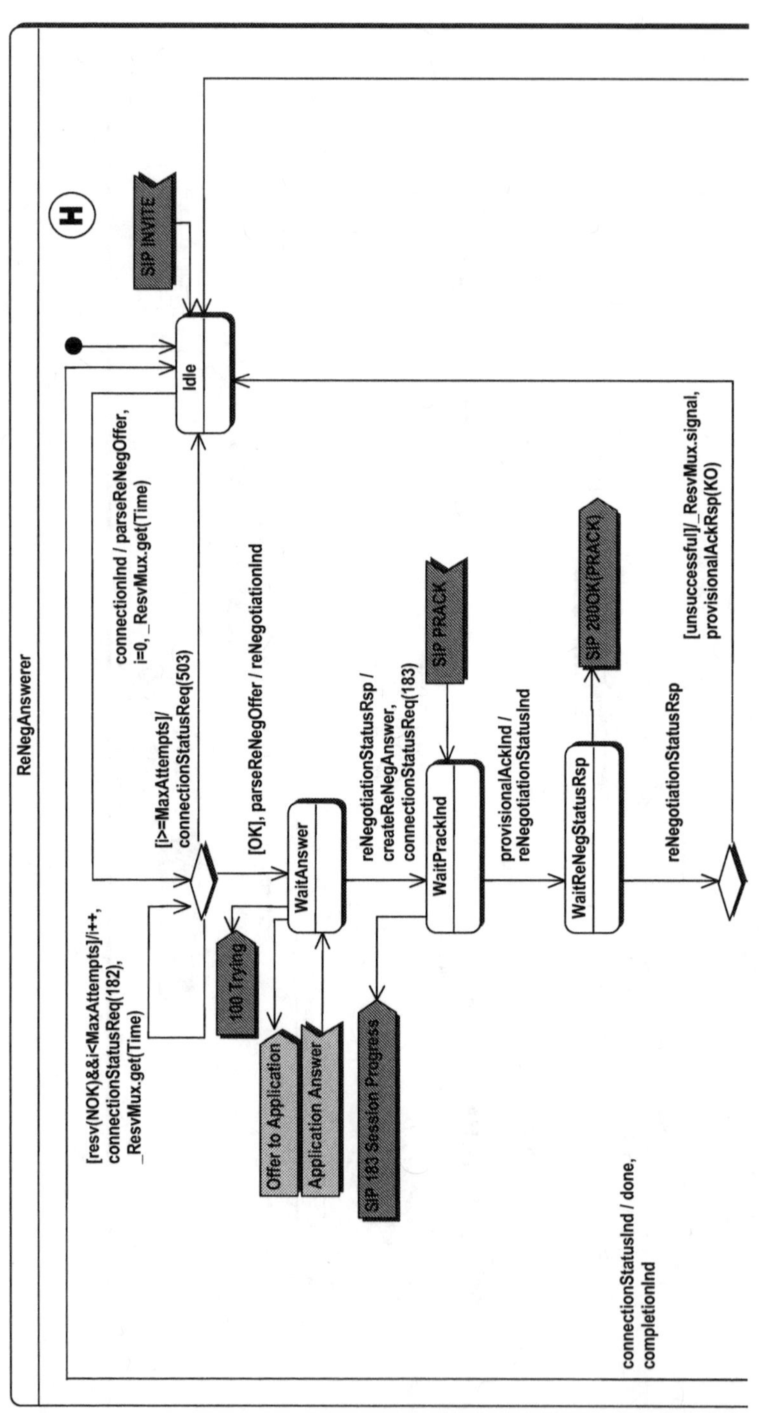

Figure 98: E2ENP UA Answerer-Re-negotiation history state (A)

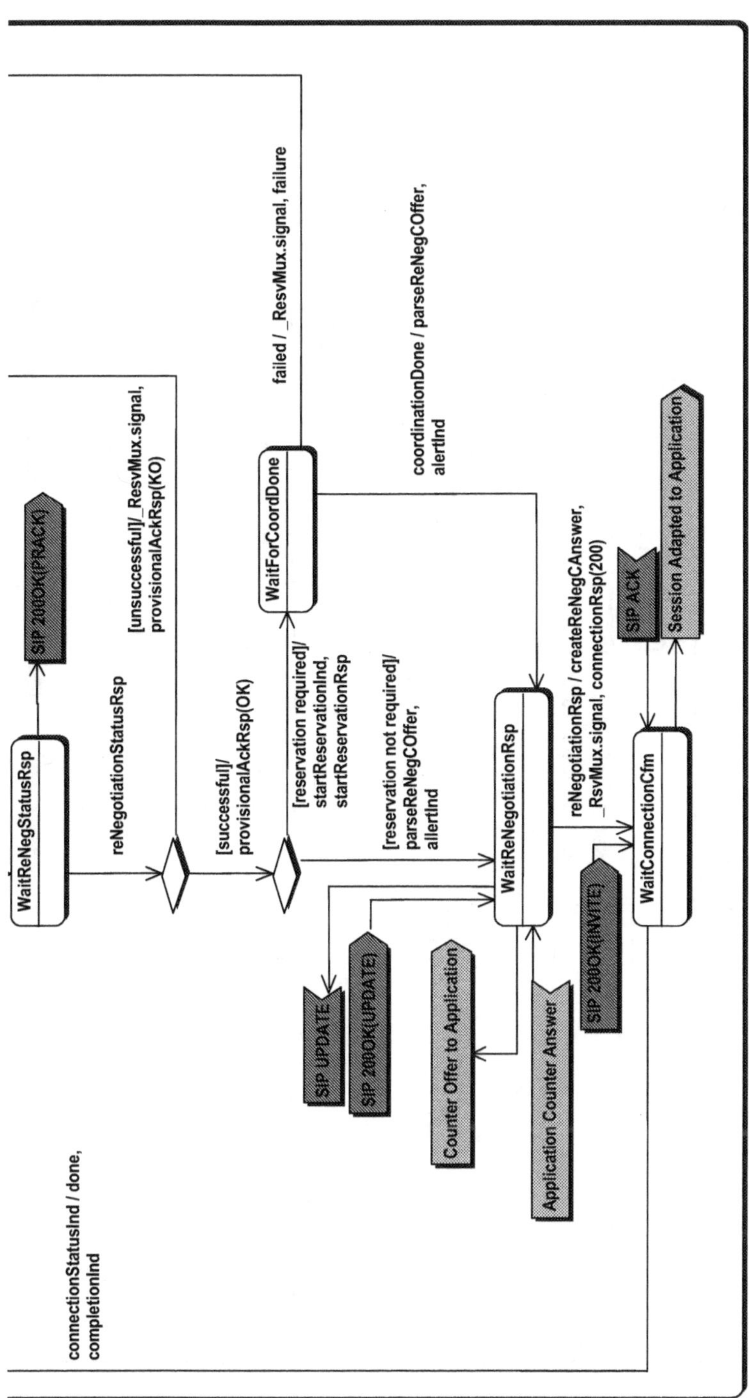

Figure 98: E2ENP UA Answerer-Re-negotiation history state (B)

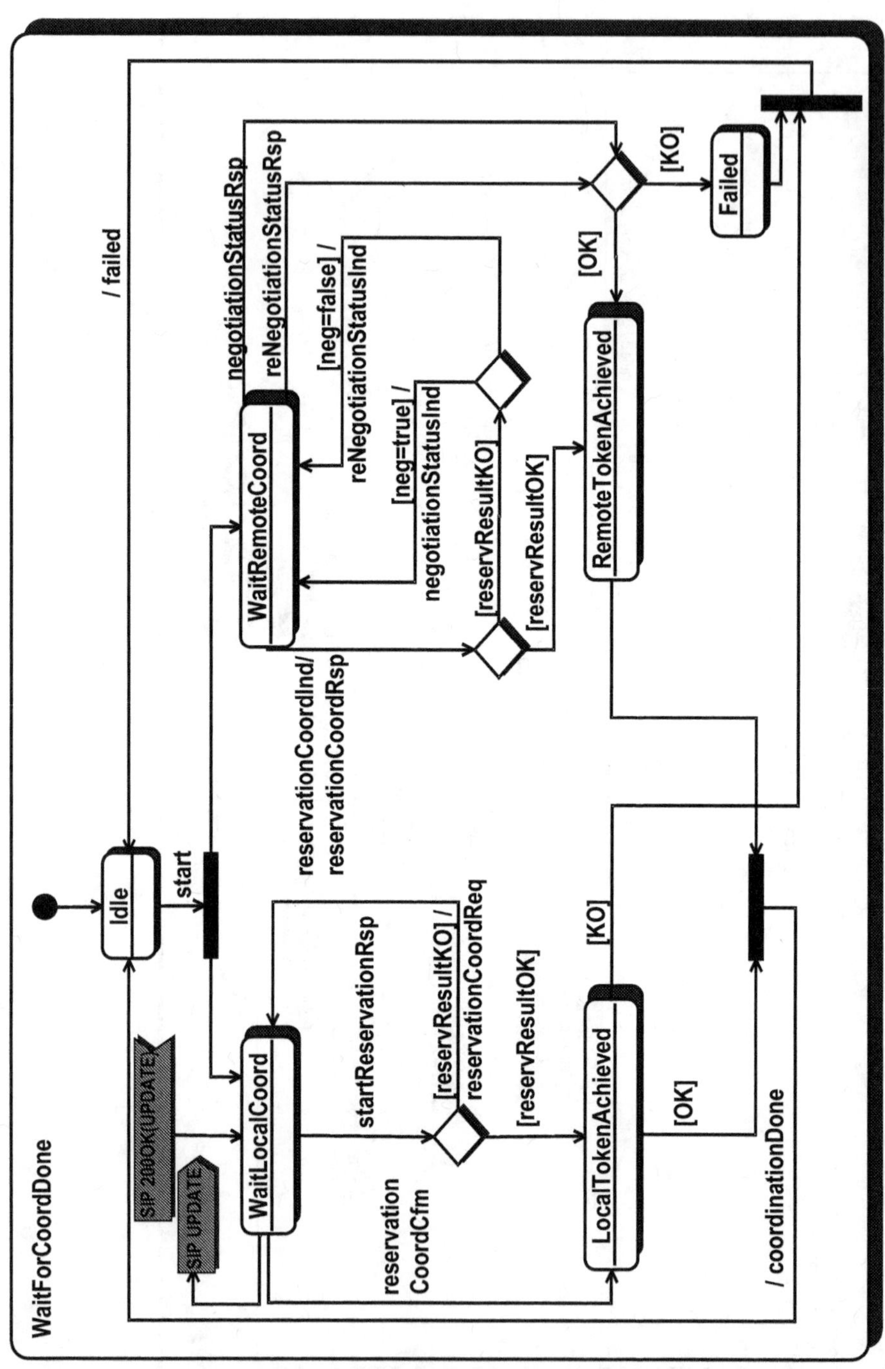

Figure 99: Answerer Sub-state *WaitForCoordDone*

5.2.2.4.2 Offerer-Answerer Finite State Machines

The realisation of the Offer/Answer Model with resources coordination is shown in Figure 94 through Figure 99. The states expressed in a coupled figure display slight overlapping of the figures for the purpose of clarity, where:

- In Figure 94AB the overlapping is in the region of the history states *NegOfferer* and *NegAnswerer*
- In Figure 95AB the overlapping is in the region of the state *WaitPrackCfm*
- In Figure 96AB the overlapping is in the region of the state *WaitNegStatusRsp*
- In Figure 97AB the overlapping is in the region of the guard condition *[successful]/provisionalAckReq* that initiates the switch to the state *WaitPrackCfm*
- In Figure 98AB the overlapping is in the region of the state *WaitReNegStatusRsp*

The difference of the E2ENP Offerer/Answerer model with resources coordination compared with the model presented in [36] is that E2ENP always enforces the coordination of resources in order to cope with the different provider policies for enabling QoS guarantees that may occur in the access networks of the affected peers (see also Sections 2.3.1.2 and 2.3.2).

Since, the E2ENP API for the application of the Offer/Answer Model with E2ENP is already shown in Section 3.1.3, this section discusses only the meaning of the states at management of the E2ENP actions. Note that the *signal* states in all of the diagrams (Figure 94 through Figure 99) indicate the communication of the E2ENP FSMs with the respective hierarchical FSMs of the carrier UA and of the application/middleware.

Figure 94A shows the initial state of the E2ENP UA after it is successfully configured (state *IDLE*). The Pre-negotiations are indicated with the states *WaitPreNegCfm* (at the Offerer side of the E2ENP UA) and *WaitPreNegRsp* (at the Answerer side), where *WaitPreNegCfm* manages the sending of offers and the waiting for answers and *WaitPreNegRsp* – the reception of offers and the sending of answers to these offers. The leasing of PDUs is managed with the states *WaitRenewLeaseCfm* (at the Offerer) and *WaitRenewLeaseRsp* (at the Answerer). The application of these states is similar to the application of Pre-negotiation states. The Pre-negotiation and the Leasing states are also associated with the acquisition (*_ResvMux.get*) and the releasing (*_ResvMux.signal*) of the MUTEX as explained in the previous Section 5.2.2.4.1. The Negotiation and Re-negotiation phases are associated with the history states *NegOfferer* and *NegAnswerer*, correspondingly, with *ReNegOfferer* and *ReNegAnswerer* (the names of the states indicate the respective peer roles and phases – see Figure 94AB). The history states for Negotiation and Re-negotiation are detailed in Figure 95 through Figure 98 (see also below and Sections 3.1.3 and 5.2.2.4.1). A successful Negotiation/Re-negotiation clears the history of the respective states and the state *SessionEstablished* indicates the successful establishment/adaptation of a multimedia session. The *SessionEstablished* state (Figure 94B) is associated with the management of the running multimedia sessions and with the maintenance of the PDU caches. Furthermore, this state serves as an indication for the application and the E2ENP UA to differentiate between Negotiations and Re-negotiations. The failure conditions associated with the Negotiation/Re-negotiation are associated with the *WaitToClear* state (Figure 94B). The end of a running session (Figure 94B) is indicated with the *WaitDisconnectionCfm* (at the Offerer) and with *WaitReleaseRsp* (at the Answerer). The transition to the states *WaitToClear*, *WaitDisconnectionCfm* and *WaitReleaseRsp* and the releasing of the *SessionEstablished* state is managed in the hierarchical state machine of the application.

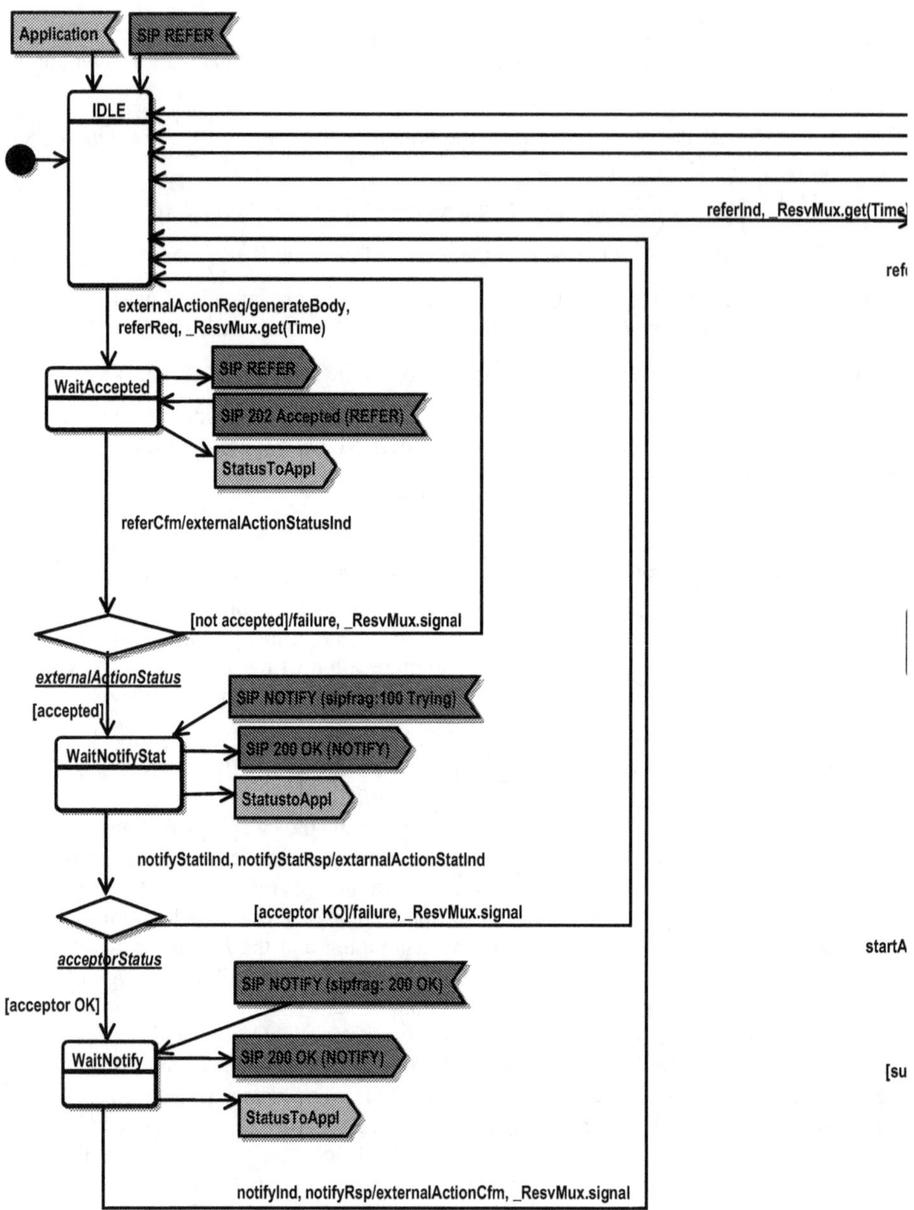

Figure 100: External action state machine
(A)

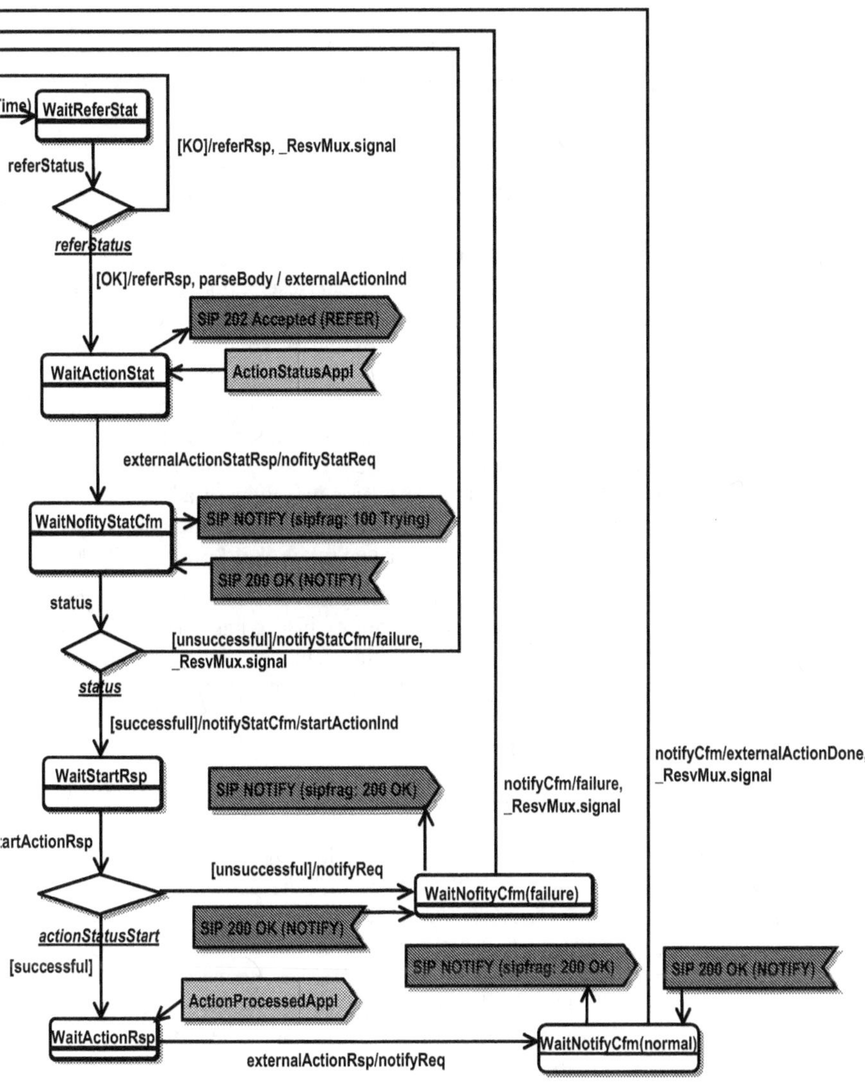

Figure 100: External-action state machine (B)

The states of the Offerer for Negotiation and Re-negotiation are similar (i.e. *NegOfferer* and *ReNegOfferer* – Figure 95 and Figure 97) as a similar SIP INVITE process is being coordinated. The same holds true for the states of the Answerer (*NegAnswerer* and *ReNegAnswerer* – Figure 96 and Figure 98). The states within the Negotiation/Re-negotiation sub-states correspond to the coordination for sending and receiving of SIP messages and the names of these states are similar to the respective SIP requests/responses (detailed explanation of the applied SIP methods and the E2ENP UA processing is given in Section 3.1.3 and in [170]). There are two special features connected with the execution of the Negotiation/Re-negotiation state with respect to the SIP coordination. First, the booking of the MUTEX is expressed with the initial decision about the number of done attempts to book the MUTEX (directly after the *IDLE* state in the A parts of Figure 95 through Figure 98). See also the respective explanation in Section 5.2.2.4.1. Secondly, the state of the Answerer *WaitForCoordinationDone* (Figure 96B and Figure 98B, and Figure 99) represents the actual resources-coordination process of the E2ENP UA. The Answerer side is the one that coordinates the state of the local reservation at the application (managed with the *WaitLocalCoord* state – Figure 99) and of the remote reservation at the Offerer (*WaitRemoteCoord*). If the required state of the reservation as seen from the application is achieved (i.e. *reservResultOK* guard condition (Figure 99) holds true for both the Offerer and the Answerer), the reservation is successfully coordinated and a media session can follow. The final switching of the state machine and the sending of an answer about the successfully done coordination is managed in the *LocalTokenAchieved* and *RemoteTokenAchieved* states (Figure 99). The Answerer may request additional round of reservation, respectively may signal weakening of the requested resource conditions with an additional UPDATE signalling round that is managed in the *WaitLocalCoord* state.

The Offerer/Answerer state machine discussed in this section represents the basis for the E2ENP negotiations and resources-coordination processing. Respectively, the performance of this state machine is evaluated within Chapter 6 to verify the E2ENP concept of full and short E2ENP phases (see Section 3.1.2) and of PDU referencing (see Sections 3.1.4 and 4.3) as means for optimising the signalling at multimedia-session establishment and at corresponding adaptations.

5.2.2.4.3 External-Action Finite State Machines

The External-Action state machine (Figure 100AB – below) is an extension of the basic Offerer/Answerer state machine. Similar to the Offerer and the Answerer Negotiation and Re-negotiation states (see Section 5.2.2.4.2), the states of the Requestor and the Acceptor bear the names of the corresponding SIP messages coordinated via the External-Action state machine (see also Section 3.3.1.2 and Figure 45AB). The Requestor part of the state machine is shown in Figure 100A and the Acceptor in Figure 100B. The two parts of Figure 100AB are slightly overlapped in the region of the message *referInd, _ResvMux.get(Time)*. The Requestor and the Acceptor apply the MUTEX (Section 5.2.2.4.1) in a similar way as the non-INVITE transactions. If they cannot acquire the lock at initial try, they simply terminate and delegate the handling to the application. Such behaviour is considered acceptable for the application playing the role of Requestor, as adaptation through session migration is actually not critical for the adaptation process of the application on a single peer if the migration does not succeed (i.e. the application simply continues on the initial device) or not relevant if the application wishes to adapt quickly (i.e. the session migration is more expensive as the simple adaptation between the peers involved in a session with the means of the Offer/Answer Model).

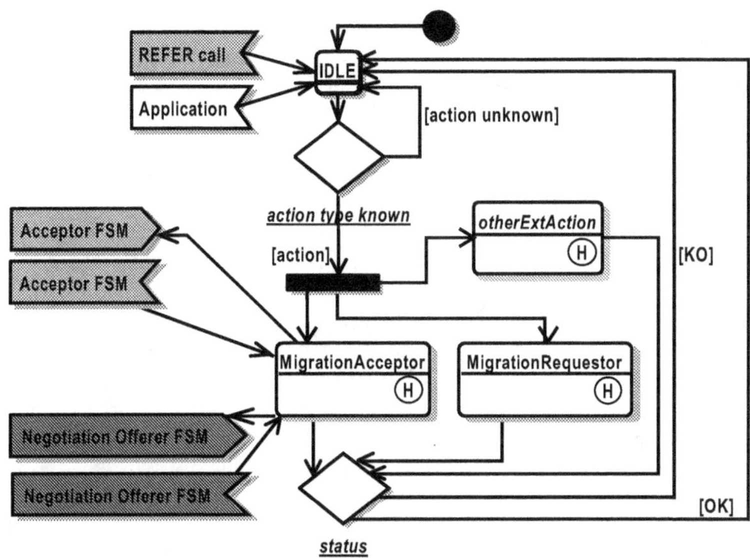

Figure 101: State machine – E2ENP O/A Model and External Actions

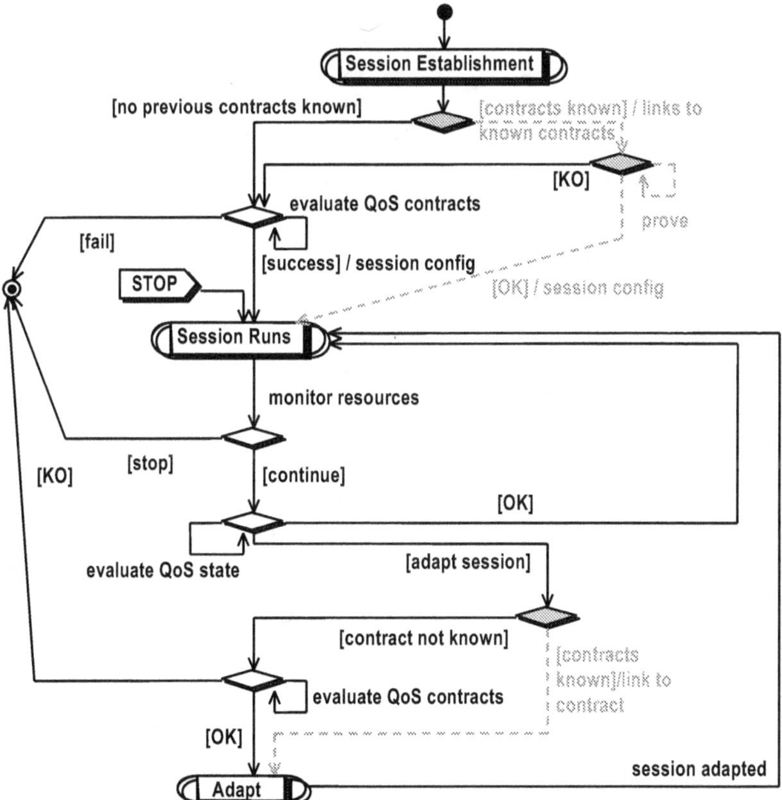

Figure 102: Service/session processing with and without memory

The application of the External-Action state machine with the MUTEX requires the delegation of the lock acquired from the Acceptor to the Negotiation-Offerer state machine. Respectively, in this case the application does not need to book the MUTEX but directly proceeds with the negotiation with the Passively Involved in the migration peer (see also Figure 45B). Correspondingly, Figure 101 shows the additional states for external actions adopted in the root state of the E2ENP UA. The communication between the Negotiation-Offerer and the Acceptor state machines is indicated with the signals *Negotiation Offerer FSM* and *Acceptor FSM.* This communication is associated with the *WaitActionRsp* state (Figure 100B) of the External-Action state machine where actually the delegation of the call from the External-Action state machine to the Negotiation-Offerer takes place. The call is delegated back to the Acceptor when the Negotiation-Offerer completes. The *MigrationAcceptor* state (Figure 101) encapsulates the External-Action state machine for the Acceptor and is a history state as it leads to resetting of resource conditions at the device playing the Acceptor role. In case the migration-coordination sequence is interrupted in between, the Acceptor peer can restore its state back to the time point where the migration was initiated. The same holds true for the history state *MigrationRequestor* used to restore the state of the Requestor device in case the migration does not succeed.

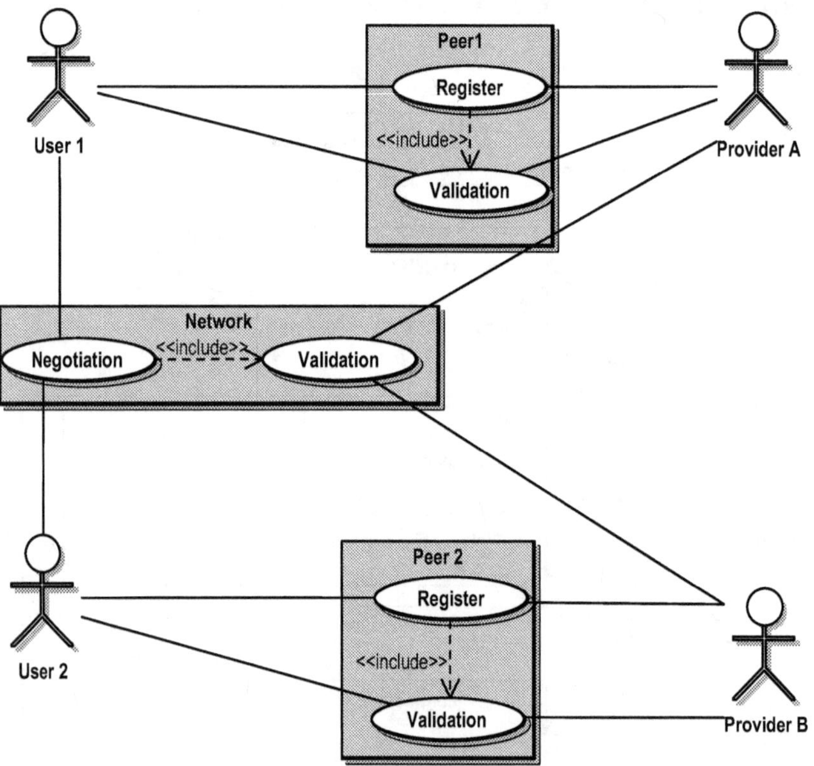

Figure 103: Registration, validation and negotiation use case

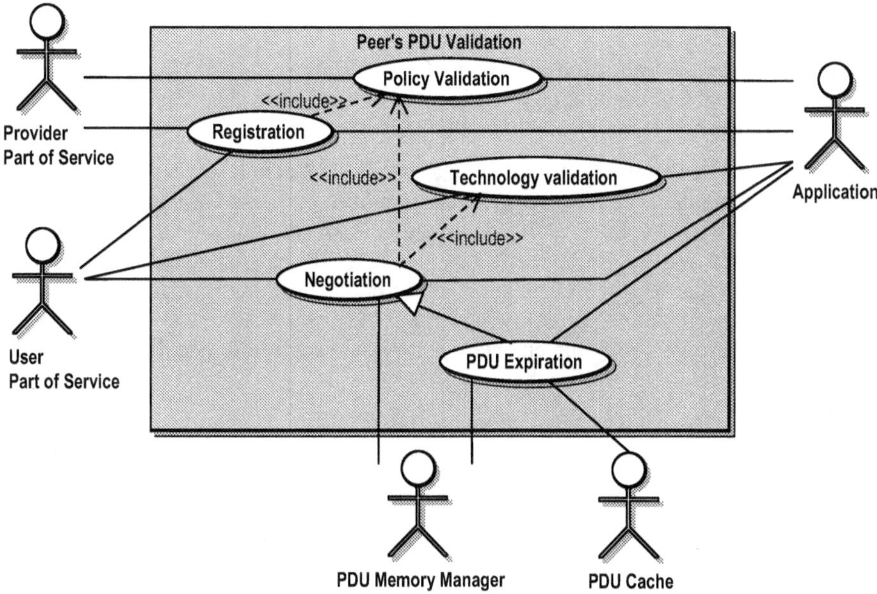

Figure 104: Actors in the validation procedure

5.3 High-Level E2ENP Management

This section presents some recommendations to the application/middleware in terms of utilization of the E2ENP concept. Detailed recommendations are also shown in Annex E.

5.3.1 Applications Using Memory at Session Management

The applications using memory to control and build hierarchical QoS specifications profit at multimedia-session management from the recollection of previous negotiations. This fact is illustrated with an Activity Diagram (Figure 102), where the short management decisions enabled through E2ENP are shown in grey shade (i.e. the shaded rhombi, the dashed lines and the associated text). In cases when the application has knowledge of earlier negotiations which outcome is applicable at time points of session establishment or of adaptation, the application simply signals the corresponding conditions and utilises the short-decision forms. Otherwise, the application has to prepare new contracts and contexts and negotiate them with its peer application, i.e. the application takes the longer decision paths in the algorithm that are indicated with *evaluate QoS contracts* decisions in Figure 102. Consequently, the actual and potential configurations of E2ENP result in *win-win* negotiations with stable results for the application, enabling the applications to optimise their session-management decisions (see also Chapters 1, 3 and 4).

5.3.2 Validation Algorithm Use Cases

The application of the validation algorithm (presented in Section 2.3.2.2.2) interplaying with the registration and negotiation procedures of the peers and the providers is illustrated with two Use Case diagrams (Figure 103 and Figure 104).

Figure 103 shows the distributed case of the validation procedure, where the terminal's registration at the provider generates a profile for constraining the performance of the applicable services. Such profiles are used in the local validation at the peers and a definition of such profiles is demonstrated in Annex B. At negotiation, the provider controls and interacts with the negotiation of the peers, resulting in manipulation of the peers technical specifications for enforcing of provider rules. The results of such network validation in terms of E2ENP implementation are also discussed in detail in Chapters 2, 3 and 4.

Figure 104 displays the interaction of the software components responsible for managing of the negotiations and the hierarchical QoS specifications at interactions with the validation. The user's registration at the provider is managed through the user and the provider application interfaces (e.g. the terminals application GUI and the administrators UI at the provider). Furthermore, the provider may sell its devices in a pre-configured form, expressed in Figure 104 with the direct interaction of the provider with the *Policy Validation* procedure. The user configures its application through the *Technology Validation* procedure and the outcome of the two validation procedures influences the *Negotiation* and the *PDU Expiration* procedures that are managed correspondingly from the *Application*, the *PDU Cache* of the E2ENP UA and the *PDU Memory Manager* of the application/middleware.

This chapter demonstrated the implementation of the E2ENP UA. Evaluations of the E2ENP UA at its basic Offerer/Answerer mode and measurement results confirming the E2ENP concepts of hierarchical-QoS-specification fragmentation and session management through recollection of past negotiations are presented in the next Chapter 6.

6 E2ENP Performance Evaluations

This chapter presents performance evaluations of the E2ENP User Agent at its Offerer/Answerer execution mode. The Offerer/Answerer mode is the primary operation mode of all negotiation scenarios (see also Chapter 3). Consequently, it is chosen for the tests of the E2ENP UA to prove the impact of the basic E2ENP performance on applications using the protocol.

In Section 6.1, the test environment for performing local and network tests of the E2ENP UA is presented. Section 6.2 discusses the negotiation and the QoS configuration scenarios chosen for the evaluations of the E2ENP UA. The operation of the major User Agent components that have primary impact on the performance of the UA (i.e. the XML Parser and Generator, the E2ENP FSMs and the Carrier Stacks – see also Chapter 5) is shown and commented in Section 6.3. Furthermore, Section 6.3 discusses the impact of the generalised signalling through the E2ENP UA in comparison to the pure SIP signalling. Section 6.4 presents the network performance of the User Agent within different emulated stationary and mobile networks. Finally, Section 6.5 discusses the relationship of E2ENP to other similar state-of-the-art protocols and demonstrates comparison results between E2ENP and SDPng applications.

Table 6: Parameters of the used test PCs

#	Host Type	CPU	RAM	OS
1	End-device	AMD Duron 700 MHz	256 MB	WinNT/SP 6
2	End-device	Intel PIII 500 MHz	256 MB	WinNT/SP 6
3	End-device	Intel PIII 500 MHz	256 MB	Win2000/SP 4
4	End-device	AMD Athlon 750 MHz	384 MB	Win2000/SP 4
5	End-device	AMD Duron 800 MHz	256 MB	Win2000/SP 4
6	End-device	Intel PIII 700 MHz	256 MB	WinXP/SP 2
7	End-device	Intel PIII 1 GHz	256 MB	WinXP/SP 2
8	Router	AMD Athlon 550 MHz	128 MB	SuSe Linux 2.4.19
9	Router	AMD Duron 1.2 GHz	128 MB	SuSe Linux 2.4.19

Table 7: Emulated network types with NIST NET

Type	Bandwidth	Delay
GSM	16 kbit/s	60 msec
GSM/GPRS	64 kbit/s	60 msec
GPRS	170 kbit/s	45 msec
UMTS	384 kbit/s	45 msec
WLAN	1500 kbit/s	10 msec
LAN	100 Mbit/s	10 msec

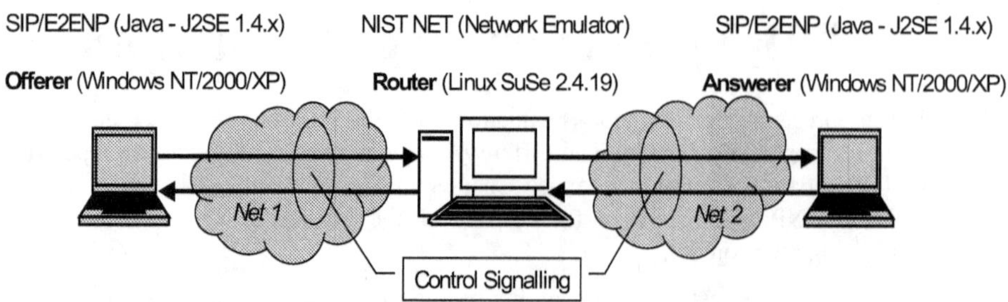

Figure 105: Network test-bed

6.1 Test Environment

The performance of the E2ENP User Agent was tested in different environments corresponding to different end-devices and different network types. The used test devices are listed in Table 6 and the investigated network types in Table 7. The tests executed with the E2ENP User Agent include single-PC tests to estimate the performance of the individual software building blocks of the UA. Furthermore, the network tests with two E2ENP UAs and network emulator were carried out to measure and compare the roundtrip times of the E2ENP negotiations in the Offerer/Answerer mode of the UAs in different environments.

In particular, the single-PC tests estimate the performance of the components associated with the major processing steps of the automatic negotiations, i.e. language interpreters, like the Parser and the Generator, and the blocks responsible for coordination and synchronisation of the communication, like the E2ENP FSMs and the Data Carriers, are evaluated (see Chapter 5). The tests of the UA components on a single PC represent idealised performance environment, i.e. they are made to exclude the possibility of distortions that may be caused through the network components. The results of the complete performance of the UA for the single-PC tests correspond approximately to the behaviour of the UA within local area network (LAN). This performance was estimated through measurements and comparison of the one-PC measurements with network LAN-based measurements. The device used for the local performance tests presented in Section 6.3 is shown as entry number one in Table 6. This device was chosen for the investigation of the operation of the individual UA components as it represents approximately the average PC performance (i.e. based on the CPU performance) of all the end-devices used in the tests (see Table 6). The CPU performance is considered to have major impact on the signalling of the end-devices due to the complexity of the E2ENP FSMs. On the contrary, the used memory and its access are per design optimised (see Section 5.2.2.2) and the management of the PDUs is rather uncomplicated due to the minimised sizes of the PDUs (see also Sections 3.1.2 and 4.3). Hence, the results achieved through the single-PC measurements correspond to some average performance of the UA components in the investigated end-system environments.

The network performance tests with the E2ENP UA include several scenarios (see the scenarios descriptions in Section 6.2). All these scenarios display however several common features corresponding to the applied networks and test-bed (see Figure 105). The tested scenarios represent the network performance of the E2ENP UA for complete negotiation rounds (i.e. Pre-negotiations, Negotiations and Re-negotiations) in different networks and using different end-devices. The end-devices used in the network measurements are listed as entries 2 to 7 in Table 6 and for the measurement purposes they are applied as Offerer and Answerer in a way that slow, middle and fast end-device performance is demonstrated (see Section 6.4). The devices used to emulate the network router performance are listed in Table 6 under 8 and 9. The measurements with the two different router devices did not show substantial differences, due to the idealisation of the test environment (i.e. only the bandwidth and the delay are considered as signification – see Table 7). Hence, the results presented here correspond to the fast router performance (i.e. entry 9 in Table 6). The emulated networks are presented in Table 7. The chosen parameters for the emulation are taken from recommendations for testing of the respective network technologies as IP networks [142] – [145] and correspond to some configuration proposals associated with the applied network emulator NIST NET [278]. Additionally, it is assumed that the transfer of the E2ENP signalling is ensured through DiffServ-like technology (e.g. combinations of DiffServ Assured and Expedited Forwarding, see [147] – [149]), so that no packet loss and no jitter occur on the network. The assumption that the application signalling shall have preferential treatment based on implicit network reservations stems from the fact that the loss and the unpredictable delays of the control data are crucial for the overall performance of the service. Furthermore, implicit reservations of network resources for the signalling assures that E2ENP can be applied efficiently in different environments and also at handovers, i.e. if the reservation is explicit (e.g. with RSVP [4]), it would increase the delays for the signalling as the reservation should be made before the signalling starts. Moreover, in ad-hoc environments, explicit reservations are not well applicable due to the fluctuations of the ad-hoc network structure that results in frequent changes of the network-resource availability. It is up to the application logic of the ad-hoc nodes to grant preferential treatment for the signalling thus ensuring QoS for the service controls (see also Section 2.1.1 and [65]). Consequently, the preferential treatment of the control data represents the quality of service for the signalling as a part of the global QoS of the service. However, there is an argument against the general application of preferential treatment for the signalling, as it can be used for service attacks thus blocking the complete service. Nevertheless, the security concern is refutable considering that E2ENP shall be treated as a premium and added-value service for which the terminals subscribe, thus getting also corresponding security controls from the providers of the network and of the services (see Sections 2.1, 2.3.2.1 and 3.2). An additional argument that preferential treatment of the control data cannot severely influence the overall network performance, so that quality-based services are disturbed, is the fact that the control information is relatively small and occurs rarely compared to the media traffic and the amount of the control data is restricted also within the terminals, e.g. per definition for the SIP performance [62]. Consequently, DiffServ-like solution for the E2ENP transfer and network treatment is a realistic solution applicable for evaluations of the E2ENP UAs.

Considering the above assumptions and conditions about the performance of the network to guarantee QoS for the control signalling of the services, the applied test-bed for proving the network operation of the E2ENP UA is designed as shown in Figure 105. The test-bed consists of two end-devices with an E2ENP UA on each of them that take the role either of an Offerer, or of an Answerer. In case on of asymmetric characteristics of the used end-devices, the more powerful terminal is chosen to be an Answerer, since the Answerer has the greater impact on the signalling due to the fact that it coordinates the reservation process at E2ENP Negotiations and Re-negotiations (see Section 5.2.2.4.2 and the respective results in Section 6.3.2). The Router (Figure 105) emulates the DiffServ-like network performance using NIST NET [278]. The emulator imitates the connections to two access networks of equal types (see Table 7 for the tested network technologies) for examining the network performance of the UAs per single network-type scenario (e.g.

GPRS scenario, UMTS scenario or LAN scenario). Each of the devices executing the E2ENP UAs (i.e. the Offerer and the Answerer) is attached to one of the access networks and these terminals are interconnected through the Router.

The evaluation of the performance of the UA components and of the network measurements is made using a software module that generates microsecond marks (see Section 5.2.1.2) at every call of the respective processing or negotiation methods of the UA (see Chapters 3 and 5). The correspondent performance times of the measured modules and of the network roundtrips are calculated through evaluations of the differences between the generated marks.

The performed measurements are the average values of 10 to 20 measurements per table entry or per measurement point on a diagram (see Sections 6.3 through 6.5). The relatively small number of measurements is however significant enough for the evaluation of the E2ENP UA performance, due to the idealised test environment (see above). Consequently, the measurements displayed only minimal dispersion (under 10%).

Note that due to specific initialisation of the Java Virtual Machine, it was observed that every initial measurement after re-starting of a terminal results in approximately two times longer delays then every subsequent performance of the UA. Consequently, in order to exclude the initialisation delays due to instantiation of persistent components, the results for the measurements shown later on in this chapter exclude all the initial measurements done after a terminal re-starting.

6.2 Test Scenarios

The applied QoS configuration for the subsequent tests corresponds to the following performance scenario:

> *The multimedia session to be established and later on adapted consists of one unidirectional audio and video stream. No more than three alternative codecs and no more than three QoS contracts per stream are used. Streams need to be synchronized, forming two different QoS contexts.*

This scenario can be associated with a VoD Service that needs to be adapted after a handover or may be adapted through session migration to another device. The only test that differs from this scenario is the XML-processors evaluation (see Section 6.3.1) that simply repeats in the XML code some of the structures associated with the above scenario to achieve specific message sizes for testing the XML processing impact on the E2ENP UA.

The choice of the above scenario for the tests was made in consideration with the following limitations:

- Due to the flexibility of the E2ENP negotiation language (see Chapter 4), it is not possible to test all potential scenarios for session configurations. Hence, a scenario presenting the major configuration options of the protocol is chosen that enables the definition of different QoS contracts and contexts and the respective referencing thereof. Thus, the profit of using different kinds of negotiation phases is demonstrated (see also Section 3.1.2 and Annex C).

- The tests with the SIP-based carriers have to follow the restrictions of the protocol processors upon the sizes of the transported messages [62]. For jSIP [184] and NIST SIP [185], the size of the complete message (i.e. SIP headers and the message body) is limited to approximately 10 kByte, which value corresponds to the size of the UDP buffer. Consequently, the chosen configuration scenario should fit this message size to cover all the tested sub-scenarios (see also below).

- The chosen scenario should be applicable for comparison with state-of-the-art protocols. Therefore, the configuration options chosen for the scenario should also be available in other protocols (in this case SDPng – see Sections 4.2, 6.5 and Annex C).

The tests presented in this work include comparison between the E2ENP phases (i.e. Pre-negotiation, Negotiation and Re-negotiation) considering short and full phases (see Section 3.1.2). The chosen E2ENP mode for the tests is the push mode (see Section 3.1.1), as it represents the shortest communication mode upon which all the other modes are based. Furthermore, the push mode is applied for purposes of comparison with state-of-the-art protocols as they perform in principle using push mode only (see the discussion in Section 3.1.1). When using SIP with E2ENP in the push mode, a Pre-negotiation consists of two E2ENP-body-containing SIP messages and each Negotiation or Re-negotiation consists of four such messages. The first two E2ENP Negotiation/Re-negotiation messages correspond to an initial configuration (i.e. an Offer and an Answer) and the other two messages are the re-configuration (i.e. a Counter-Offer and a Counter-Answer) after a resource reservation process. The Pre-negotiation is performed with the SIP MESSAGE method and the Negotiation/Re-negotiation using the INVITE transaction (see Section 3.1.3). The Offerer and the Answerer ping-pong the same massage contents (with different E2ENP headers) within one negotiation round, i.e. they exchange PDUs corresponding to the applied negotiation type Pre-negotiation, Negotiation and Re-negotiation. Thus, only the impact of the E2ENP UA performance is estimated, as the role of the application for the tests is dummy. The different tested sub-scenarios of E2ENP are:

- *Short phases test* is the case with full Pre-negotiation and with short Negotiation and Re-negotiation that corresponds to ~5kByte XML code and ~70 XML lines per Pre-negotiation document, ~ 4kByte and ~70 XML lines per Negotiation and ~2kByte and ~20 XML lines per Re-negotiation. This test is made to compare the performance of the different E2ENP phases and for the purpose of estimation of the E2ENP FSMs and carriers performance. Note that the volume size of the Pre-negotiation and Negotiation differ, though the XML lines numbers are almost equal, due to the grater number of XML attributes in the Pre-negotiation description. The number of XML lines per Offer and Answer corresponds to the number of operation events in the XML processors (see Sections 5.2.2.1 and 6.3.1), consequently, these numbers play major role at evaluation of the XML processing in E2ENP.

- *Short vs. full phases test* at session establishment and at adaptation includes XML documents describing full Negotiation and short Re-negotiation. This test corresponds to ~7kByte code and ~130 XML lines per Negotiation and ~2kByte code and ~20 XML lines per Re-negotiation. Respectively, both full Negotiation and Re-negotiation have ~7kByte code and ~130 XML lines. This test is made for the purpose of estimation how well the E2ENP short phases concept performs in comparison with application of only full phases, i.e. to be able to make a statement about the gains of E2ENP in comparison with state-of-the-art protocols for session management. Since, the most commonly used session management protocols are either too simple compared with the configuration options of E2ENP (like SDP [86]) or are designed in other then the XML technology (like H.245 [63]), an estimation about the gains of short negotiation rounds is made through comparison of E2ENP with itself (see Section 6.5).

- *SDPng vs. E2ENP test* is the simulation of the SDPng code through E2ENP (see Section 6.5) and it corresponds to an XML document sized ~4kByte and ~70 XML lines. SDPng is the only protocol that has similar contents and definition format like E2ENP. However, there are no existing processors for this protocol. Consequently and since E2ENP applies several SDPng concepts for its own descriptions (see Section 4.2), SDPng is simulated with E2ENP descriptions to make further estimations about the gains of the E2ENP. The specific features of this simulation are commented in Section 6.5.

6.3 E2ENP UA Components Performance

The major components involved in the processing of the E2ENP protocol are the Parser and the Generator for converting the PDUs from XML into UA-internal format and vice versa (see Section 5.2.2.1), the E2ENP FSMs together with the PDU caches for coordinating the negotiation process (see Sections 5.2.2.2 and 5.2.2.4) and the Data Carriers for transporting the negotiation messages (see Section 5.2.2.3). These individual E2ENP UA components are evaluated on a single PC (see Section 6.1 and the first entry in Table 6). Furthermore, the performance of the UA coordination and synchronisation components is measured using the full Pre-negotiation and short Negotiation/Re-negotiation as described in Section 6.2.

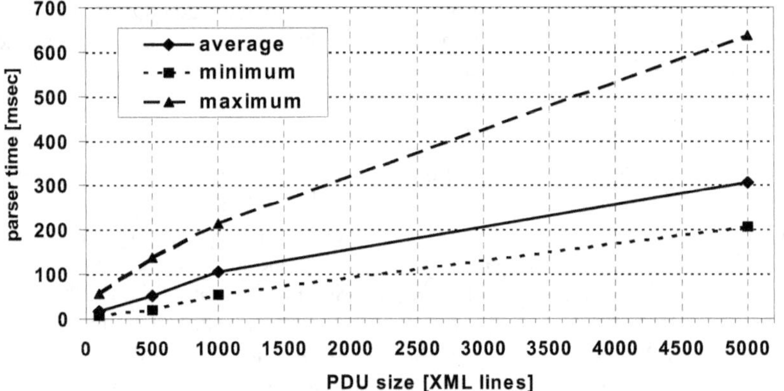

Figure 106: Parser XML Processing times

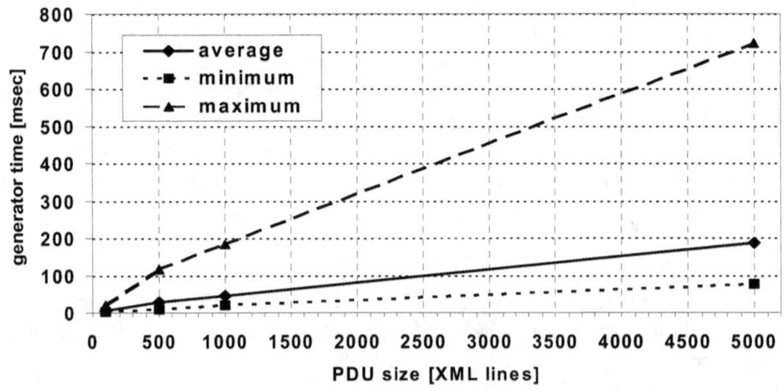

Figure 107: Generator XML Processing times

6.3.1 XML Parser and Generator Performance

The performance of the Parser and the Generator components depends on the size of the negotiated information. The basic unit used when processing the XML payloads is an *XML line*. The applied SAX parsing involves sequential XML-code processing, generating a single event per XML line when producing the UA internal representation (see Sections 5.2.2.1 and 6.2). Respectively, the serialisation of the UA internal representation into XML code is also a sequential process involving the successive visiting of the components of the object tree representing a PDU to extract the string representation of the respective XML elements (see Section 5.2.2.1).

The initial test of the Parser and the Generator involves a separate evaluation of these components, which is independent from the E2ENP UA, in order to prove the XML processing for larger PDU documents then the prescribed sizes from the SIP-based carriers (see Section 6.2). The Parser/Generator performance is proven using PDUs with approximately 100, 500, 1000 and 5000 input lines, where the 100 lines XML code corresponds to ~6kByte PDU document. The expected well-scaling behaviour of the Parser and the Generator due to the sequential processing steps of the PDUs is proven through the test and the results thereof are presented in Figure 106 and Figure 107, where the minimum the maximum and the average (of multiple measurements) processing is displayed. The results of the XML processing, compared with the results of the overall UA processing (see Sections 6.3.2 and 6.3.3), show that the XML processing is efficient enough even for large PDU documents and its impact on the overall E2ENP UA processing is relatively small.

For the tests of the Parser and the Generator within the E2ENP UA architecture, their processing times are included in the times of the E2ENP UA performance (see Section 6.3.2), as in the current implementation of the E2ENP UA the communication of the Parser and Generator with the FSMs is synchronous due to implementation's simplification purposes.

Table 8: E2ENP UA performance using different underlying carriers

Average processing time of the E2ENP UAs as Offerer and Answer in [msec] (excluding the SIP transport)		
Negotiation Phase	*jSIP Offerer*	*jSIP Answerer*
Pre-negotiation	571	244
Negotiation	58	356
Re-negotiation	9	210
Negotiation Phase	*NIST SIP Offerer*	*NIST SIP Answerer*
Pre-negotiation	393	292
Negotiation	37	606
Re-negotiation	10	171
Negotiation Phase	*RMI Offerer*	*RMI Answerer*
Pre-negotiation	818	329
Negotiation	83	297
Re-negotiation	39	172

In a real example with E2ENP PDUs (see the scenarios description is Section 6.2), the Pre-negotiation and the Negotiation payload sizes vary with the number of media streams, QoS contracts and contexts. However, the size of a short Re-negotiation for adaptation of the multimedia service is always fixed due to the specific referencing to the hierarchical QoS specification (see also Section 4.3.5). The effects of applying short and full negotiations resulting in differently sized PDUs is further detailed in Section 6.5.

6.3.2 E2ENP UA Performance

The E2ENP UA performance includes all the additional impact caused from the resources coordination and the PDU management done with the E2ENP FSMs, i.e. excluding the data transfer and the synchronisation with the transportation layer UA (e.g. SIP UA) that are part of the Generic Carrier Service Provider of the E2ENP architecture (see Sections 5.2.1.1 and 6.3.3). Consequently, the E2ENP UA evaluation represents the composite operation of the E2ENP FSMs, of the Parser and the Generator and of the PDU Caches, as the XML processors and the Caches are synchronous and run in the thread environment of the FSMs (see also Chapter 5). The evaluations at the Offerer/Answerer include the estimation and the corresponding accumulation of the measurement marks (see Section 6.1) generated when launching E2ENP UA and SIP UA interface methods (see Section 3.1.3). Considering the E2ENP UA architecture (Section 5.2.1.2), the concerned interfaces at this evaluation are IF2 and IF3. Note, that IF1 is used only once at E2ENP UA configuration and the rest of the interface methods (i.e. IF4, IF5 and IF6) are executed in between the IF2 and IF3 execution. Consequently, the E2ENP UA performance is the accumulated performance of its components excluding the performance of the carrier protocol. Although the displayed evaluation concerns the pure E2ENP UA performance, some customisation of the E2ENP FSM timers is considered due to the different reactions times of the transport layer protocols (i.e. TCP and UDP) used within the carriers (i.e. jSIP, NIST SIP and RMI-based SIP surrogate). The appropriate periods for these times were estimated within the initial debugging and tests of the E2ENP UA.

Table 8 shows the complete response times of the E2ENP modules per negotiation rounds (i.e. per Pre-negotiation, per Negotiation and per Re-negotiation phases) for the Offerer and the Answerer roles of the UA in case that full Pre-negotiation and short Negotiation/Re-negotiations are done (see also the respective scenario description in Section 6.2). Thus, the impact of the short phases performance is demonstrated confirming that the short phases can reduce the response times of the session coordination processing (see also Section 3.1.2 and the examples in Section 4.3 and Annex C).

The tests show several effects of the FSMs implementation (see Table 8). The Parser and the Generator are bound to the FSMs only at an initial call to the XML processors. This lazy initialisation results in longer Pre-negotiation time for both SIP and RMI due to the time needed to start the XML SAX parser (Table 8 – Offerer side). A binding procedure to the RMI-based carrier (i.e. the binding to the RMI remote object) is made every time a remote communication takes place and this results in longer response times for the RMI-based SIP surrogate (Table 8 – Offerer side). The re-binding to the RMI remote object is done for two purposes, i.e. to simulate re-launching of the E2ENP UA and to simulate asynchronous communication, where the communication channel exists only during the negotiation procedure. At the Answerer side, the RMI- and jSIP-based implementations are comparable, which verifies the stability of the E2ENP UA FSM, since the major amount of negotiation processing takes place at the Answerer side. However, the Answerer implementation based on NIST SIP displays longer processing times at Negotiation, due to the fact that the E2ENP FSMs simulate sleeping state between the Pre-negotiation and the Negotiation and they reconnect to the carrier at the beginning of the Negotiation. In this case, per design of the NIST SIP UA the Answerer side is re-initialised lazily which leads to the respective delay.

Considering the above results, a recommendation for future developments of the E2ENP UA is to decouple the E2ENP language processing and the Carrier Protocol binding from the negotiation process itself in order to avoid the above discussed negative effects on the negotiation process, due to lazy initialisation or re-binding of components in case they have to be restarted. This means that the E2ENP FSMs, the XML processors and the Generic Carrier Service Provider have to be individually configurable and executable, which is currently not the case. The result of such configurability would allow to execute the sleeping state of the E2ENP FSMs only, preserving the rest of the components in active state. Although the Pre-negotiation phase is not that crucial for the overall service performance, as the Pre-negotiation is always made independently from the controlled media sessions, the optimisation of the binding of the FSMs to the language processors can optimise the Pre-negotiation's performance, e.g. a dummy call to the SAX Parser at UA initialisation can start it before actual negotiations that use the Parser are made. However, the delay in the negotiations due to shortcomings in the pluggability of the carrier UAs is a non-desired effect for the coordination process of the multimedia adaptations, for the case that the E2ENP UA wishes to save terminal resources through going into a sleeping state.

Table 9: Performance of the carriers as Generic Carrier Service Provider

Negotiation Phase	**Average Roundtrip Time [msec]**		
	jSIP	*NIST SIP*	*RMI*
Pre-negotiation	691	4069	222
Negotiation	65	112	111
Re-negotiation	48	49	70

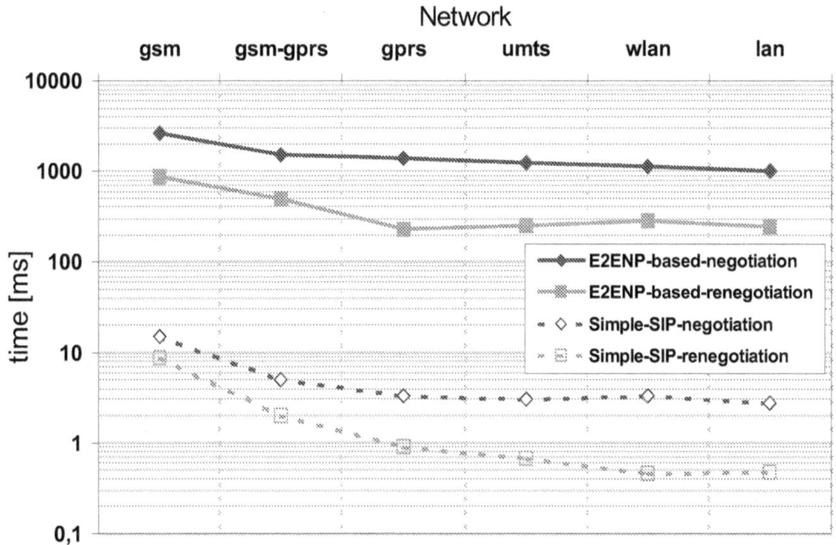

Figure 108: E2ENP UA and pure SIP UA roundtrip times

Nonetheless, the sleeping state of the E2ENP UA is an add-on functionality that depends on the type of the terminal and the applications using the UA. Consequently, the recommendation for treating of carriers that display poor pluggability is either to configure the E2ENP UA not to un-bind them in the sleeping state of the UA or in case that the application strongly depends on saving of terminal resources, the application should use only appropriately re-pluggable carrier UAs. The utilization of the RMI-based SIP surrogate is not recommendable for general-purpose usage of the E2ENP UA but only for tests of the UA, as RMI is actually not designed to perform in mobile environment due to the usage of TCP as a transportation protocol (i.e. TCP can be interrupted only with very high costs for stopping and restarting of the TCP thread, see also [180]). However, the RMI-based SIP surrogate proved to be the most stable application for the presented network tests (see Sections 6.4 and 6.5), as the open-source SIP stacks used in the tests displayed multiple synchronisation problems (see also the discussion in Section 5.2.2.3).

6.3.3 Data Carriers Performance – E2ENP UA vs. SIP UA

The E2ENP UA uses an adaptor layer to connect the E2ENP FSMs to the applied carriers and to synchronise the performance of the carriers with the E2ENP UA coordination process. The wrapping of the carriers into E2ENP's synchronisation architecture represents the E2ENP concept of Generic Carrier Service Provider (see Section 5.2.1.1).

The results shown in Table 9 summarise the performance for accomplishing negotiation roundtrips with the E2ENP's Generic Carrier Service Provider based on jSIP, NIST SIP and the SIP surrogate with RMI, i.e. the evaluation of the end-to-end SIP connectivity. These results are achieved evaluating the pure E2ENP UA performance (see Section 6.3.2) and the overall connectivity over E2ENP as seen from the application, i.e. the E2ENP UA performance times for synchronisation and XML processing are subtracted from the application-to-application response times (as measured at IF3 – Section 5.2.1.2).

Results handled in the literature [279][280] about the performance of pure SIP implementations present SIP-roundtrip times that are 10 to 100 times lesser then the operation times of the Generic Carrier Service Provider in combination with the E2ENP UA state machines (see the results in Figure 108). It should be noted that the results in the literature are made using only simple INVITE transactions without resources coordination and multi-application-management capabilities of the user agent. The existing measurements of SIP UA performance are based on pure SIP connectivity designed predominantly for high-throughput networks (e.g. LAN) and having no additional synchronisation and management features for the applications. Comparative tests with jSIP and NIST SIP based on pure SIP-connectivity were also made to prove the SIP investigations done by other authors and the results of the comparison produced similar values for the roundtrip times to those commented in the literature (see Figure 108). Consequently, it is evident how high the price is for providing generalised communication through the E2ENP UA using different carrier-protocol implementations, abstracting the application from the specific signalling on the network and providing a multi-application-management environment. The E2ENP UA supports appropriate synchronisation applicable in different kinds of networks (see also Section 6.4) and coordinates the resources reservation in the terminals and on the network, thus assisting the applications' own management and simplifying it.

The results for the NIST SIP performance in Table 9 show again the negative effect on the negotiations when having lazy initialisation of the carrier stack (see the Pre-negotiation entry for NIST SIP in Table 9 and the corresponding comments in Section 6.3.2). Furthermore, the longer roundtrip times with the SIP implementations based on UDP transfer (jSIP and NIST SIP) for the Pre-negotiation depend on the fact that the SIP UAs apply longer time-outs for the SIP non-INVITE transactions (in this case MESSAGE, see also Sections 2.3.1 and 3.1.3) in order to be able to guarantee the transfer of the Offer and the Answer with only two effective SIP-messages, i.e. MESSAGE and 200OK for MESSAGE. On the contrary, the RMI-based (respectively TCP-based) carrier is quicker at Pre-negotiation compared with the native SIP implementations, as RMI does not apply additional synchronisation of the connection at the session layer. At the level of the carrier, the application of short communication phases displays also shorter response times (compare with the E2ENP UA performance in Section 6.3.2) as the amount of data to be transferred is lesser then in the full phases (see also Section 6.5).

6.4 E2ENP UA Network Performance

The network tests of the E2ENP UA were made in several stages in accordance with the development stages of the UA and the availability of the test-bed hardware/software.[30] In the following, a summarisation of all the executed network tests as average performance results is presented, commenting the evaluations and discussing further the observed E2ENP UA behaviour.

The first network test (Figure 109 below) displays an evaluation of the E2ENP UA roundtrip performance applying the RMI-based SIP surrogate and using three different end-system setups with slow PCs (see Table 6 – entries 2 and 3), with PCs with middle CPU performance (see Table 6 – entries 4 and 5) and with fast PCs (Table 6 – entries 6 and 7). The computing power of the chosen PCs is considered predominant for the performance of the E2ENP UAs due to the application of state machines (as active components) for managing the coordination of the multimedia process. On the contrary, the application of memory is less critical due to the optimised sizes of the managed data (i.e. the E2ENP PDUs) and due to the ability of the E2ENP UA to restrict the size of the applicable memory through the cache-management mechanism of the UA (see also Section 3.1.4). Consequently, the chosen terminals for the tests have almost the same memory parameters and only the CPU power varies. Since the major processing amount is on the Answerer side (see also Section 6.3.2), the more powerful machines in the scenarios with middle and fast CPU performance are in the Answerer role. The observation from the test on different PCs (Figure 109) is that the E2ENP UA roundtrip performance is directly proportional to the CPU speed. Additionally, it is again evident that the short phases (Negotiation and Re-negotiation) optimise the session signalling (see also Section 6.3.2 and 6.3.3). The scenario analysis shows that even for relatively short session descriptions (see the scenarios in Section 6.2) the Re-negotiation has up to ~50% less signalling impact compared to the negotiation procedures for establishing of a multimedia session (Pre-negotiation, Negotiation). This difference increases in case of more complex then the chosen session descriptions as the size of the short Re-negotiation is almost fix (see also Sections 4.3.5 and 6.5). In case of slow networks (GSM), the number of transported messages dominates over the synchronisation times of the E2ENP UA. Hence, the Negotiation phase is slower then the Pre-negotiation as it has 12 SIP messages compared to the 2 SIP messages of the Pre-negotiation (see also Section 3.1.3).

---

[30] First the RMI-based SIP surrogate was available, followed by the jSIP implementation and finally by the NIST SIP implementation. Consequently, the major network test were made in parallel with the development of the E2ENP UA over three years, but some of the initial tests were not reproduced at later test phases due to limitations and hardware/software reconfigurations of the applied test-bed.

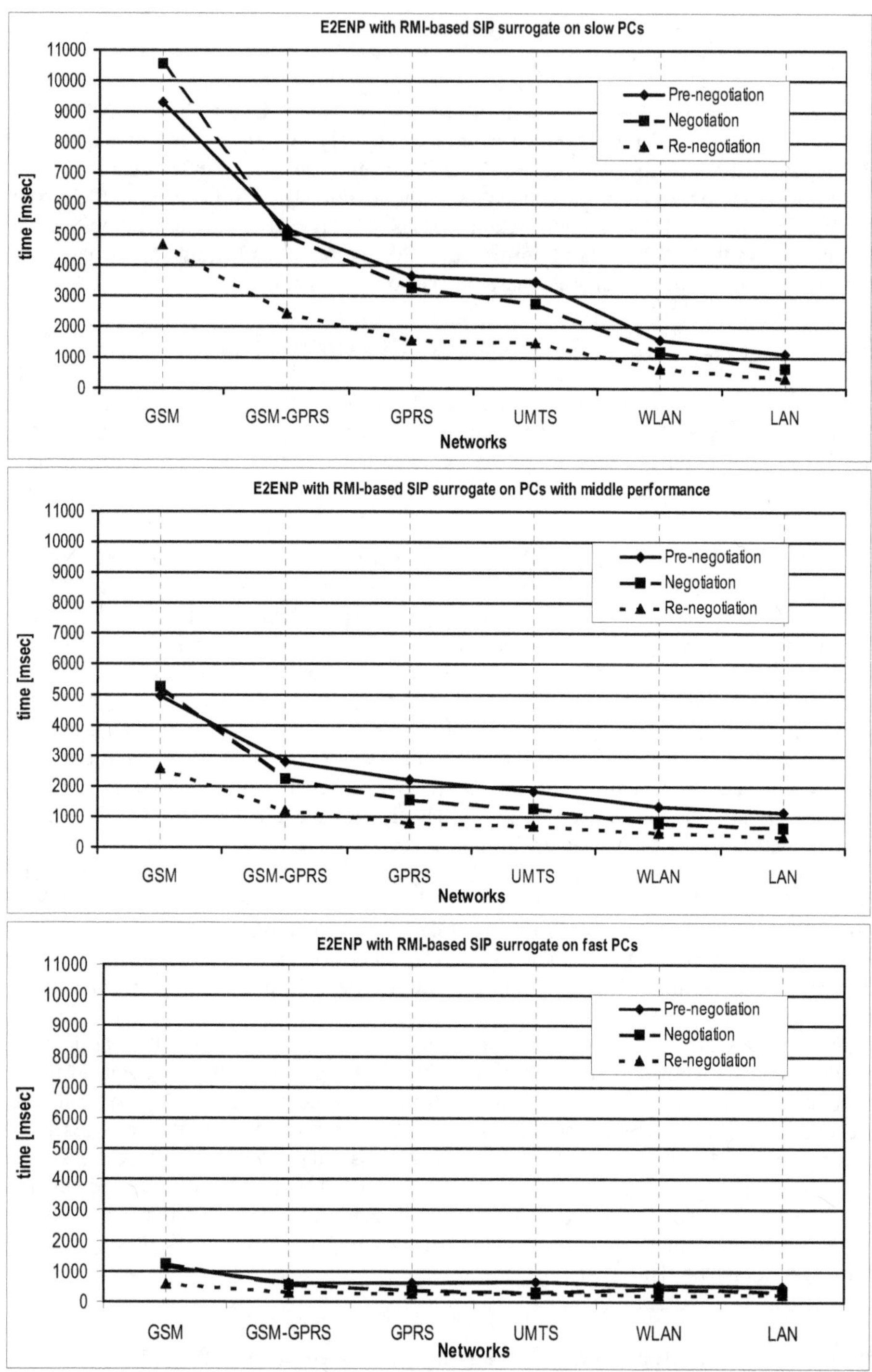

Figure 109: E2ENP UA roundtrip-times with the RMI-based SIP surrogate on different PCs and in different networks

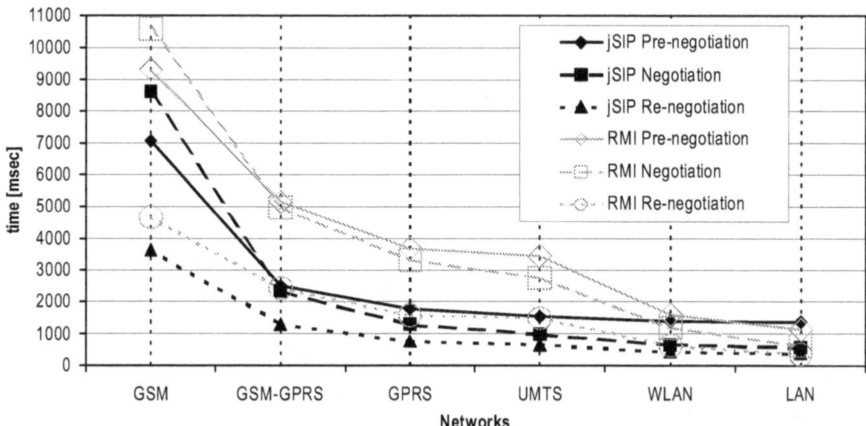

Figure 110: Comparison between E2ENP UAs with jSIP and RMI-based SIP surrogate carriers in different networks using slow PCs

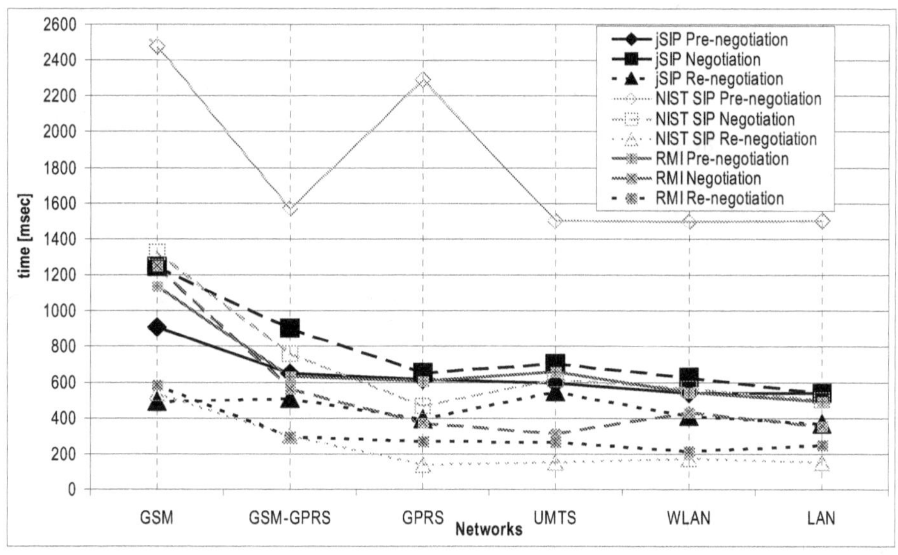

Figure 111: Comparison between E2ENP UAs with jSIP, NIST SIP and RMI-based SIP surrogate carriers in different networks using fast PCs

The second network test (Figure 110) includes the comparison of a SIP-native E2ENP implementation and the implementation based on SIP-surrogate with RMI. For this test, the jSIP-based solution was chosen because of the higher stability of this SIP implementation (see also the third test below). Figure 110 presents the results of the comparison between the SIP-native and the RMI-based SIP-surrogate implementations on the slow-terminal setup to demonstrate an approximation to the performance of mobile devices. The measurements show that due to the UDP-based transport that relies on the explicit synchronization within the SIP code, the jSIP-based E2ENP performs better in low-bandwidth networks. If more data rate is available, the complexity of the explicit SIP synchronization increases the impact of the peer performance, thus the end-to-end application signalling is slower compared to the RMI-based implementation. Within high-bandwidth networks, the RMI-based SIP surrogate shows better results, since the transport synchronization is done with TCP. This example confirms also that in slow networks (GSM) the transportation time of the Negotiations is higher then the time of the Pre-negotiations due to the greater number of messages in the negotiation rounds (see the first example above and Figure 109). However, the coordination and the synchronisation times of the E2ENP UAs are predominant in higher-bandwidth networks, hence, the Pre-negotiation is the slowest phase as it carries the greatest amount of coordination data and the carrier UAs need more time to provide successful transportation of the two negotiation messages within just two transportation calls (see also Section 6.3.3). The relation that the Re-negotiation is ~50% quicker then the rest of the negotiation phases holds also true for the SIP-native environment. Similar tests on devices with middle CPU performance confirm the observations done for the SIP-native E2ENP implementation. However, the test on fast machines (see below) displayed some negative effects with respect to the stability of the performance of the SIP-native implementations.

The third network test (Figure 111) is used to discuss some negative effects of the time-based synchronisation within the SIP-native E2ENP implementations. These effects showed most clearly on the fast-PCs terminal configuration, where the fixed time-outs and response times associated with the carrier protocol implementations and partially with the E2ENP FSMs influenced negatively the coordination process, so that the roundtrip times in higher data-rate networks increased compared with slower networks. This is due to the manual adjustment of the time-outs in the Generic Carrier Service Provider (this is most evident in the GPRS measurement in Figure 111, see also Section 5.2.2.3). The E2ENP FSMs timers were intentionally not reconfigured for the entire test series (i.e. shown in Figure 109 through Figure 111) in order to study the synchronisation effects when the E2ENP FSMs run on different terminal devices. The major problem in the final test was the coordination between the E2ENP FSMs and the Generic Carrier Service Provider that was solved using manual adaptation (see above). However, the application of the manual time-out adjustment is not that fine leading to the effects shown in Figure 111. Some better configuration of the SIP-native solutions (jSIP and NIST SIP) appeared inconvenient, as the aim of the test was to prove the suitability of available open-source SIP implementations as they are, with possibly minimal amount of debugging and possibly minimal redesigning and reconsideration of the SIP standard [62] implemented in jSIP and NIST SIP. Both jSIP and NIST SIP contain fixed synchronisation timers for the communication roundtrips, which correspond to the SIP standardisation recommendation [62]. However, these timers work only for a limited amount of configurations and networks, thus limiting the application of one single SIP implementation within different networks and terminals. Especially problematic appeared the application of NIST SIP, as it performed stably only within higher bandwidth networks. As result of these observations, it is recommendable that future versions of the E2ENP UA consider the application of event-based synchronisation (as provided recently with Java 1.5.x version of the SDK [281][282]) and apply adaptability of the UA time-outs based on detection and monitoring of the underlying network. The event-based synchronisation was not available in Java for the time of the development of the E2ENP UA. Furthermore, the monitoring of the network requires interplay between a middleware that implements the monitoring and that configures the E2ENP UA. Consequently, the rebuilding of the carrier protocol and the E2ENP state machines to adopt event-based synchronisation and time-out adaptations may need reconsideration of the design of these components, eventually leading also to additional standardisation proposals on how to more conveniently build the SIP state machines in order to achieve better applicability in variable kinds of networks and devices.

A proposal, how to apply dynamic configurations of middleware components that depend on changing parameters within the lower layer components (e.g. in the network interfaces) was made within the scope of the MIND project [20][130], where the an enhancement (named enhanced socket layer) on top of existing network devices was proposed to coordinate such dynamic reconfigurations in the middleware and to provide the middleware with corresponding information appropriate for its reconfiguration. The development of the E2ENP UA was undertaken assuming the existence of such enhanced socket layer (such configuration layer is partially implemented also within the E2ENP UA as configuration modules – see Section 5.2) and the UA evaluation confirms the necessity of such an abstraction on top of lower layer components to enable the proper re-configurability of quality-based middleware.

In addition to the observations about the synchronisation, the third network test (Figure 111) presents also some further effects of the applied SIP-native carriers:

- The effect of lazy initialisation of the NIST SIP leading to slower Pre-negotiations is again visible in this example (see also Section 6.3.3)
- NIST SIP displays the fastest Re-negotiation due to the applied *at-most-once* data transfer semantic (see also Section 5.2.2.3). Considering, this fact a topic to be handled in future versions of the E2ENP UA is the possibility to apply different transfer semantics in the SIP UA in accordance with the sizes of the applied message bodies and with the types of the underlying networks, thus optimising further the communication between the terminals. The application of different transfer semantics can also be coordinated via an enhanced socket layer as discussed above.

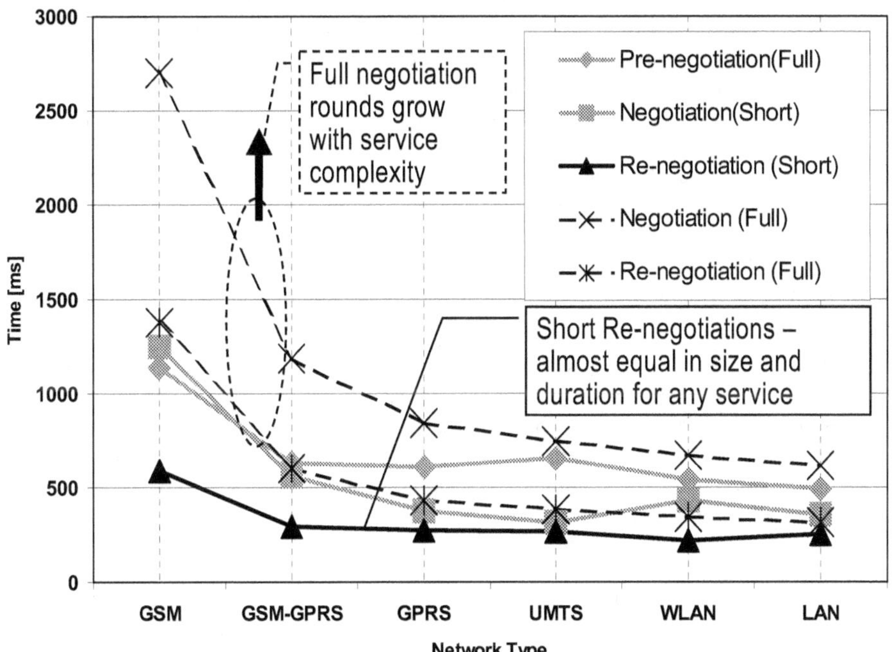

Figure 112: Gain of short negotiation rounds compared to the full negotiation rounds (RMI-based SIP surrogate with fast PCs)

Table 10: E2ENP and SDPng comparative measurements

Negotiation Phase	Average Roundtrip Time [msec]			
	jSIP		*RMI*	
	E2ENP	SDPng	E2ENP	SDPng
Negotiation	1280	1200	1094	1057
Re-negotiation	311	521	329	604

The tests presented in this section approve the E2ENP concept that short communication phases for session establishment and session adaptation represent an efficient communication process between terminals. One additional network measurement is presented in the next Section 6.5 where the application of short vs. full Negotiation and Re-negotiation phases is commented also in consideration with the state-of-the-art protocols.

6.5 Comparison with State-of-the-Art Protocols

Of special interest in this work is the comparison of the E2ENP solution (especially considering the possibility for short Re-negotiations) with state-of-the-art protocols like SDP [86], H.245 [63] and SDPng [81]. However, the common basis for comparison between E2ENP and the state-of-the-art solutions is relatively limited due to the number of innovative features proposed within the scope of E2ENP, like the different negotiation modes and phases, the application of hierarchical QoS specification and the ability to express memory of past configurations (see Section 3.1). Furthermore, the description formats of the comparable session management solutions also differ, which makes the comparison possible only upon the sizes of the exchanged messages. Finally, the comparison scenario should consider that E2ENP also coordinates network-resources reservations. Consequently, the state-of-the-art examples should also support this feature for the purpose of comparative tests. Considering these facts a scenario where all of the technologies can be compared is an evaluation between Negotiations and Re-negotiations at push mode of the protocols as the state-of-the-art solutions have no notion of Pre-negotiation phase, apply predominantly push mode at negotiations and do not support short phases.

The first test is a comparison of E2ENP with itself (Figure 112) to demonstrate the impact of the full Negotiations and Re-negotiations in relationship to the short ones and it is made on the fast-PC configuration (Table 6 – entries 6 and 7). This test demonstrates average results based on the jSIP and the NIST SIP carriers. The test is based partially on measurements and partially on approximations made in accordance with measured E2ENP behaviour. The full Pre-negotiation and the short Negotiation/Re-negotiation are measurements and the full Negotiation and Re-negotiation are smoothed to remove the effects of the manual configuration of the Generic Carrier Service Provider (see also Section 6.4). The smoothing is done in the region of the GPRS measurement point. The approximation is based on the evaluated relations between the different E2ENP negotiation phases as estimated on slower end-system configurations.

It is assumed that the language processing of E2ENP and of SDP and H.245 would not display high processing diversities, because of the fact all of them are parsed and generated sequentially[31] and due to the evaluation (see Section 6.3.1) that the language processing represents relatively small part of the overall E2ENP UA processing considering the complexity of the session coordination as a whole. As a result, the comparison evaluates the synchronisation and the coordination of the data transfer and the sizes of the negotiated data, i.e. even for the simplest representative of the state-of-the-art session description mechanisms SDP, the messages sizes can comprise documents with more then 10kByte volume (as presented in a discussion on the MMUSIC mailing list [283]), which would affect accordingly the massages transfer. Having in mind the made assumptions, the results[32] shown in Figure 112 present that the short Negotiation/Re-negotiation phases have ~30% to ~50% shorter response times compared with the full phases even for the chosen short descriptions (see Section 6.2). Especially high is the gain of short Re-negotiations as they have approximately the same size for scenarios with any complexity. Note that with SIP [62] and the Offer/Answer Model [35][36] (see Figure 112) the Re-negotiation roundtrip is always shorter then the Negotiation even in case of full phases due to the additional synchronisation time for ringing at Negotiation (see Sections 2.3.1.2 and 3.1.3). This investigation ends in a conclusion that the E2ENP's innovation with application of short phases for multimedia session establishment and adaptation can considerably improve the negotiation process of terminals.

The second test (Table 10) includes a comparison between E2ENP and SDPng, as they are very much alike also structurally (see also Sections 4.2 and 6.2). However, SDPng does not have hierarchical-QoS-specification mechanisms for defining session descriptions and it does not apply PDU headers. Furthermore, the basic SDPng description model [81] does not contain elements for QoS-specific codec parameterisation and for transport QoS. Due to these differences, the E2ENP PDU for full Negotiation is larger (~7kByte) then the SDPng one (~4kByte) according to the PDU specifications for this test (see also Annex C). However, the tested re-negotiation PDUs are for E2ENP short Re-negotiation ~2kByte and for the SDPng re-negotiation ~4kByte, since SDPng applies always full re-negotiations. The results of the SDPng and E2ENP application-to-application roundtrip times are shown in Table 10 and indicate the total time needed to establish and explicitly adapt QoS through negotiations for a multimedia service as described in Section 6.2. The measurements are taken on a single PC (Table 6 – first entry) and they correspond approximately to LAN behaviour. The session establishment times for E2ENP and SDPng are almost similar. E2ENP is slightly slower (6%) due to the larger amount of exchanged information as result of the overhead of the hierarchical QoS model. In contrast, the average response of E2ENP for the Re-negotiation is ~240 msec (or 57 %) faster than SDPng, due to the application of E2ENP short phases. In scenarios exposing higher complexity then the demonstrated test, the E2ENP short Re-negotiation is even more efficient, as the size of the E2ENP Re-negotiation payload is highly limited due to the referencing to the hierarchical QoS specification, whereas the size of SDPng re-negotiations varies with the complexity of the described application/session scenario. It should also be noted that E2ENP directly points at the changed configuration data (e.g. enforce contract X, block contract Y) whenever performing adaptations through short Re-negotiations. In contrast, SDPng Re-negotiations

---

[31] The SDP parsers and generators associated with jSIP and NIST SIP are implemented with sequential language processing. The ASN language used for the design of H.245 (see ASN.1 [210]) does not differ substantially in its structure in comparison to XML; consequently, it can be also parsed/generated sequentially like XML.

[32] The results correspond to E2ENP documents sized ~5kByte per Pre-negotiation, ~ 4kByte per short Negotiation, ~2kByte per short Re-negotiation and ~7kByte per full Negotiation/Re-negotiation – see Section 6.2

must be compared to the previously exchanged data to recognize the change. Consequently, the E2ENP referencing mechanism can reduce the time for taking adaptation decisions at the Answerer side as the Offerer prepares and signals the adaptation decision directly upon the results of the QoS monitoring (see also Section 5.3.1).

This chapter presented the major results and evaluations of the operation of the E2ENP UA and discussed the profits for the session-management performance achieved through E2ENP in comparison with the state of the art. Complete summary and conclusions on the innovations proposed through the E2ENP specification and development are discussed in the next Chapter 7.

7 Summary and Conclusions

This work provides an extensive study on existing session configuration and negotiation mechanisms to derive requirements for an optimised session-negotiation protocol and to discuss missing features of the current application-signalling approaches for multimedia services.

Consequently, the work presented the novel negotiation protocol for QoS named End-to-End Negotiation Protocol. The special feature of the protocol is the modular negotiation of different service capabilities at application and at transport layer and the specification of multiple operation abstractions in form of QoS contracts and contexts. The hierarchical QoS model and its referencing mechanisms enable the reduction of the signalling load for the applications at session establishment and adaptation scenarios. The measurement results demonstrate the scalability of the E2ENP negotiations and show that this mechanism can significantly reduce the time of QoS adaptation processes. The comparative measurements of E2ENP and SDPng operation confirm the advantages of E2ENP in comparison with the state-of-the-art service-configuration and -negotiation approaches.

The E2ENP negotiation language is designed to introduce flexible adaptation scenarios different from the currently existing sender/receiver adaptations. These novel negotiation and adaptation capabilities enable the interaction of the protocol with negotiation-mentoring entities thus enhancing existing mechanisms (like the Offer/Answer Model) with alternative application scenarios.

The implementation of E2ENP integrates standard service-description and negotiation technologies to enable the application of multimedia services in heterogeneous operation environments. The specification of E2ENP lead to multiple standardisation proposals and influenced the specifications of SDPng and MPEG-21 DIA. E2ENP integrates XML document-linking mechanism and introduces also its own logical linking to components of the hierarchical QoS specification, which is a protocol-specific approach not available in the current XML technology. The E2ENP linking and the integration of the XML Signature Technology make the modular interaction of mentoring entities with the protocol possible and represent specific E2ENP security mechanism. E2ENP enables the application of the Session Initiation Protocol in different modes and phases and allows the utilisation of E2ENP's User Agents in different negotiation- and scenario-specific roles. Furthermore, the E2ENP UA state machines represent an enhancement to the SIP state machines. They enable the concurrent sharing of the signalling environment between applications with different priorities and the integration of SIP to perform in a multi-application environment.

E2ENP includes in its negotiation process the *economy principle* thus providing coordination for the resources reservations of the negotiating peers. The considerations about the resource management of the applications as a whole lead to the development of QoS-guarantees recommendations for the signalling and to emulation of such. Unlike currently existing mechanisms that provide QoS guarantees only for the user-perceivable media, the E2ENP contribution regards the complete communication among applications for ensuring QoS.

Since E2ENP represents a completely novel approach for orchestration of the multimedia services executed in heterogeneous environments, the application of E2ENP within QoS architectures leaves still some open questions with respect to the application of provider profiles and the validation of the negotiation statements, the inter-working of the protocol with the resource-management mechanisms of the services, the utilisation of security and accounting approaches in relation to the negotiations and further research on the failure management at negotiation and resources coordination. This contribution provides where appropriate considerations and recommendations about future work to be performed in order to integrate E2ENP into global service-management architectures like 3GPP IMS or ITU H.323.

The negotiation protocol introduced in this contribution represents a novel application for service and session management. It enables the proper and optimal preparation and adaptation of application configurations to guarantee the operation of different multimedia services in heterogeneous environments and to ensure the flexible adjustment of the services to different user requirements and application scenarios.

The work on E2ENP affected the results of and produced contributions to several industry and academia research projects in the area of provisioning of next generation networks and services. The major work on E2ENP was performed within the scope of the projects AKOM, MIND and DAIDALOS.

Annex A Projects Overview – E2ENP Development Steps

This annex gives an overview on the projects that influenced the work on E2ENP and provides a short summary on the publications (in co-authorship of the author of this book) originated during the development of E2ENP and of the concepts for more efficient end-to-end QoS signalling and media controls.

The work on the End-to-End Negotiation Protocol was accomplished within the scope of several projects, i.e. BRAIN (Broadband Radio Access for IP Based Networks) / MIND (Mobile IP based Network Developments) [130], MASA (Mobility and Service Adaptation in Heterogeneous Mobile Environments) [197][198], DAIDALOS (Designing Advanced network Interfaces for the Delivery and Administration of Location independent, Optimised personal Services) [181] and AKOM (Adaptivität in heterogenen Kommunikationsnetzen mit drahtlosem Zugang) [284].

The QoS broker developed at University of Ulm (Distributed Systems Department) was initially designed to provide adaptation of the user-perceivable media only (see Section 2.4.2.1). For this purpose, the control processes within the broker included distributed negotiation and re-negotiation [17][70] similar to the Offer/Answer Model with SDP [35][36] (see also Section 2.3.1.2 and Annex E). However, during the broker's development within the projects BRAIN/MIND, MASA and the two initial phases of the AKOM project, it became increasingly important that also optimisation of the control processes within the broker should be handled. The requirements for the adaptation of the control data provided within these projects served as an impetus for initiating the work on E2ENP. A basic set of requirements for novel and more efficient end-to-end signalling was published as an IETF draft [14] within the scope of the MIND project. Additionally, these early E2ENP research was published as a conference contribution [130]. Further achievements in the MIND project (and within the phase 2 of the AKOM framework that was running in parallel to MIND) were the development of the initial E2ENP language definition [173] and of the E2ENP software architecture [170] made available as patent applications. These concepts were also presented at the MIND public workshops [171][172] at the finalisation phase of the MIND project. The results of the MIND project were further applied (also within the scope of AKOM) for refining the E2ENP negotiation models (see Section 3.1) with ideas about ad-hoc communications and provider interactions with the protocol and for implementing and testing the E2ENP User Agent based on SIP surrogate with RMI and on jSIP. The outcomes of this research were published as HICSS [65], IEEE [131][132] and WSEAS [133] journal and conference contributions. The work on E2ENP continued within the DAIDALOS project and within the final phase of the AKOM framework and lead to the development of one additional dialect of SDPng based on MPEG-21 DIA concepts. Furthermore, research on advanced adaptation-control scenarios with content adaptation nodes and session mobility was carried out. Additional studies on the provider-managed systems were also performed. These investigations lead to the optimisation of the E2ENP negotiation language also in terms of adoption of advanced MPEG and XML concepts (see Section 4.1.3). Furthermore, some of the advanced scenarios (i.e. session mobility – see Sections 3.3 and 5.2.2.4.3) were applied within the software architecture of the E2ENP UA. The concept of XML processing used within the E2ENP UA (Section 5.2.2.1) proved successful also for the implementation of the MPEG-21-DIA-based SDPng dialect. At this phase of the development of the E2ENP UA, the NIST SIP carrier was also integrated in the UA and further performance measurements were executed. The results of the research done within DAIDALOS and AKOM were published as IETF drafts, ISO/IEC MPEG contributions and as conference papers as follows:

- The content adaptation system using advanced adaptation models was presented at the first DAIDALOS Workshop [250] and as paper contribution to IPSI [92]
- Concepts of the MPEG-21-DIA-based SDPng dialect and proposals for improvement of the SDP in terms of better negotiation of provider-controlled resources were presented as contributions to IETF [96][285] and ISO/IEC MPEG [95][97] and as conference papers at EuroIMSA [202] and WIAMIS [221].
- Some additional aspects of the MPEG-21-DIA-based session controls were also provided as an IETF contribution [286]
- Results on application of SIP-only service-discovery and service-controls solution were presented as a paper contribution to ICNS [239]
- Finally, the results of the PhD thesis of the author "Coordination of Multimedia Services and Applications in Mobile, Heterogeneous Network Environment" [287] lead to three further publications in the area of peer-to-peer SIP applications [288], application of SIP in a middleware architecture [289] and application of terminal-dependent and -independent service mobility with SIP and SDP [290].

The work on E2ENP within the scope of this book was motivated through the above-mentioned projects. This work continued over four years and resulted in the investigations and the contributions within this book.

Annex B Service/Network Profiles at Registrations

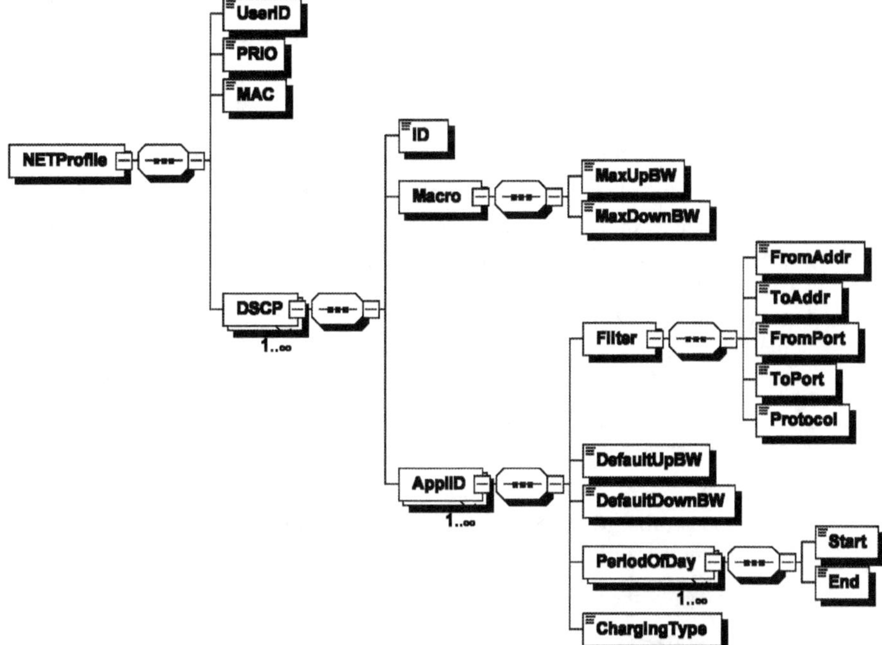

Figure 113: XML schema for provider network profile

This annex presents an example definition of network and service profiles used by the providers to restrict the access of the users to specific network and application services. This profile information can be exchanged at terminal registration to the provider-owned network or can be fixed-saved within the terminal as a provider-specific terminal pre-configuration (see also Sections 2.3.2.2.2, 2.4.1 and 3.2). The profile information is used within the terminals for properly validating requests for service configuration originated terminal-internally from the respective user-applications or from the negotiating partners of the terminal at the negotiation process. The profiles presented here stem partially from concepts developed within the DAIDALOS project for provider-controlled network access [92].

The network profile based on DiffServ controls (Figure 113) consists of the following elements:

- *NETProfile* is the root of the corresponding XML schema definition
- *UserID* is the unique identifier of the user. Upon this field the provider can recognise the user as such and prove the granted application services for the user
- *PRIO* identifies the network service priority for the respective user, e.g. the users can be differentiated as business and private users with corresponding profiles
- *MAC (Media Access Control)* identifies the physical address of the provider network of terminal's attachment

- *DSCP (Differentiated Services Code Point)* is an identifier of the DiffServ class used for the corresponding network service. One or multiple such classes can be associated with the user of the respective access network
- *ID* is the value of the *DSCP* identifier denoting the service class
- *Macro* identifies the maximal up-/down-streaming bandwidth, where *MaxUpBW* and *MaxDownBW* denote the respective bandwidth limitation levels
- *ApplID* identifies an application which QoS is being managed via DiffServ (one or multiple applications can be DiffServ controlled), where:
  - o *Filter* indicates the exact end-points within the access network (*FromAddr*, *FromPort*, *ToAddr*, *ToPort*) and the protocol (*Protocol*) being managed via DiffServ. Note that the end-points in the access network are associated with the user's terminal and the provider's proxy. However, the end-points of the application, e.g. between two terminals or between a terminal and an application server, are established dynamically during the negotiation process
  - o *DefaultUpBW* and *DefaulDownBW* are the bandwidth limitation levels per application for up-/down-channels for communication
  - o *PeriodOfDay* with *Start* and *End* denote the time validity for the defined application profile. Multiple periods of profile validity can be defined indicating times with higher or lower network load
  - o *ChargingType* is the type of charging the applications for the used network traffic

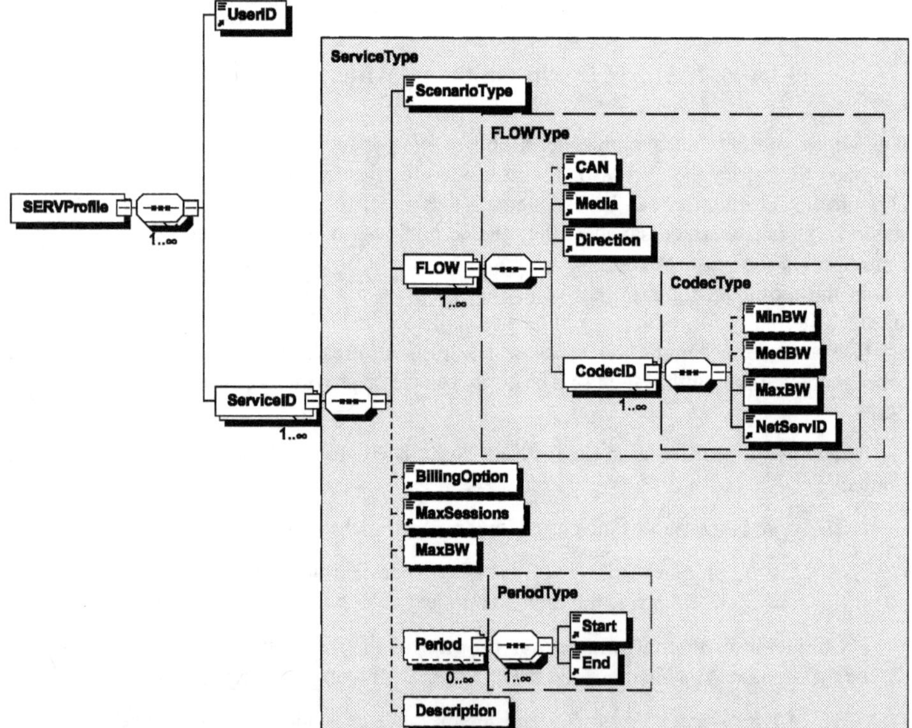

Figure 114: XML schema for provider service profile

Figure 114 presents the service profile that refines the application configuration options for the respective users and services. The fields of the service profile mean:

- *SERVProfile* is the root of the XML schema for the service definitions
- *UserID* is the unique identifier of the user. Upon this field the provider can recognise the user as such and prove the granted network services for the user
- *ServiceID* identifies the application service/-s granted for the user
- *ScenarioType* denotes the allowed application-level service-adaptation scenarios allowed for the user, e.g. proxy-controlled services, session migration, etc.
- *Flow* identifies the data-flows associated with the service, where
  - *CAN (Content Adaptation Node)* provides additional information on the art of accessing of network-situated content-adaptation services
  - *Media* identifies the type of the media associated with the data-flow
  - *Direction* denotes the flow direction (i.e. send, receive) and the activation possibilities (i.e. active, inactive) for the flow
  - *CodecID* is the allowed media coding format with indication of the bandwidth limits for network transportation (i.e. *MinBW*, *MedBW*, *MaxBW*) and the associated network service class (*NetServID*)
- *BillingOption* identifies the art of charging for the service
- *MaxSessions* indicates the number of allowed sessions per service
- *MaxBW* denotes the allowed maximal bandwidth for all of the running sessions per service
- *Period* with *Start* and *End* indicates the validity time for the service
- *Description* is any additional description associated with the service

Annex C E2ENP XML Examples Used for Measurements

The figures presented in this annex (Figure 115 – Figure 123) show E2ENP examples used for the measurements.

```
<e2enpDesc xmlns="http://www.iana.org/sdpng-e2enp" xmlns:e2enp="http://www.iana.org/sdpng/e2enp"
xmlns:sdpng="http://www.iana.org/sdpng" xmlns:rtp="http://www.iana.org/sdpng/rtp"
xmlns:xsi="http://www.w3.org/2001/XMLSchema-instance"
xsi:schemaLocation="http://www.iana.org/sdpng/e2enp e2enp-02.xsd
http://www.iana.org/sdpng/rtp rtp-pkg-sdpng-04.xsd
http://www.iana.org/sdpng-e2enp sdpng-e2enp-01.xsd">
   <e2enp:purpose>
      <e2enp:session user="Mary" session-id="2890844526" version="2890842807" nettype="IN"
      addrtype="IP4" addr="43.196.180.1">
         <e2enp:expires time="3600"/>
      </e2enp:session>
      <e2enp:description type="request" name="pre-negotiation" style="full" mode="push"/>
   </e2enp:purpose>
   <e2enp:qosdef name="capabilities">
      <e2enp:audio-codec name="PCMU" scope="applicable"/>
      <e2enp:audio-codec name="G729" scope="applicable"/>
      <e2enp:audio-codec name="G722" scope="possible"/>
      <e2enp:pt name="rtp-avp-0" pt="0" format="PCMU"/>
      <e2enp:pt name="rtp-avp-18" pt="18" format="G729"/>
      <e2enp:pt name="rtp-avp-9" pt="9" format="G722"/>
      <e2enp:video-codec name="H263" scope="applicable"/>
      <e2enp:video-codec name="H261" scope="applicable"/>
      <e2enp:video-codec name="WAVI" scope="possible"/>
      <e2enp:pt name="rtp-avp-34" pt="34" format="H263"/>
      <e2enp:pt name="rtp-avp-31" pt="31" format="H261"/>
      <e2enp:pt name="rtp-avp-100" pt="100" format="WAVI">
         <e2enp:qosdef name="capabilities">
            <e2enp:video-codec frame-rate-range="5,30" frame-size-set="QCIF,CIF"
            color-quality-set="9100,10000" overall-quality-range="9100,10000"/>
         </e2enp:qosdef>
      </e2enp:pt>
   </e2enp:qosdef>
   <e2enp:qosdef name="contracts">
      <e2enp:qos-contracts for="audio">
         <e2enp:contract type="audio" name="audio-contract-1" sampling-rate-set="4000,8000"
         channel-set="1"/>
         <e2enp:contract type="audio" name="audio-contract-2" sampling-rate-set="8000,12000"
         channel-set="1,2"/>
         <e2enp:contract type="audio" name="audio-contract-3" sampling-rate-set="12000,16000"
         channel-set="1"/>
         <e2enp:map contract="audio-contract-1" format="rtp-avp-0" peer-role="receiver"/>
         <e2enp:map contract="audio-contract-2" format="rtp-avp-18" peer-role="receiver"/>
         <e2enp:map contract="audio-contract-3" format="rtp-avp-9" peer-role="receiver"/>
      </e2enp:qos-contracts>
```

Figure 115: Full Pre-negotiation (XML file part one)

```
        <e2enp:qos-contracts for="video">
            <e2enp:contract type="video" name="video-contract-1" frame-rate-set="10,15" frame-size-set="CIF"
            color-quality-set="2,8" overall-quality-range="9500,9800"/>
            <e2enp:contract type="video" name="video-contract-2" frame-rate-set="15,20" frame-size-set="QCIF"
            color-quality-set="8,16" overall-quality-range="9800,9900"/>
            <e2enp:contract type="video" name="video-contract-3" frame-rate-set="20,25"
            frame-size-set="QCIF,CIF" color-quality-set="16,32" overall-quality-range="9900,9960"/>
            <e2enp:map contract="video-contract-1" format="rtp-avp-34" peer-role="receiver"/>
            <e2enp:map contract="video-contract-2" format="rtp-avp-31" peer-role="receiver"/>
            <e2enp:map contract="video-contract-3" format="rtp-avp-100" peer-role="receiver"/>
        </e2enp:qos-contracts>
        <e2enp:qos-contracts for="transport">
            <e2enp:contract type="transport" name="transport-contract-1" pbr="320" sbr="320" mps="72"
            mbs="72" qtp="guaranteed-service">
                <e2enp:suboption name="sensitivity-1" type="SI-parm" me2ed="80" me2edv="10" mplr="2E-2"/>
                <e2enp:suboption name="sensitivity-2" type="SI-parm" me2ed="120" me2edv="20" mplr="2E-2"/>
            </e2enp:contract>
            <e2enp:contract type="transport" name="transport-contract-2" pbr="96" sbr="96" mps="72"
            mbs="72" qtp="guaranteed-service">
                <e2enp:suboption name="sensitivity-1" type="SI-parm" me2ed="80" me2edv="10" mplr="2E-2"/>
                <e2enp:suboption name="sensitivity-2" type="SI-parm" me2ed="120" me2edv="20" mplr="2E-2"/>
            </e2enp:contract>
            <e2enp:contract type="transport" name="transport-contract-3" pbr="29" sbr="29" mps="72"
            mbs="72" qtp="guaranteed-service">
                <e2enp:suboption name="sensitivity-1" type="SI-parm" me2ed="80" me2edv="10" mplr="2E-2"/>
                <e2enp:suboption name="sensitivity-2" type="SI-parm" me2ed="120" me2edv="20" mplr="2E-2"/>
            </e2enp:contract>
        </e2enp:qos-contracts>
    </e2enp:qosdef>
</e2enpDesc>
```


Figure 116: Full Pre-negotiation (XML file part two)

```
<e2enpDesc xmlns="http://www.iana.org/sdpng-e2enp" xmlns:e2enp="http://www.iana.org/sdpng/e2enp"
xmlns:sdpng="http://www.iana.org/sdpng" xmlns:rtp="http://www.iana.org/sdpng/rtp"
xmlns:xsi="http://www.w3.org/2001/XMLSchema-instance"
xsi:schemaLocation="http://www.iana.org/sdpng/e2enp e2enp-02.xsd
http://www.iana.org/sdpng/rtp rtp-pkg-sdpng-04.xsd
http://www.iana.org/sdpng-e2enp sdpng-e2enp-01.xsd">
   <e2enp:purpose>
      <e2enp:session user="Mary" session-id="2890844526" version="2890842807" nettype="IN"
      addrtype="IP4" addr="43.196.180.1">
         <e2enp:expires time="3600"/>
      </e2enp:session>
      <e2enp:use>
         <e2enp:session user="Mary" session-id="2890843112" version="2890841393" nettype="IN"
         addrtype="IP4" addr="43.196.180.1"/>
      </e2enp:use>
      <e2enp:description type="request" name="negotiation" style="short" mode="push"/>
   </e2enp:purpose>
   <e2enp:qoscfg>
      <e2enp:component name="audio-stream-1" media="audio">
         <sdpng:alt name="AVP-audio-0">
            <rtp:session format="rtp-avp-0">
               <rtp:udp role="receive" nettype="IN" addrtype="IP4" addr="43.196.180.1" rtp-port="7800" rtcp-
               port="7801"/>
            </rtp:session>
         </sdpng:alt>
         <sdpng:alt name="AVP-audio-9">
            <rtp:session format="rtp-avp-9">
               <rtp:udp role="receive" nettype="IN" addrtype="IP4" addr="43.196.180.1" rtp-port="7840" rtcp-
               port="7851"/>
            </rtp:session>
         </sdpng:alt>
      </e2enp:component>
      <e2enp:component name="video-stream-1" media="video">
         <sdpng:alt name="AVP-video-34">
            <rtp:session format="rtp-avp-34">
               <rtp:udp role="receive" nettype="IN" addrtype="IP4" addr="43.196.180.1" rtp-port="7900" rtcp-
               port="7901"/>
            </rtp:session>
         </sdpng:alt>
         <sdpng:alt name="AVP-video-31">
            <rtp:session format="rtp-avp-31">
               <rtp:udp role="receive" nettype="IN" addrtype="IP4" addr="43.196.180.1" rtp-port="7920" rtcp-
               port="7921"/>
            </rtp:session>
         </sdpng:alt>
      </e2enp:component>
   </e2enp:qoscfg>
```


Figure 117: Short Negotiation (XML file part one)

```
<e2enp:qoscfg level="stream">
   <!-- adaptation path for single streams -->
   <e2enp:adapath name="audio1" ref_component="audio-stream-1">
      <e2enp:alt default="true" name="nominal" ref_contract="audio-contract-1" ref_transport-
      contract="transport-contract-1" ref_suboption="sensitivity-1" scope="applicable"/>
      <e2enp:alt name="choice1" ref_contract="audio-contract-3" ref_transport-contract="transport-
      contract-3" ref_suboption="sensitivity-2" scope="applicable"/>
   </e2enp:adapath>
   <e2enp:adapath name="video1" ref_component="video-stream-1">
      <e2enp:alt default="true" name="nominal" ref_contract="video-contract-1" scope="applicable"/>
      <e2enp:alt name="choice1" ref_contract="video-contract-2" scope="possible"/>
   </e2enp:adapath>
   <!-- Possible associations of streams between user A and B -->
   <e2enp:context name="association1-1" scope="applicable">
      <e2enp:comp name="element1" ref_adapath="audio1"/>
      <e2enp:comp name="element2" ref_adapath="video1"/>
      <e2enp:constraints>
         <e2enp:par name="lipsync-delay" ref_adapath="audio1" max="2"/>
         <e2enp:par name="aggregated-bw" max="64000"/>
      </e2enp:constraints>
   </e2enp:context>
   <e2enp:context name="association1-2" scope="possible">
      <e2enp:comp name="element1" ref_adapath="audio1"/>
   </e2enp:context>
   <e2enp:adapath name="associations-A-B">
      <e2enp:alt default="true" name="nominal" ref_context="association1-1"/>
      <e2enp:alt name="choice1" ref_context="association1-2"/>
   </e2enp:adapath>
</e2enp:qoscfg>
</e2enpDesc>
```

Figure 118: Short Negotiation (XML file part two)

```
<e2enpDesc xmlns="http://www.iana.org/sdpng-e2enp" xmlns:e2enp="http://www.iana.org/sdpng/e2enp"
xmlns:sdpng="http://www.iana.org/sdpng" xmlns:rtp="http://www.iana.org/sdpng/rtp"
xmlns:xsi="http://www.w3.org/2001/XMLSchema-instance"
xsi:schemaLocation="http://www.iana.org/sdpng/e2enp e2enp-02.xsd
http://www.iana.org/sdpng/rtp rtp-pkg-sdpng-04.xsd
http://www.iana.org/sdpng-e2enp sdpng-e2enp-01.xsd">
   <e2enp:purpose>
      <e2enp:session user="Mary" session-id="2890844526" version="2890842807" nettype="IN"
      addrtype="IP4" addr="43.196.180.1">
         <e2enp:expires time="3600"/>
      </e2enp:session>
      <e2enp:use>
         <e2enp:session user="Mary" session-id="2890843112" version="2890841393" nettype="IN"
         addrtype="IP4" addr="43.196.180.1"/>
      </e2enp:use>
      <e2enp:description type="request" name="re-negotiation" style="short" coordination="true"/>
   </e2enp:purpose>
   <e2enp:enforce>
      <e2enp:contract-level path="associations-A-B.association1-1.video1.video-contract-2"/>
      <e2enp:target name="$contract"/>
   </e2enp:enforce>
   <e2enp:block>
      <e2enp:contract-level path="associations-A-B.association1-1.video1.video-contract-1"/>
      <e2enp:target name="$contract"/>
   </e2enp:block>
</e2enpDesc>
```

Figure 119: Short Re-negotiation (XML file)

```
<e2enpDesc xmlns="http://www.iana.org/sdpng-e2enp" xmlns:e2enp="http://www.iana.org/sdpng/e2enp"
xmlns:sdpng="http://www.iana.org/sdpng" xmlns:rtp="http://www.iana.org/sdpng/rtp"
xmlns:xsi="http://www.w3.org/2001/XMLSchema-instance"
xsi:schemaLocation="http://www.iana.org/sdpng/e2enp e2enp-02.xsd
http://www.iana.org/sdpng/rtp rtp-pkg-sdpng-04.xsd
http://www.iana.org/sdpng-e2enp sdpng-e2enp-01.xsd">
   <e2enp:purpose>
      <e2enp:session user="Mary" session-id="2890844526" version="2890842807" nettype="IN"
      addrtype="IP4" addr="43.196.180.1">
         <e2enp:expires time="3600"/>
      </e2enp:session>
      <e2enp:description type="request" name="negotiation" style="full" mode="push"/>
   </e2enp:purpose>

   <!-- Same "e2enp:qosdef " sections as the full Pre-negotiation and same " e2enp:qoscfg" sections as the
   short Negotiation-->

</e2enpDesc>
```

Figure 120: Full Negotiation

```
<e2enp:purpose>
  <e2enp:session user="Mary" session-id="2890844526" version="2890842807" nettype="IN" addrtype="IP4"
  addr="43.196.180.1">
    <e2enp:expires time="3600"/>
  </e2enp:session>
  <e2enp:description type="request" name="re-negotiation" style="full" mode="push"/>
</e2enp:purpose>
```

Figure 121: Full Re-negotiation PDU header

```
<e2enpDesc xmlns="http://www.iana.org/sdpng-e2enp" xmlns:e2enp="http://www.iana.org/sdpng/e2enp"
xmlns:sdpng="http://www.iana.org/sdpng" xmlns:rtp="http://www.iana.org/sdpng/rtp"
xmlns:xsi="http://www.w3.org/2001/XMLSchema-instance"
xsi:schemaLocation="http://www.iana.org/sdpng/e2enp e2enp-02.xsd
http://www.iana.org/sdpng/rtp rtp-pkg-sdpng-04.xsd
http://www.iana.org/sdpng-e2enp sdpng-e2enp-01.xsd">
  <e2enp:qosdef name="capabilities">
    <e2enp:audio-codec name="PCMU" scope="applicable"/>
    <e2enp:audio-codec name="G729" scope="applicable"/>
    <e2enp:audio-codec name="G722" scope="possible"/>
    <e2enp:pt name="rtp-avp-0" pt="0" format="PCMU"/>
    <e2enp:pt name="rtp-avp-18" pt="18" format="G729"/>
    <e2enp:pt name="rtp-avp-9" pt="9" format="G722"/>
    <e2enp:video-codec name="H263" scope="applicable"/>
    <e2enp:video-codec name="H261" scope="applicable"/>
    <e2enp:video-codec name="WAVI" scope="possible"/>
    <e2enp:pt name="rtp-avp-34" pt="34" format="H263"/>
    <e2enp:pt name="rtp-avp-31" pt="31" format="H261"/>
    <e2enp:pt name="rtp-avp-100" pt="100" format="WAVI"/>
  </e2enp:qosdef>
  <e2enp:qosdef name="contracts">
    <e2enp:qos-contracts for="audio">
      <e2enp:contract type="audio" name="audio-contract-1" sampling-rate-set="4000,8000"
      channel-set="1"/>
      <e2enp:contract type="audio" name="audio-contract-2" sampling-rate-set="8000,12000"
      channel-set="1,2"/>
      <e2enp:contract type="audio" name="audio-contract-3" sampling-rate-set="12000,16000"
      channel-set="1"/>
      <e2enp:map contract="audio-contract-1" format="rtp-avp-0" peer-role="receiver"/>
      <e2enp:map contract="audio-contract-2" format="rtp-avp-18" peer-role="receiver"/>
      <e2enp:map contract="audio-contract-3" format="rtp-avp-9" peer-role="receiver"/>
    </e2enp:qos-contracts>
    <e2enp:qos-contracts for="video">
      <e2enp:contract type="video" name="video-contract-1" frame-rate-set="10,15" frame-size-set="CIF"
      color-quality-set="2,8" overall-quality-range="9500,9800"/>
      <e2enp:contract type="video" name="video-contract-2" frame-rate-set="15,20"
      frame-size-set="QCIF" color-quality-set="8,16" overall-quality-range="9800,9900"/>
      <e2enp:contract type="video" name="video-contract-3" frame-rate-set="20,25"
      frame-size-set="QCIF,CIF" color-quality-set="16,32" overall-quality-range="9900,9960"/>
      <e2enp:map contract="video-contract-1" format="rtp-avp-34" peer-role="receiver"/>
      <e2enp:map contract="video-contract-2" format="rtp-avp-31" peer-role="receiver"/>
      <e2enp:map contract="video-contract-3" format="rtp-avp-100" peer-role="receiver"/>
    </e2enp:qos-contracts>
  </e2enp:qosdef>
```

Figure 122: SDPng simulation (XML file part one)

```
<e2enp:qoscfg>
    <e2enp:component name="audio-stream-1" media="audio">
        <sdpng:alt name="AVP-audio-0">
            <rtp:session format="rtp-avp-0">
                <rtp:udp role="receive" nettype="IN" addrtype="IP4" addr="43.196.180.1" rtp-port="7800"
                rtcp-port="7801"/>
            </rtp:session>
        </sdpng:alt>
        <sdpng:alt name="AVP-audio-9">
            <rtp:session format="rtp-avp-9">
                <rtp:udp role="receive" nettype="IN" addrtype="IP4" addr="43.196.180.1" rtp-port="7840"
                rtcp-port="7851"/>
            </rtp:session>
        </sdpng:alt>
    </e2enp:component>
    <e2enp:component name="video-stream-1" media="video">
        <sdpng:alt name="AVP-video-34">
            <rtp:session format="rtp-avp-34">
                <rtp:udp role="receive" nettype="IN" addrtype="IP4" addr="43.196.180.1" rtp-port="7900"
                rtcp-port="7901"/>
            </rtp:session>
        </sdpng:alt>
        <sdpng:alt name="AVP-video-31">
            <rtp:session format="rtp-avp-31">
                <rtp:udp role="receive" nettype="IN" addrtype="IP4" addr="43.196.180.1" rtp-port="7920"
                rtcp-port="7921"/>
            </rtp:session>
        </sdpng:alt>
    </e2enp:component>
</e2enp:qoscfg>
</e2enpDesc>
```

Figure 123: SDPng simulation (XML file part two)

Annex D E2ENP Codec Classification Schemes

This annex demonstrates the audio and video classicisation schemes (CS) applied within the End-to-End Negotiation Protocol (see also Section 4.1.3.3). The CS numbers in the following two tables repeat the MPEG-7 MDS numbers and enhance the number sequence with additional popular codecs.

Audio CS (CS Name – *urn:mpeg:e2enp:cs:AudioCodingFormatCS:2001*):

CS Number	Name	Standardized by
1	Linear PCM	International Telephone and Telegraph
2	ITU-T G.7xx	ITU-T
2.1	ITU-T G.711	ITU-T
2.1.1	ITU-T G.711 PCMU	ITU-T
2.1.2	ITU-T G.711 PCMA	ITU-T
2.2	ITU-T G.721	ITU-T
2.3	ITU-T G.722	ITU-T
2.3.1	ITU-T G.722.1	ITU-T
2.3.2	ITU-T G.722.2	ITU-T
2.3.3	ITU-T G.723	ITU-T
2.3.4	ITU-T G.723.1	ITU-T
2.3.5	ITU-T G.726	ITU-T
2.3.6	ITU-T G.727	ITU-T
2.3.7	ITU-T G.728	ITU-T
2.3.8	ITU-T G.729	ITU-T
3	MPEG-1 Audio	ISO/IEC MPEG
3.1	MPEG-1 Audio Layer I	ISO/IEC MPEG
3.2	MPEG-1 Audio Layer II	ISO/IEC MPEG
3.3	MPEG-1 Audio Layer III	ISO/IEC MPEG
4	MPEG-2 Audio	ISO/IEC MPEG
4.1	MPEG-2 Audio Low Sampling Rate	ISO/IEC MPEG
4.1.1	MPEG-2 Audio Low Sampling Rate Layer I	ISO/IEC MPEG
4.1.2	MPEG-2 Audio Low Sampling Rate Layer II	ISO/IEC MPEG
4.1.3	MPEG-2 Audio Low Sampling Rate Layer III	ISO/IEC MPEG
4.2	MPEG-2 Backward Compatible Multi-Channel	ISO/IEC MPEG
4.2.1	Backward Compatible Multi-Channel Layer	ISO/IEC MPEG
4.2.2	Backward Compatible Multi-Channel Layer	ISO/IEC MPEG
4.2.3	Backward Compatible Multi-Channel Layer	ISO/IEC MPEG
4.3	MPEG-2 Audio AAC	ISO/IEC MPEG
4.3.1	MPEG-2 Audio AAC Low Complexity Profile	ISO/IEC MPEG
4.3.2	MPEG-2 Audio AAC Main Profile	ISO/IEC MPEG
4.3.3	MPEG-2 Audio AAC Sampling Rate Scalable Profile	ISO/IEC MPEG
5	MPEG-4 Audio	ISO/IEC MPEG
5.1	MPEG-4 Audio Synthetic Profile	ISO/IEC MPEG
5.1.1	MPEG-4 Audio Synthetic Profile @ Level 1	ISO/IEC MPEG
5.1.2	MPEG-4 Audio Synthetic Profile @ Level 2	ISO/IEC MPEG
5.1.3	MPEG-4 Audio Synthetic Profile @ Level 3	ISO/IEC MPEG
5.2	MPEG-4 Audio Speech Profile	ISO/IEC MPEG

5.2.1	MPEG-4 Audio Speech Profile @ Level 1	ISO/IEC MPEG
5.2.2	MPEG-4 Audio Speech Profile @ Level 2	ISO/IEC MPEG
5.3	MPEG-4 Audio Scalable Profile	ISO/IEC MPEG
5.3.1	MPEG-4 Audio Scalable Profile @ Level 1	ISO/IEC MPEG
5.3.2	MPEG-4 Audio Scalable Profile @ Level 2	ISO/IEC MPEG
5.3.2	MPEG-4 Audio Scalable Profile @ Level 3	ISO/IEC MPEG
5.3.3	MPEG-4 Audio Scalable Profile @ Level 4	ISO/IEC MPEG
5.4	MPEG-4 Main Audio Profile	ISO/IEC MPEG
5.4.1	MPEG-4 Main Audio Profile @ Level 1	ISO/IEC MPEG
5.4.2	MPEG-4 Main Audio Profile @ Level 2	ISO/IEC MPEG
5.4.3	MPEG-4 Main Audio Profile @ Level 3	ISO/IEC MPEG
5.4.4	MPEG-4 Main Audio Profile @ Level 4	ISO/IEC MPEG
5.5	MPEG-4 High Quality Audio Profile	ISO/IEC MPEG
5.5.1	MPEG-4 High Quality Audio Profile @ Level 1	ISO/IEC MPEG
5.5.2	MPEG-4 High Quality Audio Profile @ Level 2	ISO/IEC MPEG
5.5.3	MPEG-4 High Quality Audio Profile @ Level 3	ISO/IEC MPEG
5.5.4	MPEG-4 High Quality Audio Profile @ Level 4	ISO/IEC MPEG
5.5.5	MPEG-4 High Quality Audio Profile @ Level 5	ISO/IEC MPEG
5.5.6	MPEG-4 High Quality Audio Profile @ Level 6	ISO/IEC MPEG
5.5.7	MPEG-4 High Quality Audio Profile @ Level 7	ISO/IEC MPEG
5.5.8	MPEG-4 High Quality Audio Profile @ Level 8	ISO/IEC MPEG
5.6	MPEG-4 Low Delay Audio Profile	ISO/IEC MPEG
5.6.1	MPEG-4 Low Delay Audio Profile @ Level 1	ISO/IEC MPEG
5.6.2	MPEG-4 Low Delay Audio Profile @ Level 2	ISO/IEC MPEG
5.6.3	MPEG-4 Low Delay Audio Profile @ Level 3	ISO/IEC MPEG
5.6.4	MPEG-4 Low Delay Audio Profile @ Level 4	ISO/IEC MPEG
5.6.5	MPEG-4 Low Delay Audio Profile @ Level 5	ISO/IEC MPEG
5.6.6	MPEG-4 Low Delay Audio Profile @ Level 6	ISO/IEC MPEG
5.6.7	MPEG-4 Low Delay Audio Profile @ Level 7	ISO/IEC MPEG
5.6.8	MPEG-4 Low Delay Audio Profile @ Level 8	ISO/IEC MPEG
5.7	MPEG-4 Natural Audio Profile	ISO/IEC MPEG
5.7.1	MPEG-4 Natural Audio Profile @ Level 1	ISO/IEC MPEG
5.7.2	MPEG-4 Natural Audio Profile @ Level 2	ISO/IEC MPEG
5.7.3	MPEG-4 Natural Audio Profile @ Level 3	ISO/IEC MPEG
5.7.4	MPEG-4 Natural Audio Profile @ Level 4	ISO/IEC MPEG
5.8	MPEG-4 Mobile Audio Internetworking Profile	ISO/IEC MPEG
5.8.1	MPEG-4 Mobile Audio Internetworking Profile @ Level 1	ISO/IEC MPEG
5.8.2	MPEG-4 Mobile Audio Internetworking Profile @ Level 2	ISO/IEC MPEG
5.8.3	MPEG-4 Mobile Audio Internetworking Profile @ Level 3	ISO/IEC MPEG
5.8.4	MPEG-4 Mobile Audio Internetworking Profile @ Level 4	ISO/IEC MPEG
5.8.5	MPEG-4 Mobile Audio Internetworking Profile @ Level 5	ISO/IEC MPEG
5.8.6	MPEG-4 Mobile Audio Internetworking Profile @ Level 6	ISO/IEC MPEG
6	GSM	3GPP
6.1	GSM 06.10	3GPP, ETSI
6.2	GSM 06.60	3GPP
7	AMR	3GPP

8	AC3	ATSC A/52
9	DTS	Digital Theater Systems Inc.
10	ATRAC	SONY
10.1	ATRAC 1	SONY
10.2	ATRAC 2	SONY
10.3	ATRAC 3	SONY
11	WMA	Microsoft
12	DVI	Intel, IMA
12.1	DVI4	Intel, IMA
12.2	VDVI	Intel, IMA
13	LPC10e (FIPS 1015)	US Government

Video CS (CS Name – *urn:mpeg:e2enp:cs:VisualCodingFormatCS:2001*):

CS Number	Name	Standardized by
1	MPEG-1 Video	ISO/IEC MPEG
2	MPEG-2 Video	ISO/IEC MPEG
2.1	MPEG-2 Video Simple Profile	ISO/IEC MPEG
2.1.1	MPEG-2 Video Simple Profile @ Main Level	ISO/IEC MPEG
2.2	MPEG-2 Video Main Profile	ISO/IEC MPEG
2.2.1	MPEG-2 Video Main Profile @ Low Level	ISO/IEC MPEG
2.2.2	MPEG-2 Video Main Profile @ Main Level	ISO/IEC MPEG
2.2.3	MPEG-2 Video Main Profile @ High 1440 Level	ISO/IEC MPEG
2.2.4	MPEG-2 Video Main Profile @ High Level	ISO/IEC MPEG
2.3	MPEG-2 Video SNR Scalable Profile	ISO/IEC MPEG
2.3.1	MPEG-2 SNR Scalable Profile @ Low Level	ISO/IEC MPEG
2.3.2	MPEG-2 Video SNR Scalable Profile @ Main Level	ISO/IEC MPEG
2.4	MPEG-2 Video Spatial Scalable Profile	ISO/IEC MPEG
2.4.1	MPEG-2 Video Spatial Scalable Profile @ Main Level	ISO/IEC MPEG
2.4.2	MPEG-2 Video Spatial Scalable Profile @ High 1440 Level	ISO/IEC MPEG
2.4.3	MPEG-2 Video Spatial Scalable Profile @ High Level	ISO/IEC MPEG
2.5	MPEG-2 Video High Profile	ISO/IEC MPEG
2.5.1	MPEG-2 Video High Profile @ Main Level	ISO/IEC MPEG
2.5.2	MPEG-2 Video High Profile @ High 1440 Level	ISO/IEC MPEG
2.5.3	MPEG-2 Video High Profile @ High Level	ISO/IEC MPEG
2.6	MPEG-2 Video Multiview Profile	ISO/IEC MPEG
2.6.1	MPEG-2 Video Multiview Profile @ Main Level	ISO/IEC MPEG
2.7	MPEG-2 Video 4:2:2 Profile	ISO/IEC MPEG
2.7.1	MPEG-2 Video 4:2:2 Profile @ Main Level	ISO/IEC MPEG
3	MPEG-4 Visual	ISO/IEC MPEG
3.1	MPEG-4 Visual Simple Profile	ISO/IEC MPEG
3.1.1	MPEG-4 Visual Simple Profile @ Level 0	ISO/IEC MPEG
3.1.2	MPEG-4 Visual Simple Profile @ Level 1	ISO/IEC MPEG
3.1.3	MPEG-4 Visual Simple Profile @ Level 2	ISO/IEC MPEG
3.1.4	MPEG-4 Visual Simple Profile @ Level 3	ISO/IEC MPEG
3.2	MPEG-4 Visual Simple Scalable Profile	ISO/IEC MPEG
3.2.1	MPEG-4 Visual Simple Scalable Profile @ Level 1	ISO/IEC MPEG
3.2.2	MPEG-4 Visual Simple Scalable Profile @ Level 2	ISO/IEC MPEG
3.3	MPEG-4 Visual Advanced Simple Profile	ISO/IEC MPEG

3.3.1	MPEG-4 Visual Advanced Simple Profile @ Level 0	ISO/IEC MPEG
3.3.2	MPEG-4 Visual Advanced Simple Profile @ Level 1	ISO/IEC MPEG
3.3.3	MPEG-4 Visual Advanced Simple Profile @ Level 2	ISO/IEC MPEG
3.3.4	MPEG-4 Visual Advanced Simple Profile @ Level 3	ISO/IEC MPEG
3.3.5	MPEG-4 Visual Advanced Simple Profile @ Level 4	ISO/IEC MPEG
3.3.6	MPEG-4 Visual Advanced Simple Profile @ Level 5	ISO/IEC MPEG
3.4	MPEG-4 Visual Core Profile	ISO/IEC MPEG
3.4.1	MPEG-4 Visual Core Profile@ Level 1	ISO/IEC MPEG
3.4.2	MPEG-4 Visual Core Profile@ Level 2	ISO/IEC MPEG
3.5	MPEG-4 Visual Core-Scalable Profile	ISO/IEC MPEG
3.5.1	MPEG-4 Visual Core-Scalable Profile @ Level 1	ISO/IEC MPEG
3.5.2	MPEG-4 Visual Core-Scalable Profile @ Level 2	ISO/IEC MPEG
3.5.3	MPEG-4 Visual Core-Scalable Profile @ Level 3	ISO/IEC MPEG
3.6	MPEG-4 Visual Advanced Core Profile	ISO/IEC MPEG
3.6.1	MPEG-4 Visual Advanced Core Profile @ Level 1	ISO/IEC MPEG
3.6.2	MPEG-4 Visual Advanced Core Profile @ Level 2	ISO/IEC MPEG
3.7	MPEG-4 Visual Main Profile	ISO/IEC MPEG
3.7.1	MPEG-4 Visual Main Profile @ Level 2	ISO/IEC MPEG
3.7.2	MPEG-4 Visual Main Profile @ Level 3	ISO/IEC MPEG
3.7.3	MPEG-4 Visual Main Profile @ Level 4	ISO/IEC MPEG
3.8	MPEG-4 Visual N-bit Profile	ISO/IEC MPEG
3.8.1	MPEG-4 Visual N-bit Profile @ Level 2	ISO/IEC MPEG
3.9	MPEG-4 Visual Advanced Real Time Simple Profile	ISO/IEC MPEG
3.9.1	MPEG-4 Visual Advanced Real Time Simple Profile @ Level 1	ISO/IEC MPEG
3.9.2	MPEG-4 Visual Advanced Real Time Simple Profile @ Level 2	ISO/IEC MPEG
3.9.3	MPEG-4 Visual Advanced Real Time Simple Profile @ Level 3	ISO/IEC MPEG
3.9.4	MPEG-4 Visual Advanced Real Time Simple Profile @ Level 4	ISO/IEC MPEG
3.10	MPEG-4 Visual Advanced Coding Efficiency Profile	ISO/IEC MPEG
3.10.1	MPEG-4 Visual Advanced Coding Efficiency Profile @ Level 1	ISO/IEC MPEG
3.10.2	MPEG-4 Visual Advanced Coding Efficiency Profile @ Level 2	ISO/IEC MPEG
3.10.3	MPEG-4 Visual Advanced Coding Efficiency Profile @ Level 3	ISO/IEC MPEG
3.10.4	MPEG-4 Visual Advanced Coding Efficiency Profile @ Level 4	ISO/IEC MPEG
3.11	MPEG-4 Visual Simple Studio Profile	ISO/IEC MPEG
3.11.1	MPEG-4 Visual Simple Studio Profile @ Level 1	ISO/IEC MPEG
3.11.2	MPEG-4 Visual Simple Studio Profile @ Level 2	ISO/IEC MPEG
3.11.3	MPEG-4 Visual Simple Studio Profile @ Level 3	ISO/IEC MPEG
3.11.4	MPEG-4 Visual Simple Studio Profile @ Level 4	ISO/IEC MPEG
3.12	MPEG-4 Visual Core Studio Profile	ISO/IEC MPEG
3.12.1	MPEG-4 Visual Core Studio Profile @ Level 1	ISO/IEC MPEG
3.12.2	MPEG-4 Visual Core Studio Profile @ Level 2	ISO/IEC MPEG
3.12.3	MPEG-4 Visual Core Studio Profile @ Level 3	ISO/IEC MPEG
3.12.4	MPEG-4 Visual Core Studio Profile @ Level 4	ISO/IEC MPEG
3.13	MPEG-4 Visual Fine Granularity Scalable Profile	ISO/IEC MPEG
3.13.1	MPEG-4 Visual Fine Granularity Scalable Profile @ Level 0	ISO/IEC MPEG
3.13.2	MPEG-4 Visual Fine Granularity Scalable Profile @ Level 1	ISO/IEC MPEG
3.13.3	MPEG-4 Visual Fine Granularity Scalable Profile @ Level 2	ISO/IEC MPEG
3.13.4	MPEG-4 Visual Fine Granularity Scalable Profile @ Level 3	ISO/IEC MPEG
3.13.5	MPEG-4 Visual Fine Granularity Scalable Profile @ Level 4	ISO/IEC MPEG
3.13.6	MPEG-4 Visual Fine Granularity Scalable Profile @ Level 5	ISO/IEC MPEG
3.14	MPEG-4 Visual Simple Face Animation Profile	ISO/IEC MPEG
3.14.1	MPEG-4 Visual Simple Face Animation Profile @ Level 1	ISO/IEC MPEG

3.14.2	MPEG-4 Visual Simple Face Animation Profile @ Level 2	ISO/IEC MPEG
3.15	MPEG-4 Visual Simple FBA Profile	ISO/IEC MPEG
3.15.1	MPEG-4 Visual Simple FBA Profile @ Level 1	ISO/IEC MPEG
3.15.2	MPEG-4 Visual Simple FBA Profile @ Level 2	ISO/IEC MPEG
3.16	MPEG-4 Visual Basic Animated Texture Profile	ISO/IEC MPEG
3.16.1	MPEG-4 Visual Basic Animated Texture Profile @ Level 1	ISO/IEC MPEG
3.16.2	MPEG-4 Visual Basic Animated Texture Profile @ Level 2	ISO/IEC MPEG
3.17	MPEG-4 Visual Scalable Texture Profile	ISO/IEC MPEG
3.17.1	MPEG-4 Visual Scalable Texture Profile @ Level 1	ISO/IEC MPEG
3.18	MPEG-4 Visual Advanced Scalable Texture Profile	ISO/IEC MPEG
3.18.1	MPEG-4 Visual Advanced Scalable Texture Profile @ Level 1	ISO/IEC MPEG
3.18.2	MPEG-4 Visual Advanced Scalable Texture Profile @ Level 2	ISO/IEC MPEG
3.18.3	MPEG-4 Visual Advanced Scalable Texture Profile @ Level 3	ISO/IEC MPEG
3.19	MPEG-4 Visual Hybrid Profile	ISO/IEC MPEG
3.19.1	MPEG-4 Visual Hybrid Profile @ Level 1	ISO/IEC MPEG
3.19.2	MPEG-4 Visual Hybrid Profile @ Level 2	ISO/IEC MPEG
4	JPEG	ISO/IEC ITU-T
5	MJPEG	IETF
6	JPEG2000	ISO/IEC, ITU-T
6.1	JPEG2000 jp2p Profile	ISO/IEC, ITU-T
6.1.1	JPEG2000 jp2p Profile – Level 0	ISO/IEC, ITU-T
6.1.2	JPEG2000 jp2p Profile – Level 2	ISO/IEC, ITU-T
6.1.3	JPEG2000 jp2p Profile – Level 3	ISO/IEC, ITU-T
6.2	MJPEG2000 mjp2 Profile	ISO/IEC, ITU-T
6.3	MJPEG2000 mj2s Profile	ISO/IEC, ITU-T
7	H261	ITU-T
8	H263	ITU-T
8.1	H263 baslineProfile	ITU-T
8.1.1	H263 baslineProfile, Level 10	ITU-T
8.1.2	H263 baslineProfile, Level 20	ITU-T
8.1.3	H263 baslineProfile, Level 30	ITU-T
8.1.4	H263 baslineProfile, Level 40	ITU-T
8.1.5	H263 baslineProfile, Level 45	ITU-T
8.1.6	H263 baslineProfile, Level 50	ITU-T
8.1.7	H263 baslineProfile, Level 60	ITU-T
8.1.8	H263 baslineProfile, Level 70	ITU-T
8.2	H263 h320Profile	ITU-T
8.2.1	H263 h320Profile, Level 10	ITU-T
8.2.2	H263 h320Profile, Level 20	ITU-T
8.2.3	H263 h320Profile, Level 30	ITU-T
8.2.4	H263 h320Profile, Level 40	ITU-T
8.2.5	H263 h320Profile, Level 45	ITU-T
8.2.6	H263 h320Profile, Level 50	ITU-T
8.2.7	H263 h320Profile, Level 60	ITU-T
8.2.8	H263 h320Profile, Level 70	ITU-T
8.3	H263 backwardCompatibleProfile	ITU-T
8.3.1	H263 backwardCompatibleProfile, Level 10	ITU-T
8.3.2	H263 backwardCompatibleProfile, Level 20	ITU-T
8.3.3	H263 backwardCompatibleProfile, Level 30	ITU-T
8.3.4	H263 backwardCompatibleProfile, Level 40	ITU-T
8.3.5	H263 backwardCompatibleProfile, Level 45	ITU-T

8.3.6	H263 backwardCompatibleProfile, Level 50	ITU-T
8.3.7	H263 backwardCompatibleProfile, Level 60	ITU-T
8.3.8	H263 backwardCompatibleProfile, Level 70	ITU-T
8.4	H263 v2WirelessProfile	ITU-T
8.4.1	H263 v2WirelessProfile, Level 10	ITU-T
8.4.2	H263 v2WirelessProfile, Level 20	ITU-T
8.4.3	H263 v2WirelessProfile, Level 30	ITU-T
8.4.4	H263 v2WirelessProfile, Level 40	ITU-T
8.4.5	H263 v2WirelessProfile, Level 45	ITU-T
8.4.6	H263 v2WirelessProfile, Level 50	ITU-T
8.4.7	H263 v2WirelessProfile, Level 60	ITU-T
8.4.8	H263 v2WirelessProfile, Level 70	ITU-T
8.5	H263 v3WirelessProfile	ITU-T
8.5.1	H263 v3WirelessProfile, Level 10	ITU-T
8.5.2	H263 v3WirelessProfile, Level 20	ITU-T
8.5.3	H263 v3WirelessProfile, Level 30	ITU-T
8.5.4	H263 v3WirelessProfile, Level 40	ITU-T
8.5.5	H263 v3WirelessProfile, Level 45	ITU-T
8.5.6	H263 v3WirelessProfile, Level 50	ITU-T
8.5.7	H263 v3WirelessProfile, Level 60	ITU-T
8.5.8	H263 v3WirelessProfile, Level 70	ITU-T
8.6	H263 conversationalProfile	ITU-T
8.6.1	H263 conversationalProfile, Level 10	ITU-T
8.6.2	H263 conversationalProfile, Level 20	ITU-T
8.6.3	H263 conversationalProfile, Level 30	ITU-T
8.6.4	H263 conversationalProfile, Level 40	ITU-T
8.6.5	H263 conversationalProfile, Level 45	ITU-T
8.6.6	H263 conversationalProfile, Level 50	ITU-T
8.6.7	H263 conversationalProfile, Level 60	ITU-T
8.6.8	H263 conversationalProfile, Level 70	ITU-T
8.7	H263 conversationInternetProfile	ITU-T
8.7.1	H263 conversationInternetProfile, Level 10	ITU-T
8.7.2	H263 conversationInternetProfile, Level 20	ITU-T
8.7.3	H263 conversationInternetProfile, Level 30	ITU-T
8.7.4	H263 conversationInternetProfile, Level 40	ITU-T
8.7.5	H263 conversationInternetProfile, Level 45	ITU-T
8.7.6	H263 conversationInternetProfile, Level 50	ITU-T
8.7.7	H263 conversationInternetProfile, Level 60	ITU-T
8.7.8	H263 conversationInternetProfile, Level 70	ITU-T
8.8	H263 conversationInterlaceProfile	ITU-T
8.8.1	H263 conversationInterlaceProfile, Level 10	ITU-T
8.8.2	H263 conversationInterlaceProfile, Level 20	ITU-T
8.8.3	H263 conversationInterlaceProfile, Level 30	ITU-T
8.8.4	H263 conversationInterlaceProfile, Level 40	ITU-T
8.8.5	H263 conversationInterlaceProfile, Level 45	ITU-T
8.8.6	H263 conversationInterlaceProfile, Level 50	ITU-T
8.8.7	H263 conversationInterlaceProfile, Level 60	ITU-T
8.8.8	H263 conversationInterlaceProfile, Level 70	ITU-T
8.9	H263 highLatencyProfile	ITU-T
8.9.1	H263 highLatencyProfile, Level 10	ITU-T
8.9.2	H263 highLatencyProfile, Level 20	ITU-T

8.9.3	H263 highLatencyProfile, Level 30	ITU-T
8.9.4	H263 highLatencyProfile, Level 40	ITU-T
8.9.5	H263 highLatencyProfile, Level 45	ITU-T
8.9.6	H263 highLatencyProfile, Level 50	ITU-T
8.9.7	H263 highLatencyProfile, Level 60	ITU-T
8.9.8	H263 highLatencyProfile, Level 70	ITU-T
9	H264	ITU-T
9.1	H264 Baseline Profile	ITU-T
9.1.1	H264 Baseline Profile Level 1	ITU-T
9.1.1.1	H264 Baseline Profile Level 1.1	ITU-T
9.1.1.2	H264 Baseline Profile Level 1.2	ITU-T
9.1.1.3	H264 Baseline Profile Level 1.3	ITU-T
9.1.2	H264 Baseline Profile Level 2	ITU-T
9.1.2.1	H264 Baseline Profile Level 2.1	ITU-T
9.1.2.2	H264 Baseline Profile Level 2.2	ITU-T
9.1.3	H264 Baseline Profile Level 3	ITU-T
9.1.3.1	H264 Baseline Profile Level 3.1	ITU-T
9.1.3.2	H264 Baseline Profile Level 3.2	ITU-T
9.1.4	H264 Baseline Profile Level 4	ITU-T
9.1.4.1	H264 Baseline Profile Level 4.1	ITU-T
9.1.4.2	H264 Baseline Profile Level 4.2	ITU-T
9.1.5	H264 Baseline Profile Level 5	ITU-T
9.1.5.1	H264 Baseline Profile Level 5.1	ITU-T
9.2	H264 Main Profile	ITU-T
9.2.1	H264 Main Profile Level 1	ITU-T
9.2.1.1	H264 Main Profile Level 1.1	ITU-T
9.2.1.2	H264 Main Profile Level 1.2	ITU-T
9.2.1.3	H264 Main Profile Level 1.3	ITU-T
9.2.2	H264 Main Profile Level 2	ITU-T
9.2.2.1	H264 Main Profile Level 2.1	ITU-T
9.2.2.2	H264 Main Profile Level 2.2	ITU-T
9.2.3	H264 Main Profile Level 3	ITU-T
9.2.3.1	H264 Main Profile Level 3.1	ITU-T
9.2.3.2	H264 Main Profile Level 3.2	ITU-T
9.2.4	H264 Main Profile Level 4	ITU-T
9.2.4.1	H264 Main Profile Level 4.1	ITU-T
9.2.4.2	H264 Main Profile Level 4.2	ITU-T
9.2.5	H264 Main Profile Level 5	ITU-T
9.2.5.1	H264 Main Profile Level 5.1	ITU-T
9.3	H264 Extended Profile	ITU-T
9.3.1	H264 Extended Profile Level 1	ITU-T
9.3.1.1	H264 Extended Profile Level 1.1	ITU-T
9.3.1.2	H264 Extended Profile Level 1.2	ITU-T
9.3.1.3	H264 Extended Profile Level 1.3	ITU-T
9.3.2	H264 Extended Profile Level 2	ITU-T
9.3.2.1	H264 Extended Profile Level 2.1	ITU-T
9.3.2.2	H264 Extended Profile Level 2.2	ITU-T
9.3.3	H264 Extended Profile Level 3	ITU-T
9.3.3.1	H264 Extended Profile Level 3.1	ITU-T
9.3.3.2	H264 Extended Profile Level 3.2	ITU-T
9.3.4	H264 Extended Profile Level 4	ITU-T

9.3.4.1	H264 Extended Profile Level 4.1	ITU-T
9.3.4.2	H264 Extended Profile Level 4.2	ITU-T
9.3.5	H264 Extended Profile Level 5	ITU-T
9.3.5.1	H264 Extended Profile Level 5.1	ITU-T
10	Cell-B	SUN
11	WAVI	ETHZ
12	Sorenson Video	Sorenson Media
13	Cinepak	Compression Technologies, Inc
14	Indeo 3	Intel
15	Indeo Video Interactive	Intel
15.1	Indeo Video Interactive 4	Intel
15.2	Indeo Video Interactive 5	Intel
16	RealVideo	RealNetworks
16.1	Real Video 8	RealNetworks
16.2	Real Video 9	RealNetworks
16.3	Real Video 10	RealNetworks

Annex E Resources Coordination through E2ENP – Recommendations to E2ENP Applications

This annex provides additional information on the resources coordination using E2ENP as a recommendation for the applications of E2ENP.

The formal definition of the validation procedure (see section 2.3.2.2.2) is expressed in form of operators upon sets of application and transport parameters. An application parameter (e.g. frame-rate, frame-size etc. for video, see also Section 2.2.1) is defined as a set of or a continuous range of simple values:

$$apar = \overset{y}{\underset{i=x}{G}}(saval_i)$$

Where 'apar' is an application parameter and 'saval' are the values of this parameter within the set or the range [x, y] (x, y are real numbers where x≤y and are directly correspondent to the values an application parameter can take, e.g. $saval_x$=x). 'G' indicates a grouping function for generating the set or the range. The application of sets or ranges for monitoring the application performance provides stability for the monitoring algorithm at deciding when QoS changes occur. This is due to the fact that the human users do not notice or do not rate as unacceptable certain service performance changes, hence such changes as not evaluated as QoS changes [134] – [138]. The definition of a transport parameter (e.g. bandwidth, packet loss, etc., see also Section 2.2.1) is specified in a similar way to the definition of an application parameter:

$$tpar = \overset{z}{\underset{i=t}{G}}(stval_i)$$

Where 'tpar' is a transport parameter and 'stval' are the values of this parameter within the set or the range [t, z] (t, z are real numbers where t≤z and are directly correspondent to the values a transport parameter can take, e.g. $stval_t$=t). 'G' here is also a grouping function (for the transport parameters). As for the application parameters, the stability of the QoS monitoring is defined also through sets or ranges. There are situations where the transport performance changes of the service do not affect the application performance in a way that such transport performance shall be evaluated as QoS change [17], [139] – [141]. The complete application and transport configurations corresponding specific perceivable QoS performance are formally defined as:

$$AP = \{apar_1, apar_2, \ldots, apar_n\}$$

$$TP = \{tpar_1, tpar_2, \ldots, tpar_m\}$$

Where 'AP' is the complete application configuration of 'n' parameters 'apar' and the 'TP' is the complete transport configuration of 'm' parameters 'tpar', where 'm' and 'n' are whole numbers. The result of associating of the transport with the application parameters that correspond certain QoS performance leads to the following definition:

$$Contract = AFP(AP, TP)$$

Where the 'contract' is a single QoS contract and 'AFP' (association function for the parameters) is an operator for associating the application and the transport parameters. The specific application of this operator for generating a real QoS contract depends on the type of media (e.g. audio, video) and on the type of the applicable coding mechanisms for the media (see the respective discussion in Section 2.2.1). The exact definition of AFP can be achieved using empiric results (see [134] – [138]). The contracts are associated into contexts using the following operation:

$$\text{Context} = \text{AFQ}\{\text{Contract}_1, \text{Contract}_2, \ldots, \text{Contract}_K\}$$

Where the 'Context' is a single QoS context and 'AFQ' (association function for QoS contracts) is an operator applied upon a set of 'K' mutually dependent and associable contracts, where 'K' is a whole number. The specific application of 'AFQ' depends on the media and service features (e.g. association of video stream with audio to correlate the lip movements of a speaker with the respective speech). Details on such associations are described in Section 2.2.2.2 (see also [14][17][20][70][132][173]). Furthermore, an association between contexts is described as:

$$\text{CContext} = \text{AFC}\{\text{Context}_1, \text{Context}_2, \ldots, \text{Context}_L\}$$

Where 'CContext' indicates a complex context build through the association operator 'AFC' (association function for contexts) out of 'L' simple contexts, where 'L' is a whole number. If no complex contexts are defined for the service the operator 'AFC' is a zero operator and in this case:

$$\text{Context} = \text{CContext}$$

The complex contexts are usually grouped in accordance with specific monitoring features of the service to produce respective context switches for enabling QoS changes within the system in cases that the system resources change (see Section 2.2.2.2 and [14][17][20][70][132][173]). The operations for evaluating the applicable contexts within the scope of the service performance are then specified as:

$$\text{LCnd} = \underset{\text{LC}}{\text{LIM}}(\text{LF}(\sum_{i=1}^{U} \text{CContext}_i))$$

$$\text{PCnd} = \underset{\text{PC}}{\text{LIM}}(\text{PF}(\sum_{i=1}^{W} \text{LCnd}_i))$$

$$\text{DCnd} = \underset{\text{DC}}{\text{LIM}}(\text{NRF}(\sum_{i=1}^{V} \text{PCnd}_i))$$

These operations correspond to the single decision steps within the validation algorithm and have the following meaning:

- QoS Conditions – 'LCnd' is a locally applicable QoS condition generated out of 'U' (whole number) specific service contexts, 'PCnd' is a provider applicable QoS condition that limits the number of all applicable local conditions (this number is indicated as 'W' – whole number) and 'DCnd' is the QoS condition of the complete distributed service that is generated out of all applicable provider conditions (the number of the applicable provider conditions is indicated with 'V' – whole number).
- Decision functions – 'LF' (local function), 'PF' (provider function) and 'NRF' (negotiation and resource function) are operators indicating a decision taking method within the local system, the provider system and the remote respectively distributed resource negotiation and management system. These operators are placeholders for actual decision taking algorithms as described in the previous Section 2.3.2.2.1.
- Limitation operator – 'LIM' is a delimiting operator that indicates the application of constraints ('LC' – local constraints, 'PC' – provider constraints and 'DC' – distributed system constraints) within the decision-taking algorithm.

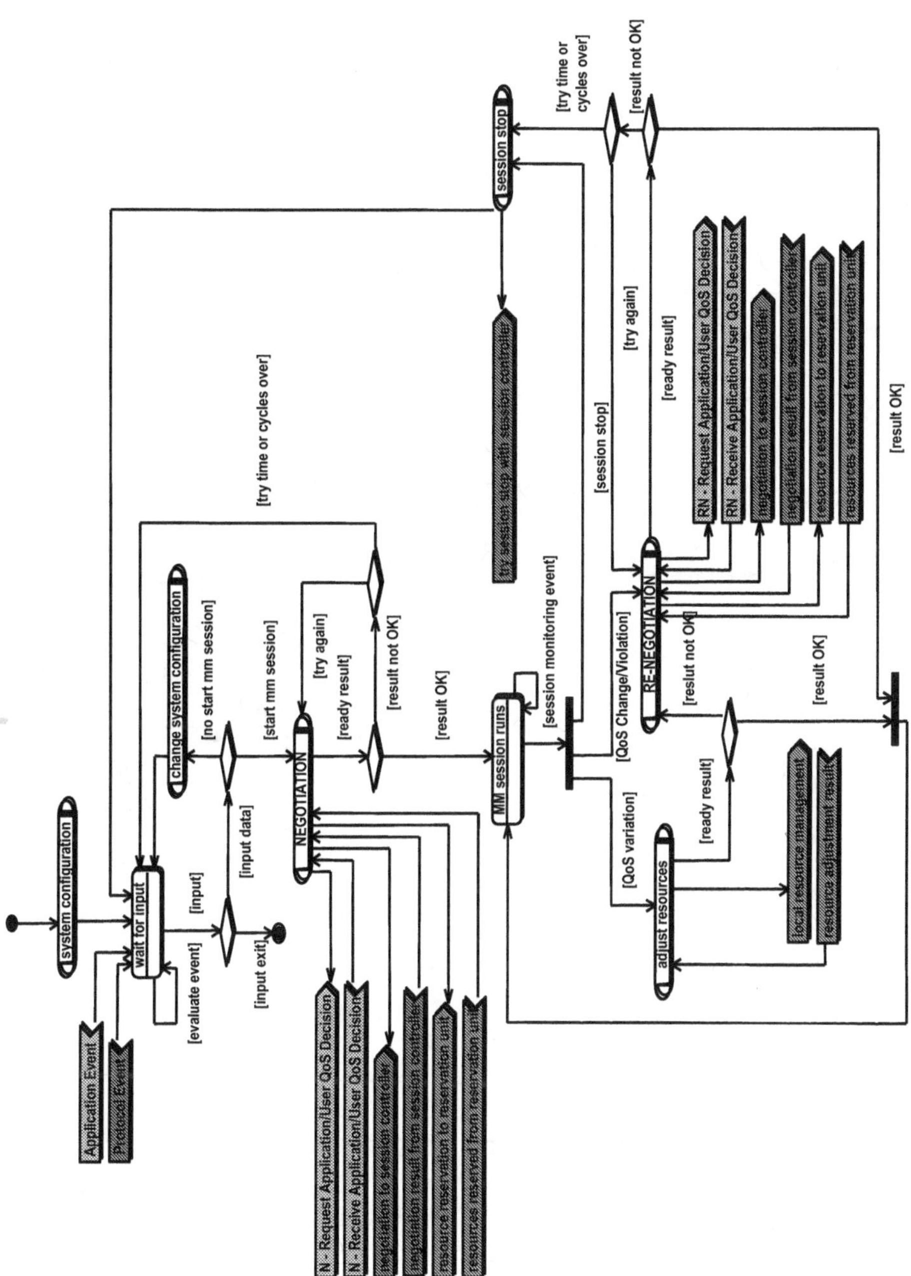

Figure 124: Session management without memory

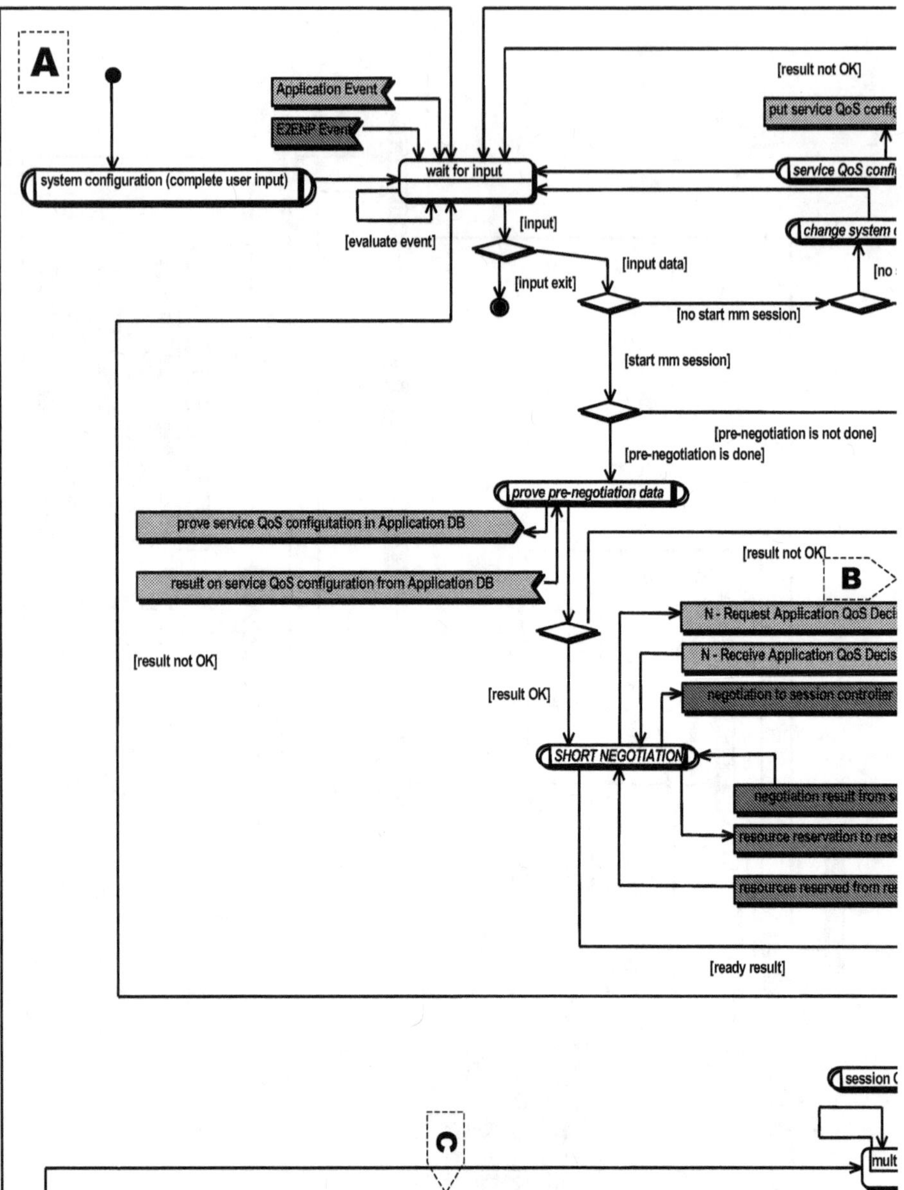

Figure 125: Session/service management with memory (A)

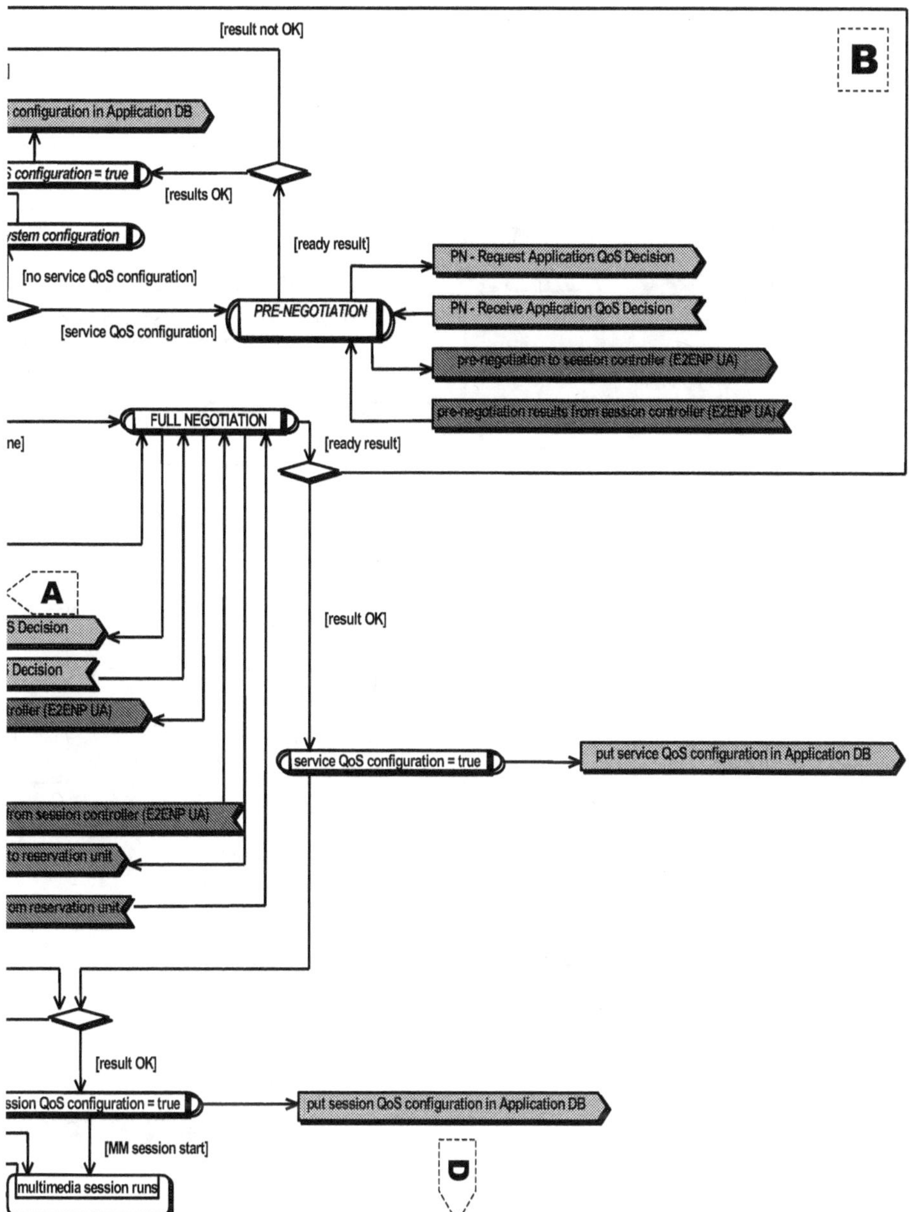

[result not OK]
B
configuration in Application DB
configuration = true
[results OK]
stem configuration
[ready result]
PN - Request Application QoS Decision
[no service QoS configuration]
PRE-NEGOTIATION
PN - Receive Application QoS Decision
[service QoS configuration]
pre-negotiation to session controller (E2ENP UA)
FULL NEGOTIATION
pre-negotiation results from session controller (E2ENP UA)
ne]
[ready result]
A
S Decision
Decision
troller (E2ENP UA)
[result OK]
service QoS configuration = true
put service QoS configuration in Application DB
from session controller (E2ENP UA)
to reservation unit
om reservation unit
[result OK]
ssion QoS configuration = true
put session QoS configuration in Application DB
[MM session start]
multimedia session runs
D

Figure 125: Session/service management with memory (B)

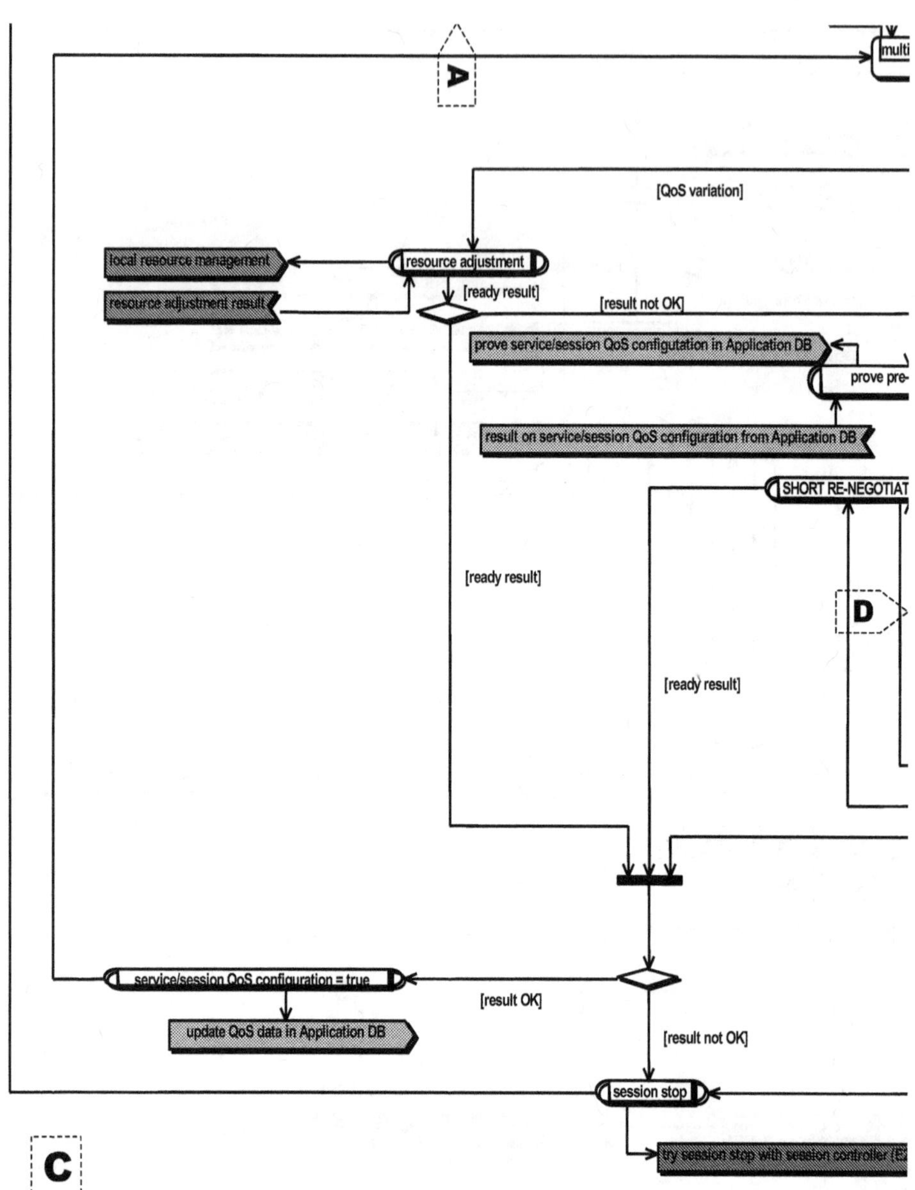

Figure 125: Session/service management with memory (C)

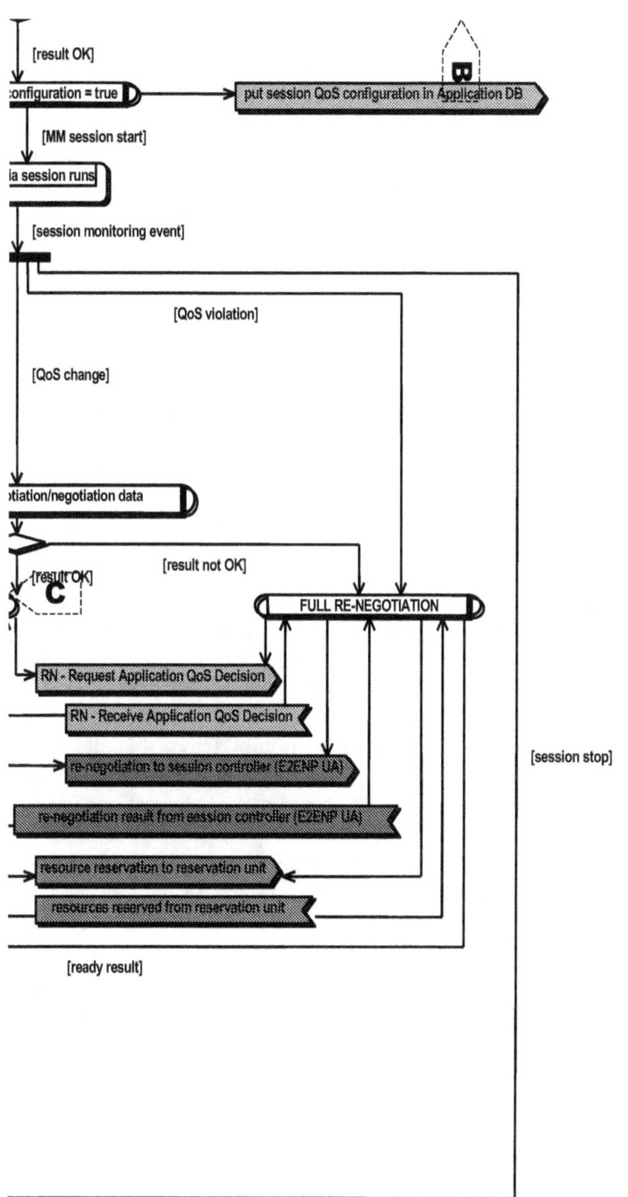
[result OK]
configuration = true
put session QoS configuration in Application DB
B
[MM session start]
ia session runs
[session monitoring event]
[QoS violation]
[QoS change]
tiation/negotiation data
[result not OK]
[result OK]
C
FULL RE-NEGOTIATION
RN - Request Application QoS Decision
RN - Receive Application QoS Decision
re-negotiation to session controller (E2ENP UA)
re-negotiation result from session controller (E2ENP UA)
resource reservation to reservation unit
resources reserved from reservation unit
[ready result]
[session stop]
D

Figure 125: Session/service management with memory (D)

The negotiation process in association with the resources management process of the services is modelled with Activity Diagrams [216][269] to express the negotiations of applications that do not use E2ENP PDU memory (Figure 124) and that use PDU memory (Figure 125) utilizing or not utilizing the possibility of E2ENP Pre-negotiations. The signals and the conditions expressed in Figure 124 and Figure 125 indicate the communication of the application with the E2ENP UA, with its internal database (indicated with DB in Figure 125) for the management of memorised PDU contents and with the system's resource management units. The signals associated with Pre-negotiations, Negotiations and Re-negotiations steering the respective negotiation rounds of the applications are denoted (in Figure 124 and Figure 125) correspondingly with PN, N and RN.

Figure 124 presents the negotiations as typical for the classical applications of session negotiations (like the Offer/Answer Model with SDP [35][36]) where no application of PDU referencing is possible. Hence, when performing session adaptations the application does not differentiate in its logic between negotiation and re-negotiation upon their contents (PDUs), but only upon the fact if a multimedia session runs or not. The respective actions in Figure 124 reflect this art of logic.

Figure 125 presents the model of an E2ENP-enabled application for performing session adaptations based on PDU memory. Figure 125 consists of four sub-figures that shall be interconnected on the places where the respective arrows point at the adjacent part of the figure, e.g. the arrows in part A point at B and C, part B points at A and D, part C at A and D and part D at B and C. The arrows indicate the exact connection points between the sub-figures. In case that the logic of the multimedia service does not support Pre-negotiations, the actions indicated in Figure 125 with names in italic letters and all of their associated conditions, signals and decisions shall be removed from the corresponding model. Figure 125 reflects the fact that E2ENP supports preparative configurations in form of Pre-negotiations, applies memory and memory-management for session configurations and E2ENP differentiates at resources management between three cases (indicated as conditions in Figure 125), namely, QoS variation, QoS change and QoS violation. The QoS variation is the case where no explicit E2ENP signalling is necessary as the resources change remains within the limits of the pre-defined QoS contracts/contexts. The QoS change is associated with an explicit short Re-negotiation and a QoS violation with an explicit full Re-negotiation, in case that the application needs to change its QoS specification as the pre-defined and initially negotiated one is no more valid. Details on the E2ENP-specific resources management are presented in Section 2.3.2.

References

[1] H. Schulzrinne et al., "RTP: A Transport Protocol for Real-Time Applications", IETF RFC 3550, July 2003

[2] H. Schulzrinne, S. Casner, "RTP Profile for Audio and Video Conferences with Minimal Control", IETF RFC 3551, July 2003

[3] R. Braden et al., "Resource ReSerVation Protocol (RSVP) -- Version 1 Functional Specification", IETF RFC 2205, September 1997

[4] A. Mankin et al., "Resource ReSerVation Protocol (RSVP) -- Version 1 Applicability Statement Some Guidelines on Deployment", IETF RFC 2208, September 1997

[5] S. Blake et al., "An Architecture for Differentiated Services", IETF RFC 2475, December 1998

[6] E. Rosen, A. Viswanathan, R. Callon, "Multiprotocol Label Switching Architecture", IETF RFC 3031, January 2001

[7] D. Durham et al., "The COPS (Common Open Policy Service) Protocol", IETF RFC 2748, January 2000

[8] M. Kalman, B. Girod, E. Steinbach, "Adaptive Playout for Real-time Media Streaming" International Symposium on Circuits and Systems, ISCAS 2002, Scottsdale (Arizona, USA), May 2002.

[9] Yi J. Liang, E. Steinbach, B. Girod, "Real-time Voice Communication over the Internet Using Packet Path Diversity", ACM Multimedia 2001, Ottawa (Canada), Sept./Oct. 2001.

[10] P. Karn et al., "Advice for Internet Subnetwork Designers", IETF RFC 3819, July 2004

[11] A. Kassler, A. Neubeck, P. Schulthess, "Real-time Filtering of Wavelet Coded Videostreams for Meeting QoS Constraints and User Priorities", Packet Video Workshop 2000, Cagliari (Italy), May 2000

[12] A. Kassler, A. Neubeck, P. Schulthess, "Filtering Wavelet based Video Streams for Wireless Inter-working", ICME2000, New York (NJ, USA), August 2000

[13] J. Koistinen, A. Seetharaman, "Worth-Based Multi-Category Quality-of-Service Negotiation in Distributed Object Infrastructures", Technical Report HPL-98-51 (R.1), July 1998

[14] T. Guenkova-Luy, A. Kassler, J. Eisl, D. Mandato, "Efficient End-to-End QoS Signaling - concepts and features", IETF Work-in-progress, draft-guenkova-mmusic-e2enp-sdpng-00, March 2002

[15] J.C. Rojas et al., "Requirements for the QoS negotiation at the Application Layer", IETF Work-in-progress, draft-rojas-mmusic-qosreq-00.txt, December 2002

[16] X. Gu, K. Nahrstedt et al., "An XML-based Quality of Service Enabling Language for the Web", Project Report - National Science Foundation, 2001

[17] A. Kassler, "Video Adaptation within a Quality of Service Architecture", PhD Thesis at University of Ulm, 2001

[18] S. Frolund, J. Koistinen, "QML: A Language for Quality Specification", HP Research Report, February 1998.

[19] K. Nahrstedt, J. M. Smith, “The QoS Broker”, IEEE Multimedia Magazine, Vol. 2, No. 1, pp. 53-67, Spring 1995.

[20] MIND Project, “Top-level architecture for providing seamless QoS, security, accounting and mobility to applications and services”, Deliverable D1.2, November 2002

[21] 3GPP TS 23.228, “IP Multimedia Subsystem (IMS)”, Technical Specification V7.0.0, June 2005

[22] 3GPP TS 24.228, “Signalling flows for the IP multimedia call control based on SIP and SDP”, Technical Specification, considered versions: from V5.1.0 (June 2002) till V5.13.0, June 2005

[23] W. Almesberger et al., “Combining IntServ and DiffServ under Linux”, I&C Technical Report, Technical Reports in Computer and Communication Sciences, September 1999

[24] D. D. Clark, D. L. Tennenhouse, “Architectural Considerations for a New Generation of Communication Protocols”, ACM SIGCOMM 1990, Philadelpia (PA, USA), September 1990

[25] H. Lu, I. Faynberg, “An Architectural Framework for Support of Quality of Service in Packet Networks”, IEEE Communications Magazine, Vol. 41, pp. 98- 105, June 2003

[26] M. Buckley, “ETSI Workshop on QoS in Next Generation Networks – Workshop Conclusions”, 21 February 2002, <http://www.etsi.org/frameset/home.htm?/QOSWORKSHOP/>

[27] A. Alwan et al., “Adaptive Mobile Multimedia Networks”, IEEE Personal Communications, pp. 34-51, April 1996

[28] D. Sisalem, “End-to-end Quality of Service Control using Adaptive Applications”, IFIP International Workshop on QoS IWQOS '97, New York (NJ, USA), May 1997

[29] R. Katz, “Adaptation and Mobility in Wireless Information Systems”, IEEE Personal Communication, Vol 1, No. 1, pp. 6–17, 1994.

[30] M. Mirhakkak, N. Schult, D. Thomsom, “Dynamic Bandwidth Management and Adaptive Applications for a Variable Bandwidth Wireless Environment”, IEEE JSAC, Vol. 19, No. 10, pp. 1984 - 1997,October 2001

[31] R. Ramanathan, R. Hain, “An Ad Hoc Wireless Testbed for Scalable, Adaptive QoS Support”, IEEE WCNC 2000, Chicago (IL, USA), November 2000

[32] G. Dobson, “Quality of Service in Service-Oriented Architectures”, (un-reviewed paper), September 2004, <digs.sourceforge.net/papers/qos.html>

[33] D. Isenberg, “The dawn of the stupid network”, ACM Networker 2.1, pp. 24-31, Feb./March 1998

[34] D. Isenberg, “The end of the middle”, IEEE Spectrum Online, Jan. 2003

[35] J. Rosenberg, H. Schulzrinne, “An Offer/Answer Model with SDP”, IETF RFC 3264, June 2002

[36] G. Camarillo et al., “Integration of Resource Management and Session Initiation Protocol (SIP)”, IETF RFC 3312, October 2002

[37] D. Wichadakul, K. Nahrstedt, “Distributed QoS Compiler”, Project Report - National Science Foundation, 2001

[38] BBN TECHNOLOGIES, “Quality Objects (QuO)”, <http://quo.bbn.com/>

[39] F. Hauck, “Dienstqualität in objektbasierten Verteilten Systemen”, digit4, 2004

[40] J.P. Loyall et al., “Specifying and Measuring Quality of Service in Distributed Object Systems”, First International Symposium on Object-Oriented Real-Time Distributed Computing (ISORC '98), Kyoto (Japan), April 1998

[41] R. Vanegas et al. “QuO's Runtime Support for Quality of Service in Distributed Objects”, Middleware'98, Lake District (England), September 1998.

[42] J. K. Cross, D. C. Schmidt, “Quality Connector - A Pattern Language for Provisioning and Managing Quality-Constrained Services in Distributed Real-time and Embedded Systems”, Patterns in Distributed Real-Time and Embedded Systems (OOPSLA’02), Seattle (WA, USA), November 2002

[43] D. Mandato, A. Kassler, T. Robles, G. Neureiter, “Concepts for Service adaptation, Scalability and QoS concepts on mobility enabled networks”, IST Global Summit 2001, Barcelona (Spain), September 2001

[44] D. Mandato, A. Kassler, T. Robles, G. Neureiter, “Handling End-to-End QoS in Mobile Heterogeneous Networking Environments”, PIMRC 2001, San Diego (CA, USA), September 2001

[45] R. Fisher, W. Ury, “Getting to YES – Negotiating Agreements without Giving In”, Penguin Books, 1991

[46] R. B. Myerson, “Game Theory – Analysis of Conflict”, First Harvard University Press (1997), Fifth printing, 2002

[47] S. Kraus, K. Sycara, A. Evenchik, “Reaching agreements through argumentation: a logical model and implementation”, Artificial Intelligence Journal, Vol. 104, No. 1-2, pp. 1 - 69, September 1998.

[48] Lectric Law Library, “Negotiation”, viewed 27 May 2005, <http://www.lectlaw.com/def2/n048.htm>

[49] Wikipedia, the free encyclopedia, “Negotiation”, last modified 13 May 2005, viewed 27 May 2005, <http://en.wikipedia.org/wiki/Negotiation>

[50] M. Weber, “Verteilte Systeme”, Spektrum Akad. Verl. 1998

[51] A. Tanenbaum, “Distributed Operating Systems”, Prentice Hall Inc., 1995

[52] A. Tanenbaum, M. van Steen, “Distributed Systems: Principles and Paradigms”, Prentice Hall Inc., 2001

[53] K. Sycara, “Resolving goal conflicts via negotiation”, Seventh National Conference on Artificial Intelligence, St. Paul (MN, USA), August 1988

[54] International Organization for Standardization, International Electrotechnical Commission and International Telecommunication Union, “Information Processing Systems - OSI Reference Model - The Basic Model”, International Standard ISO/IEC 7498-1:1994 and ITU-T Recommendation X.200, 1994

[55] Wikipedia, the free encyclopedia, “Protocol (computing)”, last modified 26 October 2005, viewed 9 November 2005, <http://en.wikipedia.org/wiki/Protocol_%28computing%29 >

[56] Wikipedia, the free encyclopedia, “Conciliation”, last modified 19 November 2004, viewed 1 June 2005, <http://en.wikipedia.org/wiki/Conciliation>

[57] Wikipedia, the free encyclopedia, “Mediation”, last modified 22 May 2005, viewed 1 June 2005, <http://en.wikipedia.org/wiki/Mediation>

[58] Wikipedia, the free encyclopedia, “Arbitration”, last modified 18 May 2005, viewed 1 June 2005, <http://en.wikipedia.org/wiki/Arbitration>

[59] J. C. Harsanyi, "Games With Incomplete Information", Nobel Lecture, December 1994, <http://www.nobel.se/economics/laureates/1994/harsanyi-lecture.pdf>

[60] M. Brunner, "Requirements for Signaling Protocols", IETF RFC 3726, April 2004.

[61] K. Nichols et al., "Definition of the Differentiated Services Field (DS Field) in the IPv4 and IPv6 Headers", IETF RFC 2474, December 1998

[62] J. Rosenberg et al., "SIP: Session Initiation Protocol", IETF RFC 3261 (obsoletes RFC 2543), June 2002

[63] International Telecommunication Union, "Control protocol for multimedia communication", ITU-T Recommendation H.245, Approved in 2005-01

[64] FIPA, "FIPA Nomadic Application Support Specification", Standard document SI00014H, December 2002

[65] P. Ruiz, J. Sanchez, E. Garcia, A. Gomez-Skarmeta, J. Botia, A. Kassler, T. Guenkova-Luy, "Adaptive Multimedia Multi-party Communication in Ad Hoc Environments", HICSS-37, Software Technology Track, Hawaii (USA), January 2004

[66] G. Klyne, "Protocol-independent Content Negotiation Framework", IETF RFC 2703, September 1999

[67] G. Huston, "Next Steps for the IP QoS Architecture", IETF RFC 2990, November 2000

[68] M. Eder, H. Chaskar, "Considerations from the Service Management Research Group (SMRG) on Quality of Service (QoS) in the IP Network", IETF RFC 3387, September 2002

[69] J. Ott, C. Perkins, "SDPng Transition", IETF Work-in-progress, draft-ietf-mmusic-sdpng-trans-04, May 2003

[70] D. Mandato, A. Kassler, "Method for achieving end-to-end quality of service negotiations for distributed multi-media applications", European Patent Application 01 107 626.2, March 2001.

[71] World Wide Web Consortium, "Web Ontology Language (OWL): Use Cases and Requirements", W3C Specification, February 2004

[72] International Organization for Standardization and International Electrotechnical Commission, "Information Technology - Multimedia Framework (MPEG-21) - Part 7: Digital Item Adaptation", ISO/IEC 21000-7:2004, 2004

[73] World Wide Web Consortium, "Extensible Markup Language (XML) 1.0 (Third Edition)", W3C Recommendation, February 2004

[74] Object Management Group, "Common Object Request Broker Architecture: Core Specification", Version 3.0.3, March 2004

[75] K. Raatikainen, "Wireless Access and Terminal Mobility in CORBA", Middleware Standards for Mobile Computing – Course at University of Helsinki, Department of Computer Science, Autumn 2003, <http://www.cs.helsinki.fi/u/kraatika/Courses/mwstd2003.html>

[76] R. S. Cardoso, F. Kon, "A Mobile Agent Infrastructure for QoS Negotiation of Adaptive Distributed Applications", International Symposium on Distributed Objects and Applications (DOA), Agia Napa (Cyprus), October 2004

[77] FIPA, "FIPA Quality of Service Ontology Specification", Standard document SC00094A, December 2002

[78] FIPA, "Foundation for Intelligent Physical Agents", <http://www.fipa.org>

[79] FIPA, "FIPA ACL Message Structure Specification", Standard document SC00061G, December 2002

[80] W. Brenner, R. Zarnekow, H. Wittig, "Intelligent Software Agents", Springer-Verlag, 1998

[81] D. Kutscher et al., "Session description and capability negotiation", IETF Work-in-progress, Considered versions: from draft-ietf-mmusic-sdpng-04, March 2002 till draft-ietf-mmusic-sdpng-08, February 2005

[82] M. Handley, C. Perkins, E. Whelan, "Session Announcement Protocol", IETF RFC 2974, October 2000

[83] H. Schulzrinne, A. Rao, R. Lanphier, "Real Time Streaming Protocol (RTSP)", IETF RFC 2326, April 1998

[84] F. Andreasen, B. Foster, "Media Gateway Control Protocol (MGCP) Version 1.0", IETF RFC 3435, January 2003

[85] C. Groves et al., "Gateway Control Protocol Version 1", IETF RFC 3525, June 2003

[86] M. Handley, V. Jacobson, C. Perkins, "SDP: Session Description Protocol", IETF RFC 4566 (obsoletes RFC 2327), July 2006

[87] L. Bos et al., "SDPng Extensions for Quality of Service Negotiation", IETF Work-in-progress, draft-bos-mmusic-sdpng-qos-00, March 2002

[88] Sun Microsystems & jGuru, "jGuru: Remote Method Invocation (RMI)", Copyright 1994-2005 Sun Microsystems, Inc., <http://java.sun.com/developer/onlineTraining/rmi/RMI.html>

[89] C. Perkins et al., "RTP Payload for Redundant Audio Data", IETF RFC 2198, September 1997.

[90] A. Kassler, A. Schorr, "Generic QoS aware Media Stream Transcoding and Adaptation", Packet Video Workshop, Nantes (France), April 2003

[91] A. Schorr, A. Kassler, G. Petrovic, "Adaptive Media Streaming in Heterogeneous Wireless Networks", IEEE International Workshop on Multimedia Signal Processing (MMSP2004), Siena (Italy), September/October 2004

[92] T. Guenkova-Luy, A. Schorr, A. Kassler, J. A. Botia Blaya, T. Inzerilli, M. Gomez, T. Mota, "Multimedia Service Provisioning in B3G Service Creation Platform", IPSI-2005, Pescara (Italy), July 2005

[93] World Wide Web Consortium, "XML Schema: Primer", "XML Schema: Structures", and "XML Schema: Datatypes", W3C Recommendations, October 2004

[94] Ch. Valentine, "XML Schemas", SYBEX, 2002

[95] I. Wolf, B. Feiten, T. Guenkova-Luy, A. Schorr, F. Hauck, A. Kassler, "MPEG-21 DIA based delivery using SDPng and RTP", ISO/IEC JTC1/SC29/WG11, Coding of Moving Pictures and Audio, MPEG2004/M10996, Redmond (WA, USA), July 2004

[96] T. Guenkova-Luy, A. Schorr, F. Hauck, A. Kassler, I. Wolf, B. Feiten, C. Timmerer, Willem Romijn, "Harmonization of Session and Capability Descriptions between SDPng and MPEG-21 Digital Item Adaptation", IETF Work-in-progress, draft-guenkova-mmusic-mpeg21-sdpng-00, February 2005

[97] T. Guenkova-Luy, A. Schorr, F. Hauck, A. Kassler, I. Wolf, B. Feiten, Ch. Timmerer, H. Hellwagner, "MPEG-21 DIA based content delivery using SDPng

controls and RTP transport", ISO/IEC JTC1/SC29/WG11, Coding of Moving Pictures and Audio, MPEG2004/M12002, Busan (Korea), April 2005

[98] World Wide Web Consortium, "XML Linking Language (XLink)", W3C Recommendation, June 2001.

[99] World Wide Web Consortium, "XML Path Language", W3C Recommendation, November 1999.

[100] World Wide Web Consortium, "XPointer", W3C Recommendation, March 2003

[101] R. Sparks, "The Session Initiation Protocol (SIP) Refer Method", IETF RFC 3515, April 2003

[102] R. Sparks, "The Session Initiation Protocol (SIP) Referred-By Mechanism", IETF RFC 3892, September 2004

[103] J. Soininen, "Transition Scenarios for 3GPP Networks", IETF RFC 3574, August 2003

[104] A. Johnston et al., "Session Initiation Protocol (SIP) Basic Call Flow Examples", IETF RFC 3665, December 2003

[105] J. Rosenberg et al., "Best Current Practices for Third Party Call Control (3pcc) in the Session Initiation Protocol (SIP)", IETF RFC 3725, April 2004

[106] G. Camarillo et al., "Transcoding Services Invocation in the Session Initiation Protocol (SIP) Using Third Party Call Control (3pcc)", IETF RFC 4117, June 2005

[107] International Telecommunication Union, "Packet-based multimedia communications systems", ITU-T Recommendation H.323, Approved in 2003-07

[108] J. S. Rosenschein, G. Zlotkin, "Rule of Encounter: Design conventions for automated negotiation among computers", The MIT Press, 1994

[109] International Organization for Standardization and International Electrotechnical Commission, "Information Technology - Multimedia Content Description Interface (MPEG-7) - Part 5: Multimedia Description Schemes", ISO/IEC 15938-5:2003, 2003

[110] J. H. Saltzer, D. P. Reed, D. D. Clark, "End-to-End Arguments in System Design", ACM Transactions on Computer Systems, pp. 277 - 288, Vol. 2, No.4, November 1984

[111] Th. Vu Duong, G. Fromentoux, J.-L. Le Roux, "A global framework for architecture analysis in telecommunications", IEEE GLOBECOM 2003, San Francisco (CA, USA), December 2003

[112] E. Bertin, E. Bury, P. Lesieur, "Intelligence distribution in next-generation networks: An architectural framework for multimedia services", IEEE International Conference on Communications (ICC2004), Paris (France), June 2004

[113] M. Welzl, L. Franzens, M. Mühlhäuser, "Scalability and Quality of Service: A Trade-off?", IEEE Communications Magazine, Vol. 41, June 2003

[114] GSM 02.17 V8.0.0 (1999-11), "Digital cellular telecommunications system (Phase 2+); Subscriber Identity Modules (SIM); Functional characteristics", Technical Specification, Nov. 1999

[115] F. Trosby, "SMS, the strange duckling of GSM", Telektronikk telecommunications journal, Vol. 2004, No. 3, 2004

[116] C. Mills, D. Hirsh, G. Ruth, "Internet Accounting: Background", IETF RFC 1272, November 1991

[117] N. Brownlee, A. Blount, "Accounting Attributes and Record Formats", IETF RFC 2924, September 2000

[118] B. Aboba, J. Arkko, D. Harrington, "Introduction to Accounting Management", IETF RFC 2975, October 2000

[119] J. Postel, "Internet Protocol", IETF RFC 791, September 1981

[120] R. Hinden, S. Deering, "Internet Protocol Version 6 (IPv6) Addressing Architecture", IETF RFC 3513, April 2003

[121] S. Deering, R. Hinden, "Internet Protocol, Version 6 (IPv6) Specification", IETF RFC 2460, December 1998

[122] J. Postel, "Transmission Control Protocol", IETF RFC 793, September 1981

[123] J. Postel, "User Datagram Protocol", IETF RFC 768, August 1980

[124] C. Rigney et al., "Remote Authentication Dial In User Service (RADIUS)", IETF RFC 2865, June 2000

[125] P. Calhoun et al., "Diameter Base Protocol", IETF RFC 3588, September 2003

[126] D. Mitton, "Authentication, Authorization, and Accounting: Protocol Evaluation", IETF RFC 3127, June 2001

[127] S. Corson, J. Macker, "Mobile Ad hoc Networking (MANET): Routing Protocol Performance Issues and Evaluation Considerations", IETF RFC 2501, January 1999

[128] K. Farooqui, L. Logrippo, J. de Meer, "The ISO Reference Model for Open Distributed Processing: An Introduction", Computer Networks and ISDN Systems Journal, Vol. 27, No. 8, pp. 1215-1229, July 1995

[129] International Organization for Standardization, International Electrotechnical Commission and International Telecommunication Union, "Information Processing Systems - Open Distributed Processing (ODP)", International Standard ISO/IEC 10746-x and ITU-T Recommendation Series X.9xx, 1995-1999

[130] A. Kassler, T. Guenkova-Luy, D. Mandato, P. Schoo, I. Armuelles, T. Robles, P. Ruiz, A. Bascuñana, "Enabling Mobile Heterogeneous Networking Environments with End-to-End QoS - The BRAIN vision and the MIND approach", Invited Paper, European Wireless Conference, Florence (Italy), February 2002

[131] T. Guenkova-Luy, A. Kassler, "End-To-End Quality of Service Coordination for Multimedia Applications in Heterogeneous, Mobile Networks", IEEE International Conference on Communications (ICC2004), Paris (France), June 2004

[132] T. Guenkova-Luy, A. Kassler, D. Mandato, "End-to-End Quality of Service Coordination for Mobile Multimedia Applications", IEEE JSAC - Advanced Mobility Management and QoS Protocols for Wireless Internet, Vol. 22, No. 5, pp. 889 - 903, June 2004

[133] T. Guenkova-Luy, A. Kassler, "End-to-End Quality of Service Coordination Models for Mobile Networks", Invited Paper, WSEAS ISA '04, Florida (USA), April 2004 and in WSEAS TRANSACTIONS on COMPUTERS, Vol. 3, No. 5, November 2004

[134] M. Alfano, N. Radouniklis, "A Cooperative Multimedia Environment with QoS Control: Architectural and Implementation Issues", ICSI Technical Report TR-96-040, International Computer Science Institute, Berkeley (CA, USA), Sept. 1996

[135] A. Watson, M. A. Sasse, "Measuring Perceived Quality of Speech and Video in Multimedia Conferencing Applications", Sixth ACM international conference on Multimedia, Bristol (UK), September 1998

[136] A. Schrader, "Audio Perception and Coding", Lecture on Multimedia Communication at University of Ulm, Summer Term, 2003

[137] C. H. Muntean, J. McManis, "A QoS-Aware Adaptive Web-Based System", IEEE International Conference on Communications (ICC2004), Paris (France), June 2004

[138] Y. Ito, Sh. Tasaka, Y. Fukuta, "Psychometric Analysis of the Effect of End–to–End Delay on User–Level QoS in Live Audio-Video Transmission", IEEE International Conference on Communications (ICC2004), Paris (France), June 2004

[139] P. Florissi, "QoSME: QoS Management Environment", PhD Thesis at Columbia University, 1996

[140] A. Smith, R. Jacobs, "Quality of Service Management within a Middleware for Large Scale Multicast Applications", Workshop on QoS Support for Real-Time Internet Applications (RTAW99), Vancouver (Canada), June 1999

[141] MIND Project, "MIND Trials Final Report", Deliverable D6.4, November 2002

[142] H. Honkasalo et al., "WCDMA and WLAN for 3G and Beyond", IEEE Wireless Communications Magazine, Vol. 9, pp. 14 - 18, April 2002

[143] J. De Vriendt et al., "Mobile Networking Evolution: A Revolution on the Move", IEEE Communications Magazine, Vol. 40, pp. 104 - 111, April 2002

[144] A. Kassler, A. Schrader, "Computer Networks – provide connectivity among a set of autonomous computers", Lecture on Multimedia Communication at University of Ulm, Summer Term, 2003

[145] International Telecommunication Union, "Broadband Mobile Communications towards a Converged World – Emerging Technology Scenario: What Are the Future Broadband Mobile Services?", ETRI Background Paper, ITU/MIC Workshop on Shaping the Future Mobile Information Society, Seoul (Korea), March 2004, <http://www.itu.int/osg/spu/ni/futuremobile/Broadbandmobile.pdf>

[146] M. Grand, "Patterns in Java: A Catalog of Reusable Design Patterns Illustrated with UML – Volume 1", 2nd Edition, Wiley Publishing, Inc., 2002

[147] J. Heinanen et al., "Assured Forwarding PHB Group", IETF RFC 2597, June 1999

[148] V. Jacobson et al., "An Expedited Forwarding PHB", IETF RFC 2598, June 1999

[149] Cisco Systems, "DiffServ – The Scalable End-to-End QoS Model", Cisco Systems Inc., White Paper, August 2005, <http://www.cisco.com/application/pdf/en/us/guest/tech/tk766/c1550/ccmigration_09186a00800a3e2f.pdf >

[150] R. Kippenhahn, "Verschlüsselte Botschaften – Geheimschrift, Enigma und Chipkarte", Rowohlt Taschenbuch Verlag GmbH, 1999

[151] A. Kassler, "Quality of Service – Network Support for Multimedia", Lecture on Multimedia Communication at University of Ulm, Summer Term, 2003

[152] International Telecommunication Union, "7 kHz audio-coding within 64 kbit/s", ITU-T Recommendation G.722, Approved in 1988-11

[153] International Telecommunication Union, "Low-complexity coding at 24 and 32 kbit/s for hands-free operation in systems with low frame loss", ITU-T Recommendation G.722.1, Approved in 2005-05

[154] International Telecommunication Union, "Wideband coding of speech at around 16 kbit/s using Adaptive Multi-Rate Wideband (AMR-WB)", ITU-T Recommendation G.722.2, Approved in 2003-07

[155] International Telecommunication Union, "40, 32, 24, 16 kbit/s adaptive differential pulse code modulation (ADPCM)", ITU-T Recommendation G.726, Approved in 1990-12

[156] International Telecommunication Union, "5-, 4-, 3- and 2-bit/sample embedded adaptive differential pulse code modulation (ADPCM)", ITU-T Recommendation G.727, Approved in 1990-12

[157] International Telecommunication Union, "Coding of speech at 16 kbit/s using low-delay code excited linear prediction", ITU-T Recommendation G.728, Approved in 1992-09

[158] International Telecommunication Union, "Coding of speech at 8 kbit/s using conjugate-structure algebraic-code-excited linear-prediction (CS-ACELP)", ITU-T Recommendation G.729, Approved in 1996-03

[159] International Telecommunication Union, "Video codec for audiovisual services at p x 64 kbit/s", ITU-T Recommendation H.261, Approved in 1993-03

[160] International Telecommunication Union, "Video coding for low bit rate communication", ITU-T Recommendation H.263, Approved in 2005-01

[161] International Telecommunication Union, "Advanced video coding for generic audiovisual services", ITU-T Recommendation H.264, Approved in 2005-03

[162] G. Fankhauser et al., "WaveVideo – An Integrated Approach to Adaptive Wireless Video," ACM Monet, Special Issue on Adaptive Mobile Networking and Computing, Vol. 4., No. 4, pp. 255 - 271, 1999.

[163] International Organization for Standardization and International Electrotechnical Commission, "Information technology – Generic coding of moving pictures and associated audio information (MPEG-2)", ISO/IEC 13818, 2000

[164] International Organization for Standardization and International Electrotechnical Commission, "Information technology – Coding of audio-visual objects (MPEG-4)", ISO/IEC 14496, 2004

[165] G. Camarillo et al., "Grouping of Media Lines in the Session Description Protocol (SDP)", IETF RFC 3388, December 2002

[166] E. Steinbach, "Receiver-based Adaptation Mechanisms for Realtime Media Delivery", Lecture on Multimedia Communication at University of Ulm, Summer Term, 2003

[167] R. Fielding et al., "Hypertext Transfer Protocol – HTTP/1.1", IETF RFC 2616, June 1999

[168] T. Berners-Lee, R. Fielding, L. Masinter, "Uniform Resource Identifier (URI): Generic Syntax", IETF RFC 3986, January 2005

[169] D. Crocker, P. Overell, "Augmented BNF for Syntax Specifications: ABNF", IETF RFC 2234, November 1997

[170] D. Mandato, A. Kassler, T. Guenkova-Luy, M. Nosse, "Software architecture for capability and quality-of-service negotiations and session establishment for distributed multimedia applications", European Patent Application EP1414211, December 2002

[171] A. Kassler, T. Guenkova-Luy, D. Mandato, T. Robles, "E2ENP End-to-End QoS Negotiation Protocol", Mobile Networking Beyond 3G - International Workshop on Mobile IP-based Network Developments, Budapest (Hungary), November 2002

[172] A. Kassler, T. Guenkova-Luy, D. Mandato, T. Robles, "An End-to-End QoS Negotiation Protocol", International Workshop on Mobile IP-based Network Developments, London (GB), October 2002

[173] D. Mandato, A. Kassler, T. Guenkova-Luy, "A method for enabling the negotiation of end-to-end qos by using the end-to-end negotiation protocol (e2enp)", European Patent Application EP1331785, January 2002

[174] IETF, "International Engineering Task Force", <http://www.ietf.org>

[175] G. Dermler et al., "A Negotiation and Resource Reservation Protocol (NRP) for Distributed Multimedia Applications", IEEE International Conference on Multimedia Computing and Systems (ICMCS '96), Hiroshima (Japan), June 1996

[176] W. Fiederer, K. Rothermel, G. Dermler, "QoS Negotiation and Resource Reservation for Distributed Multimedia Applications", International Conference on Multimedia Computing and Systems (ICMCS ′97), Ottawa (Canada), June 1997

[177] M. Boger, "Java in Verteilten Systemen: Nebenläufigkeit, Verteilung, Persistenz", dpunkt.verlag GmbH, 1999

[178] Maintained by java.net, JAIN SIP, <https://jain-sip.dev.java.net/>

[179] Internet Engineering Task Force, "Active IETF Working Groups", <http://www.ietf.org/html.charters/wg-dir.html>

[180] Ch.-W. Chen et al., "Efficient Support of Java RMI over Heterogeneous Wireless Networks", IEEE International Conference on Communications (ICC2004), Paris (France), June 2004

[181] DAIDALOS project, "Designing Advanced network Interfaces for the Delivery and Administration of Location independent, Optimised personal Services", Daidalos Flyer, June 2004, <http://www.ist-daidalos.org/daten/publications/publications.htm>

[182] P. Ting et al., "An Adaptive Hardware Platform for SDR", Software Defined Radio Forum Contribution, August 2001, <http://www.sdrforum.org/2001_docs.html>

[183] Wipro Technologies, "Software-Defined Radio", White Paper, August 2002, <http://www.broadcastpapers.com/broadband/WiproSDRadio.pdf>

[184] Sourceforge, "jSIP Home Page", tested versions v0.6 (2001) and v0.8 (2002), <http://jsip.sourceforge.net/>

[185] National Institute of Standards and Technology, "NIST-SIP 1.2 -- SIP Libraries and Tools for the People!", tested versions: v0.9 (2001) till v1.2 (2003), <http://www-x.antd.nist.gov/proj/iptel/src/nist-sip/jain-sip/docs/>

[186] Wikipedia, the free encyclopedia, "Public Switched Telephone Network", last modified 9 December 2005, viewed 9 December 2005, <http://en.wikipedia.org/wiki/Pstn>

[187] P. Srisuresh et al., "Middlebox communication architecture and framework", IETF RFC 3303, August 2002

[188] G. Klyne, "A Syntax for Describing Media Feature Sets", IETF RFC 2533, March 1999

[189] G. Klyne, "Corrections to 'A Syntax for Describing Media Feature Sets'", IETF RFC 2738, Dec. 1999

[190] G. Klyne, "Identifying Composite Media Features", IETF RFC 2938, Sept.2000

[191] M. Lohse, Ph. Slusallek, P. Wambach, “Extended Format Definition and Quality-driven Format Negotiation in Multimedia Systems”, Multimedia 2001 – EUROGRAPHICS Workshop, Manchester (UK), Sept. 2001

[192] Th. Frühwirth, S. Abdennadher, “Essentials of Constraint Programming”, Springer-Verlag, 2003

[193] R. Myerson, “Harsanyi's Games with Incomplete Information”, A review of Harsanyi's three-part paper on Bayesian games with incomplete information, written for the 50th anniversary of the journal Management Science, August 2004, <http://home.uchicago.edu/~rmyerson/research/harsinfo.pdf>

[194] F.W. Taylor, “The Principles of Scientific Management”, originally prepared for presentation to The American Society of Mechanical Engineers, 1911, <http://www.fordham.edu/halsall/mod/1911taylor.html> and <http://melbecon.unimelb.edu.au/het/taylor/sciman.htm>

[195] D. Gürtler, “Die Dagoberts – Eine Weltgeschichte des Reichtums – von Krösus bis Bill Gates”, Eichborn Verlag, 2004

[196] Th. Cormen, Ch. Leiserson, R. Rivest, “Introduction to Algorithms”, MIT Press (1990), Eighth printing, 1992

[197] A. Kassler et al., “MASA - A scalable QoS Framework”, Internet and Multimedia Systems and Applications (IMSA2003), Honolulu (USA), August 2003

[198] MASA, “Mobility and Service Adaptation in Heterogeneous Mobile Environments”, Project of Siemens, NEC Europe and University of Ulm, <http://masa.ccrle.nec.de/>

[199] M. Garcia-Martin, “Input 3rd-Generation Partnership Project (3GPP) Release 5 Requirements on the Session Initiation Protocol (SIP)”, IETF RFC 4083, May 2005

[200] H. Schulzrinne, C. Agboh, “Session Initiation Protocol (SIP) – H.323 Interworking Requirements”, IETF RFC 4123, July 2005

[201] DAIDALOS Project, “Service Creation Platform Design Specification”, Deliverable D351, August 2004

[202] T. Guenkova-Luy, A. Schorr, F. Hauck, M. Gomez, Ch. Timmerer, I. Wolf, A. Kassler, “Advanced Multimedia Management – Control Model and Content Adaptation”, IASTED International Conference on Internet and Multimedia Systems and Applications (EuroIMSA 2006), Innsbruck (Austria), February 2006

[203] Lucent Technologies, “IP Multimedia Subsystem (IMS) – Service Architecture”, White Paper, February 2005, <http://www.lucent.com/livelink/090094038005df2f_White_paper.pdf>

[204] Y. Boger, “Fine-tuning Voice over Packet services”, VOIP White Paper, November 2004, <http://www.radcom.com/radcom/WhitePapers/voip2.htm>

[205] J. Postel, J. Reynolds, “File Transfer Protocol (FTP)”, IETF RFC 959, October 1985

[206] H. Schnepp, Th. Müller, J.-F. Luy, “Channel Diversity in a Software Autoradio”, IEEE Vehicular Technology Conference, Rhodes (Greece), May 2001

[207] H. Schnepp, Th. Müller, J.-F. Luy, P. Russer, “The Implementation of Channel Diversity in Mobile Software Radio Receivers”, IEEE Microwave and Wireless Components Letters, Vol. 13, No. 8, pp.323-325, 2003

[208] 3GPP & International Telecommunication Union, “LS on Technical Report on Mobility between H.323 Multimedia Systems and GPRS/IMT2000 Networks”,

Liaison Statement, ITU-T SG 16, January 2004, <http://www.3gpp.org/ftp/tsg_sa/TSG_SA/TSGS_23/Docs/PDF/SP-040026.pdf>

[209] 3GPP & International Telecommunication Union, "Technical Report on mobility management", Liaison Statement, Q.2/SSG Rapporteur, August 2004, <http://www.3gpp.org/ftp/tsg_sa/TSG_SA/TSGS_25/Docs/PDF/SP-040538.pdf>

[210] International Telecommunication Union, "Information technology – Abstract Syntax Notation One (ASN.1): Specification of basic notation", ITU-T Recommendation X.680, Approved in 2002-07

[211] International Telecommunication Union, "Information technology – ASN.1 encoding rules: XML encoding rules", ITU-T Recommendation X.693, Approved in 2001-12

[212] International Telecommunication Union, "Information technology – ASN.1 encoding rules: Mapping W3C XML schema definitions into ASN.1", ITU-T Recommendation X.694, Approved in 2004-01

[213] B. Campbell et al., "Session Initiation Protocol (SIP) Extension for Instant Messaging", IETF RFC 3428, December 2002

[214] S. Floyd, "Congestion Control Principles", IETF RFC 2914, September 2000

[215] Cisco Systems, "Introduction to Internetworking", Cisco Systems Inc., Internetworking Technology Overview Manual, June 1999, <http://www.cisco.com/warp/public/3/at/solutions/ito.pdf>

[216] Borland Software Corporation, "Together ControlCenter", version v6.2 (2003) and "Together Architect 2006 for Eclipse", version 8.0 (2005), <http://www.borland.com/us/products/together/index.html>

[217] N. Freed, N. Borenstein, "Multipurpose Internet Mail Extensions (MIME) Part Two: Media Types", IETF RFC 2046, November 1996

[218] H. Schulzrinne, "Requirements for Resource Priority Mechanisms for the Session Initiation Protocol (SIP)", IETF RFC 3487, February 2003

[219] H. Schulzrinne, J. Polk, "Communications Resource Priority for the Session Initiation Protocol (SIP)", IETF RFC 4412, February 2006

[220] J. Rosenberg, "The Session Initiation Protocol (SIP) UPDATE Method", IETF RFC 3311, September 2002

[221] A. Kassler, T. Guenkova-Luy, A. Schorr, H. Schmidt, F. Hauck, I.Wolf, "Network-Based Content Adaptation of Streaming Media Using MPEG-21 DIA and SDPng", 7th International Workshop on Image Analysis for Multimedia Interactive Services (WIAMIS06), Special Session on (UMA), Seoul (Korea), April 2006

[222] Sun Microsystems Inc., "Java TM 2 Platform, Standard Edition (J2SE) v 1.4.2", <http://java.sun.com/j2se/1.4.2/>

[223] J. Galvin et al, "Security Multiparts for MIME: Multipart/Signed and Multipart/Encrypted", IETF RFC 1847, October 1995

[224] B. Ramsdell, "Secure/Multipurpose Internet Mail Extensions (S/MIME) Version 3.1 Message Specification", IETF RFC 3851, July 2004.

[225] T. Dierks, C. Allen, "The TLS Protocol Version 1.0", IETF RFC 2246, January 1999

[226] P. Chown, "Advanced Encryption Standard (AES) Cipher-suites for Transport Layer Security (TLS)", IETF RFC 3268, June 2002

[227] S. Kent, R. Atkinson, "Security Architecture for the Internet Protocol", IETF RFC 2401, November 1998

[228] C. Jennings, J. Peterson, "Certificate Management Service for The Session Initiation Protocol (SIP)", IETF Work-in-progress, draft-ietf-sipping-certs-02, July 2005

[229] D. Eastlake et al., "XML-Signature Syntax and Processing", IETF RFC 3275, March 2002 and W3C Recommendation, February 2002

[230] Internet Assigned Numbers Authority, "Multipart Media-Types", 2 January 2002, <http://www.iana.org/assignments/media-types/multipart/>

[231] E. Levinson, "The MIME Multipart/Related Content-Type", IETF RFC 2387, August 1998

[232] J. Rosenberg, H. Schulzrinne, "Session Initiation Protocol (SIP): Locating SIP Servers", IETF RFC 3263, June 2002

[233] H. Schulzrinne, E. Wedlund, "Application-Layer Mobility Using SIP", ACM Mobile Computing and Communications Review, Vol. 4, No. 3, pp. 47-57, July 2000

[234] R. Shacham et al., "Session Initiation Protocol (SIP) Session Mobility", IETF Work-in-progress, draft-shacham-sipping-session-mobility-01, July 2005

[235] A. B. Roach, "Session Initiation Protocol (SIP)-Specific Event Notification", IETF RFC 3265, June 2002

[236] E. Guttman et al., "Service Location Protocol, Version 2", IETF RFC 2608, June 1999

[237] Bluetooth SIG, Inc., "Specification of the Bluetooth System, version 1.2", November 2003, <http://bluetooth.org/foundry/adopters/document/Bluetooth_Core_Specification_v1.2>

[238] United States Coast Guard's Navigation Center, "Global Positioning System Standard Positioning Service Signal Specification", 2nd Edition, June 1995, <http://www.navcen.uscg.gov/pubs/gps/sigspec/gpssps1.pdf>

[239] H. Schmidt, T. Guenkova-Luy, F. Hauck, "Service Location using the Session Initiation Protocol", International Conference on Networking and Services (ICNS'06), Silicon Valley (USA), July 2006

[240] World Wide Web Consortium, "Resource Description Framework (RDF) Model and Syntax Specification", W3C Recommendation, 1999

[241] World Wide Web Consortium, "Web Ontology Language (OWL): Guide", W3C Specification, February 2004

[242] R. Mahy, B. Biggs, R. Dean, "The Session Initiation Protocol (SIP) "Replaces" Header", IETF RFC 3891, September 2004

[243] A. Johnston, O. Levin, "Session Initiation Protocol Call Control - Conferencing for User Agents", IETF Work-in-progress, draft-ietf-sipping-cc-conferencing-07, June 2005

[244] D. Komiya, X. Mingqiang, E. Shim, "Use Cases for Session Mobility in Multimedia Applications", IETF Work-in-progress, draft-komiya-mmusic-session-mobility-usecases-00, February 2006

[245] R. Mahy et al, "A Call Control and Multi-party Usage Framework for the Session Initiation Protocol (SIP)", IETF Work-in-progress, draft-ietf-sipping-cc-framework-06, March 2006

[246] R. Sparks, A. Johnston, D. Petrie, "Session Initiation Protocol Call Control - Transfer", IETF Work-in-progress, draft-ietf-sipping-cc-transfer-06, March 2006

[247] 3GPP TR 22.978, "All-IP Network (AIPN) feasibility study", Technical Report, V7.1.0, June 2005

[248] R. Sparks, "Internet Media Type message/sipfrag", IETF RFC 3420, November 2002

[249] G. Camarillo, "Framework for Transcoding with the Session Initiation Protocol (SIP)", IETF Work-in-progress, draft-ietf-sipping-transc-framework-03, November 2005

[250] A. Kassler, M. Gomez, T. Guenkova-Luy, B. Feiten, A. Schorr, I. Wolf, "Content Adaptation, Service Discovery and Enhanced Session Coordination (poster)", First DAIDALOS Public Workshop, Stuttgart (Germany), December 2004

[251] B. Marchal, "XML by Example", QUE, 2002

[252] Altova GmbH and Altova Inc., used version 4.4 (1998-2002), "XML Spy", <http://www.altova.com/>

[253] Pixware, "XMLmind XML Editor", used version 3.1.0 (2006), <http://www.pixware.fr/> and <http://www.xmlmind.com/>

[254] International Organization for Standardization, International Electrotechnical Commission and International Telecommunication Union, "Information technology - Multimedia content description interface - Part 1: Systems", ISO/IEC 15938-1:2002, 2002

[255] International Telecommunication Union, "Internet protocol data communication service - IP packet transfer and availability performance parameters", ITU-T Recommendation Y.1540, Approved in 2002-12

[256] International Telecommunication Union, "Network performance objectives for IP-based services", ITU-T Recommendation Y.1541, Approved in 2006-2

[257] VESA, "Video Electronics Association", <http://www.vesa.org>

[258] ATSC, "Advanced Television Standards Committee", <http://www.atsc.org/>

[259] Matrox Graphics Inc., "Matrox Display Resolution Guide", Technical Guide, Copyright 2006 Matrox Graphics Inc., <http://www.matrox.com/mga/workstation/3dws/products/resolution.cfm>

[260] MARANTZ, "What is HDTV?", White Paper, viewed on 24.03.2006, <http://www.marantz.com/hifi/america/hdtv/info_what_is_HDTV.html>

[261] IANA, "Internet Assigned Numbers Authority", <http://www.iana.org>

[262] Fraunhofer-Gesellschaft, "Audio & Multimedia Realtime Systems - MP3: MPEG Audio Layer-3", Audio & Multimedia Project Overview, Copyright 1998-2006 Fraunhofer-Gesellschaft, <http://www.iis.fraunhofer.de/amm/projects/mp3/index.html>

[263] DivX, Inc. (DivXNetworks), "The Official DivX 5.2 Guide", Technical Guide, Copyright 2003-2004 DivX, Inc. (DivXNetworks), <http://download.divx.com/divx/DivXUserGuide521-en.exe>

[264] W. Van Lancker et al., "A Framework for Transformations of XML within the Binary Domain", IASTED International Conference on Internet and Multimedia Systems and Applications (EuroIMSA 2006), Innsbruck (Austria), February 2006

[265] World Wide Web Consortium, "XSL Transformations (XSLT) Version 1.0", W3C Recommendation, November 1999

[266] J. Goyvaerts, "Regular Expressions: The Complete Tutorial", Lulu Press, 2006 and <http://www.regular-expressions.info/tutorial.html>

[267] D. Megginson, "SAX – Simple API for XML", version v2.0.2 (27 April 2004), <http://www.saxproject.org/>

[268] G. Booch, J. Rumbaugh, I. Jacobson, "The Unified Modeling Language User Guide", Addison Wesley Longman, 1999

[269] Object Management Group, "Unified Modeling Language: Superstructure", Version 2.0, August 2005

[270] Oracle Technology Network, "Oracle XML Developers Kit", version 27.03.2002, <http://otn.oracle.com/tech/xml/xdk/content.html>

[271] The Apache Software Foundation, "Xerces Java Parser", version v1.4.4 (1999-2000), <http://xml.apache.org/xerces-j/>

[272] R. van der Pas, "Memory Hierarchy in Cache-Based Systems", White Paper, Sun Microsystems, November 2002, <http://www.sun.com/blueprints/1102/817-0742.pdf >

[273] British Telecom & MIND Project, "SIPInterface DLL Documentation", MIND Project Internal Document and Source Code, 2001

[274] Fraunhofer FOKUS, "Project: BonePhone - Summary", version v0.8 (2001) <http://developer.berlios.de/projects/bonephone/>

[275] Aymeric Moizard, "The GNU oSIP library", (2005) <http://www.gnu.org/software/osip/>

[276] Testing Technologies, "TTsuite-SIP", (2006), <http://www.testingtech.de/products/voip_sip.php>

[277] Sun Microsystems Inc., "Java TM 2 Remote Method Invocation (RMI)", <http://java.sun.com/j2se/1.4.2/docs/api/index.html> and java.rmi

[278] National Institute of Standards and Technology, "NIST NET Home Page", <http://snad.ncsl.nist.gov/itg/nistnet/> (used version: 10 June 2002 – v.2.0.12, latest version: 20 July 2005 – v. 2.0.12 final versions of NIST Net (2.0.12b and 2.0.12c))

[279] J. Bilien, E. Eliasson, J.-O. Vatn, "Call establishment delay for secure VoIP", Workshop on Modeling and Optimization in Mobile, Ad Hoc and Wireless Networks (WiOpt'04), Cambridge (UK), March 2004

[280] K. Murakami, O. Haase, J.-Sh. Shin, Th. La Porta, "Mobility Management Alternatives for Migration to Mobile Internet Session-Based Services", IEEE JSAC - Advanced Mobility Management and QoS Protocols for Wireless Internet, Vol. 22, No. 5, pp. 818 - 833, June 2004

[281] Sun Microsystems Inc., Java TM 2 Platform, Standard Edition (J2SE) v. 5.0, <http://java.sun.com/j2se/1.5.0/>

[282] Sun Microsystems Inc., "Explicit Locks and Condition Variables", Java Tutorial 1995-2005, <http://java.sun.com/docs/books/tutorial/essential/threads/explicitlocks.html>

[283] IETF MMUSIC Mail Archive, "[MMUSIC] How big can SDP get", Discussion Subject on IETF MMUSIC mailing list, mmusic Digest, Vol 25, May 2006, <http://www1.ietf.org/mail-archive/web/mmusic/current/msg04205.html>

[284] DFG, "AKOM – Adaptivität in heterogenen Kommunikationsnetzen mit drahtlosem Zugang", DFG Project, 2000-2006, <http://www.lkn.ei.tum.de/AKOM/>

[285] T. Guenkova-Luy, A. Schorr, F. Hauck, I. Wolf, B. Feiten, M. Gomez, F. Galan, T. Mota, W. Romijn, "Stream Tracking Description for Resource Management Guarantees in the Network", IETF Work-in-progress, draft-guenkova-mmusic-sdp-ng-streamtrack-00, February 2005

[286] B. Feiten, I. Wolf, T. Guenkova-Luy, A. Schorr, A. Kassler, "New mode for rfc3640: AAC-BSAC with MPEG-21 gBSD", IETF Work-in-progress, draft-feiten-avt-BSACmodefor3640-00, February 2005

[287] T. Guenkova-Luy, "Coordination of Multimedia Services and Applications in Mobile, Heterogeneous Network Environment", PhD Thesis at University of Ulm, 2007

[288] H. Schmidt, T. Guenkova-Luy, F. J. Hauck, "A Decentral Architecture for SIP-based Multimedia Networks", Kommunikation in Verteilten Systemen (KIVS'07), Bern (Switzerland), Feb-Mar. 2007

[289] T. Guenkova et al., "A Session-Initiation-Protocol-based Middleware for Multi-Application Management", IEEE International Conference on Communications (ICC2007), Glasgow (UK), June 2007

[290] T.Guenkova-Luy et al., "Service Mobility with SIP, SDP and MPEG-21", 9th International Conference on Telecommunications (ConTEL 2007), Zagreb (Croatia), June 2007

Index

A

B

C

I

L

M

N

O

P

T

U

V

W

X

www.ingramcontent.com/pod-product-compliance
Lightning Source LLC
LaVergne TN
LVHW062308060326
832902LV00013B/2112